AF358658

Cooperative Microactuator
Devices and Systems

Cooperative Microactuator Devices and Systems

Editors

Manfred Kohl
Stefan Seelecke
Stephan Wulfinghoff

Basel • Beijing • Wuhan • Barcelona • Belgrade • Novi Sad • Cluj • Manchester

Editors

Manfred Kohl
Institute of Microstructure
Technology, Smart Materials
and Devices
Karlsruhe Institute of Technology
Eggenstein-Leopoldshafen
Germany

Stefan Seelecke
Intelligent Material
Systems Lab
Saarland University
Saarbrücken
Germany

Stephan Wulfinghoff
Institute of Materials Science
University of Kiel
Kiel
Germany

Editorial Office
MDPI
St. Alban-Anlage 66
4052 Basel, Switzerland

This is a reprint of articles from the Special Issue published online in the open access journal *Actuators* (ISSN 2076-0825) (available at: www.mdpi.com/journal/actuators/special_issues/Cooperative_Microactuator_Devices_Systems).

For citation purposes, cite each article independently as indicated on the article page online and as indicated below:

Lastname, A.A.; Lastname, B.B. Article Title. *Journal Name* **Year**, *Volume Number*, Page Range.

ISBN 978-3-7258-0902-8 (Hbk)
ISBN 978-3-7258-0901-1 (PDF)
doi.org/10.3390/books978-3-7258-0901-1

Cover image courtesy of Gowtham Arivanandhan

Contents

About the Editors

Manfred Kohl

Manfred Kohl received his Ph.D. in physics from the University of Stuttgart, Germany, in 1989. Later on, he worked as an IBM postdoctoral fellow at the T.J. Watson Research Center in Yorktown Heights, USA, and subsequently joined the Karlsruhe Institute of Technology (KIT), Germany, in 1992. He is currently a professor at the Faculty of Mechanical Engineering and Head of the Department of Smart Materials and Devices at the Institute of Microstructure Technology of KIT. He is a spokesperson of the Priority Program KOMMMA (Cooperative Multistage Multistable Microactuator Systems) of the German Science Foundation. His current research focuses on ferroelastic and ferromagnetic shape memory alloys, multimaterial micro- and nanotechnologies, as well as corresponding smart devices.

Stefan Seelecke

Stefan Seelecke received a Ph.D. in engineering science from Technical University Berlin, Berlin, Germany, in 1995. After his habilitation in 1999, he joined the Department of Mechanical and Aerospace Engineering, North Carolina State University, Raleigh, NC, USA, in 2001. He is currently a professor of systems engineering and materials science and engineering at Saarland University, Saarbrücken, Germany, where he directs the Intelligent Material Systems Lab. His research interests include the development of smart-material-based actuator and sensor systems, particularly (magnetic) shape memory alloys, piezoelectrics, and electroactive polymers.

Stephan Wulfinghoff

Stephan Wulfinghoff received his Ph.D. in engineering science from Karlsruhe University (KIT), Germany, in 2014. Subsequently, he worked as Engineer-in-Chief at the Institute for Applied Mechanics, RWTH Aachen, Germany, until his call for the professorship for Computational Materials Science at Kiel University, Kiel, Germany, in 2018. His main research interests include multiphysics modeling, continuum mechanics, shape memory alloys, microactuators, and multiscale approaches.

Preface

The smart coupling of distributed microactuators to a cooperative synergetic actuation system creates new functionalities to address complex tasks comprising combinations of force, displacement, and dynamics that have not been possible until now. Combining similar microactuators in microactuator arrays enables the control of time and spatially resolved actuation patterns, while the combination of microactuators based on different transducer principles even allows for novel process chains across different functional levels, as well as several length scales. Thus, operating many actuators in a coordinated way will generate new synergy effects like bi-/multistability and multistage performance. Thereby, different functional levels could be triggered selectively depending on external stimuli like temperature, as well as electrical and magnetic fields. Further synergies will open up through the intrinsic sensing of various system parameters.

This Special Issue collects selected research papers in the field of cooperative microactuator devices and systems. Different contributions originate from the Priority Program KOMMMA (Cooperative Multistage Multistable Microactuator Systems) of the German Science Foundation. This interdisciplinary program addresses the various challenges of cooperative microactuator systems by bringing together research groups in the different research areas of microactuators, microsystems, material science, system simulation, control, and systems engineering. The Special Issue covers a broad range of research topics, including piezoelectric, electrostatic, magnetic, shape memory, and dielectric elastomer principles, as well as combinations thereof. New concepts for the modeling and simulation of the multiphysical properties of the microactuator systems are presented with an emphasis on techniques of compact modeling via model order reduction.

Manfred Kohl, Stefan Seelecke, and Stephan Wulfinghoff
Editors

Review

Inchworm Motors and Beyond: A Review on Cooperative Electrostatic Actuator Systems

Almothana Albukhari [1,2] and Ulrich Mescheder [1,3,*]

1. Mechanical and Medical Engineering Faculty, and Institute for Microsystems Technology (iMST), Furtwangen University, 78120 Furtwangen, Germany; alb@hs-furtwangen.de
2. Department of Microsystems Engineering (IMTEK), Faculty of Engineering, University of Freiburg, 79110 Freiburg, Germany
3. Associated to the Faculty of Engineering, University of Freiburg, 79110 Freiburg, Germany
* Correspondence: mes@hs-furtwangen.de

Abstract: Having benefited from technological developments, such as surface micromachining, high-aspect-ratio silicon micromachining and ongoing miniaturization in complementary metal–oxide–semiconductor (CMOS) technology, some electrostatic actuators became widely used in large-volume products today. However, due to reliability-related issues and inherent limitations, such as the pull-in instability and extremely small stroke and force, commercial electrostatic actuators are limited to basic implementations and the micro range, and thus cannot be employed in more intricate systems or scaled up to the macro range (mm stroke and N force). To overcome these limitations, cooperative electrostatic actuator systems have been researched by many groups in recent years. After defining the scope and three different levels of cooperation, this review provides an overview of examples of weak, medium and advanced cooperative architectures. As a specific class, hybrid cooperative architectures are presented, in which besides electrostatic actuation, another actuation principle is used. Inchworm motors—belonging to the advanced cooperative architectures—can provide, in principle, the link from the micro to the macro range. As a result of this outstanding potential, they are reviewed and analyzed here in more detail. However, despite promising research concepts and results, commercial applications are still missing. The acceptance of piezoelectric materials in some industrial CMOS facilities might now open the gate towards hybrid cooperative microactuators realized in high volumes in CMOS technology.

Keywords: cooperative actuators; electrostatic actuator; inchworm motor; electrostatic motor; microsystems; micromachining; MEMS; gap-closing actuator

Citation: Albukhari, A.; Mescheder, U. Inchworm Motors and Beyond: A Review on Cooperative Electrostatic Actuator Systems. *Actuators* **2023**, *12*, 163. https://doi.org/10.3390/act12040163

Academic Editor: Qingan Huang

Received: 1 February 2023
Revised: 30 March 2023
Accepted: 31 March 2023
Published: 4 April 2023

1. Introduction

Electrostatic actuation is an actuation principle that is widely used in billions of commercial devices and products today, e.g., the actuation of the so-called primary excitation in MEMS (microelectromechanical systems) angular rate sensors (gyros), which are based on the Coriolis effect, is provided in most commercial devices electrostatically [1,2]. The successful emergence of electrostatic actuators is based on several key factors: the actuation principle does not rely on specific materials, and thus, electrostatic actuators can be realized with standard CMOS-processes, while taking full advantage of the technological advancement therein, such as the ongoing miniaturization and creation of high-aspect-ratio (HAR) microstructures, e.g., by using deep reactive-ion etching (DRIE) [3]. The feasibility of electrostatic actuators also benefited from the development of the so-called surface micromachining [4].

Additionally, the basic actuation cell is a simple capacitor, usually with one electrode fixed and the other electrode anchored via a tethered suspension to the substrate. The energy density of an electrostatic actuator and the corresponding forces can be simply derived from the change of energy stored in that capacitance. For a parallel-plate arrangement, the stored energy W in the capacitance C is given by Equation (1), where U is the voltage potential applied across the electrodes, ε_r and ε_o are the relative and vacuum permittivities, respectively, d is the distance between the electrodes, and a and b are the overlapping dimensions of the capacitor's electrodes. The acting forces can be calculated by the corresponding derivatives in respect to distance and actual overlap.

$$W = \frac{1}{2}CU^2 = \frac{1}{2}\varepsilon_r\varepsilon_o U^2\frac{ab}{d} \tag{1}$$

However, the so-called z-movement capability (change of d) is very limited due to decreasing forces for larger distances d ($\propto 1/d^2$), and due to the pull-in effect for small distances. Furthermore, without considering special effects, e.g., provided by fringe fields, electrostatic actuation is unidirectional, i.e., the forces in respect to Equation (1) are always directed to maximize the electrostatic energy stored in the capacitance. In the case of a comb-drive-like setup, this means that the force is directed to achieve a complete overlap of opposing "fingers" of the interlaced combs. Therefore, upon electrostatic actuation, a counter force starts building up in the mechanical suspensions of the movable or deformable electrode, which will drive the movable part back to the starting position once the potential between the electrodes is switched off. Based on Hooke's law, these restoring forces are linearly increasing with the deflection of the suspending beams. Thus, the restoring actuation forces are depending on the beam geometry and material properties (Young's modulus), and therefore are fixed for a given design. As discussed in several examples in Section 3, an important feature of cooperation is to achieve bidirectional operation, where the movement back to the starting position can be actively controlled.

Further advantages of electrostatic actuation are high energy transfer efficiency (low power consumption), fast response time and high resonance frequencies of most miniaturized spring-mass systems, allowing a fast dynamic operation [5]. Additionally, the electrostatic effect shows low temperature dependence and allows high operation temperatures when using suitable materials such as poly-Si and metals.

An early implementation of electrostatic actuation was the force feedback loop and self-test function integrated in the 1990s in the MEMS accelerometer ADXL50 from Analog Devices [6]. In consumer electronics, an example is projectors that are realized by the so-called digital micromirror device (DMD) from Texas Instruments (TI), which can be even integrated into smart phones such as Samsung's I8530 Galaxy Beam [7].

Different to most other actuation principles, miniaturization increases the performance of electrostatic actuation by virtue of the scaling [8] and Paschen's laws [9], whereby the former allows efficient actuation with relatively large force and energy density on the microscale (in respect to dimensions of electrodes and distance between them), and the latter provides larger breakdown voltages, which allows greater electric field strengths, in smaller gaps between electrodes.

However, electrostatic actuation also has certain limitations and drawbacks, such as low (absolute) force (typically µN), small stroke (typically µm), inclination to stiction, vulnerability to dielectric charging and—based on a planar design architecture and resulting mostly uniaxial movements—limited aptitude for "3D-operation" and more complex actuation tasks and movements, such as motors.

To overcome the inherent displacement limitation of electrostatic actuation, over the past few decades, many researchers have proposed various mechanisms that accumulate the repeated, short actuation steps of an electrostatic actuator into much longer displacements. Such mechanisms are typically called motors, which are in principle capable of indefinite motion, at least in one degree of freedom. For this purpose, some new motor concepts specifically suited for electrostatic actuation have been proposed, e.g., the so-called scratch drive actuator (SDA) [10], as well as other, already known concepts, such as the inchworm motor, have been adapted to electrostatic actuation [11]. Typically, these electrostatically driven motors make use of cooperation between different actuators and/or structures within the system.

In this review, we first provide definitions for cooperative electrostatic actuators and for different levels of cooperation. We will then present examples of cooperative electrostatic actuation principles and systems for each cooperation level. As a specific class, hybrid cooperative systems are presented, where electrostatic actuators are cooperating with actuators using other actuation principles. After a discussion, where we compare different cooperative architectures, we conclude the findings and try to present an outlook for future developments in this field. Being part of the Special Issue on "Cooperative Microactuator Devices and Systems", this review focusses on electrostatic cooperative actuation systems. Other principles will be considered in other contributions in that Special Issue.

2. Methodical Approach: Definition of Cooperative Electrostatic Microactuators

So far, no generally accepted definition or classification of cooperative microactuators is available. This holds especially true for cooperative actuator systems using electrostatic actuation as their base. In this review paper, cooperative electrostatic microactuator systems are defined by the following conditions or properties:

- Different actuators—at least one using electrostatic actuation—are integrated to form a cooperative, synergistically operating system, which generates new functionalities, and thus can also fulfill complex tasks that are impossible to achieve by the individual actuators found within the system (e.g., a larger force, a longer stroke, or a complex travel path).
- The scale of functional structures in these actuators is in the order of μm.
- The actuators are placed in a closely spaced arrangement to allow interaction between them and provide synergy of the individual microactuators and integration on a single chip, e.g., monolithic integration.

We distinguish "pure" electrostatic cooperative microactuator systems, i.e., those made up exclusively of electrostatic microactuators, from hybrid cooperative microactuator systems, in which, in addition to electrostatic actuation, other actuation principles or effects are used, e.g., to accomplish bistability or even multistability.

To structure this review paper, we are proposing, within the given definition, a classification based on system architecture and degree of cooperation. Three levels of cooperation are defined and presented:

Weakly cooperative architectures (Section 3.1):

- systems, where the cooperation is derived by the integration of independent, not coupled actuators through the electronic control system;
- a very limited number of cooperating actuators (2–4), where the cooperation is based on simple mechanical coupling structures of these actuators.

Medium cooperative architectures (Section 3.2):

- some actuators (2–5) with strong interaction using smart mechanical coupling (e.g., coupling activated by actuation itself) to allow, e.g., complex 3D-trajectories;
- architectures which are already providing by cooperation the base for integration in even more complex cooperative actuation systems.

Advanced/strongly cooperative architectures (Section 3.3):

- cooperative sub-architectures are integrated and combined into complex systems;
- huge numbers of coupled actuators integrated into a functional system.

This structure can also be viewed as somewhat indicative of the chronological order, given that early and even commercialized examples of weakly cooperative electrostatic actuator systems were introduced in the 1990s, whereas strongly cooperative systems are still a topic of fundamental research.

In a further section, Section 3.4, we present concepts, where electrostatic actuators are combined with other actuation principles. This section is called 'Hybrid System Architectures'.

An early example of a cooperative microactuator system with weak cooperation between the actuators is the DMD [12]. Here, a typical system features more than a million independent mirrors that are working in parallel to cooperatively create an image. The cooperative image creation is provided by the control logic, which individually addresses each single micromirror (pixel) without any coupling of the independent micromirrors. An example of medium cooperative architecture is xyz-stages, which allow out-of-plane movement even with a planar structural design, by using specific hinges. Finally, cooperative microactuator systems with advanced cooperation between distributed actuators are, e.g., electrostatic inchworm motors, whereby different actuators are working cooperatively to create a single step in a long sequence of a step-like motion, thus allowing very large travel ranges (several mm instead of μm) and, by combining many driving units in a cascaded system, also large forces (mN instead of μN). The quality of such cooperative electrostatic microactuators is therefore called, in this review, "strong", because the successful functioning of such systems needs not only addressability of each actuator within the system, but also coordinated action of the different coupled actuators, since they depend on each other. Additionally, such devices are able to carry loads or create relatively large forces and provide larger travel ranges.

Electrostatically driven microactuators using direct cooperation between the actuators, or cooperation at a system level, are described in several research papers. However, compared to the field of electrostatic actuation in general, the number of research papers is rather limited, and for the most part, even the papers listed in this review are not focusing on "cooperation" or cooperative aspects specifically, and do not pronounce the cooperative or coupling character of the described actuator systems. For this reason, a systematic search of related publications by keywords is difficult, and the search results provided are unreliable. For instance, a search on Google Scholar combining the single words and the full phrase "cooperative electrostatic actuation" provides only 10 results, most of them fulfilling the given definition. However, combining the keywords actuator, electrostatic and cooperative with the phrase "cooperative electrostatic", gives 74 results in Google Scholar, from which only 11 are relevant in respect to the given definition (and even not providing the papers presented in this review!). On the other hand, a general search with the keywords "electrostatic", "microactuators" and "cooperative" provides 1720 results, but many of them are not relevant in respect to the given definition of cooperation.

Similar results are found for a search on science direct: When using the three keywords "electrostatic", "actuator" and "cooperative", 1198 results are provided. However, only 104 results are found, when using the keywords "electrostatic", "microactuators" and "cooperative". These results are almost uniformly spread over the last 20 years. In contrast, an advanced search combining those three keywords in the text with the criteria that these are also used in the title or abstract fields provides only one result. By reducing the keywords needed to be found in either of those fields to "electrostatic" and "actuator", 22 search results are found.

Therefore, a simple keyword-based search on "cooperative microactuators" is not reliable. Instead, we analyzed papers dealing with miniaturized electrostatic actuators in general, but with some special features, for example, in respect to motion (e.g., feasibility to provide not only unidirectional movement), stroke (>>μm) and force (>>μN). For this review, about 140 papers were identified according to the above definition of cooperative electrostatic actuation systems. From these, 82 are presented in more detail or referenced in this review. The selected papers are representing basic principles and concepts of cooperation.

Whenever possible, we try to introduce, as examples of a given cooperation category or method, those publications where that specific idea was first presented before referencing more recent papers.

3. Design Principles and Architectures of Cooperative Electrostatic Microactuators

3.1. Weakly Cooperative Architectures

As we established in the aforementioned classification, this Section reviews actuator systems that exhibit a form of cooperation, but on a weak level, from one perspective or another. Examples of such systems are those that have actuators with little to no coupling or physical dependence among them, such as the independent optical actuators that collectively create an image or drive a multi-channel switch. Additionally, systems that consist of only few actuators and that have only simple and passive coupling elements, and those that cascade identical actuators to increase their displacement range or force, will also be reviewed in this section.

3.1.1. Cooperation by Control Logic

As already mentioned, a well-known example of this kind is the DMD [12], in which more than a million mirrors can be individually actuated and work in parallel to cooperatively create an image. The cooperative image creation is provided by the control logic, which individually addresses each single micromirror (pixel) and drive it to assume one of two states: on or off. Thus, the cooperative quality is similar to a microelectronic device, such as static or dynamic random-access memory (SRAM or DRAM) chips, where millions to even billions of basic unit cells (typically consisting of six transistors) are controlled for particular functions. In that respect, we call the quality of cooperation of such types of cooperative microactuator systems "weak".

Another early example belonging to this type of cooperative actuation systems was presented by Liu et al. [13]. In this solution, an array of mirrors is arranged vertically by micro latches on micro platforms that can be tilted electrostatically. Hence, upon selective excitation, the micro platforms are pivoted to bring their respective mirrors into the paths of light beams. In this way, a 4 × 4 optical switch is realized. The arrangement of the 16 mirror platforms can be seen in Figure 1. An exceptional feature of the surface micromachined device is the pronounced out-of-plane movement, which is created by means of electrostatic actuation in combination with suitable pivoting joints for the tilting plates on the substrate (draw-bridge beams). In the absence of a driving voltage, the tilting plate is in the up-position, and it moves downwards by applying a voltage between the plate and an electrode on the substrate beneath it. In 2002, a similar solution for a 16 × 16 switch was presented by the American company OMM Inc. (San Diego, CA, USA) [14].

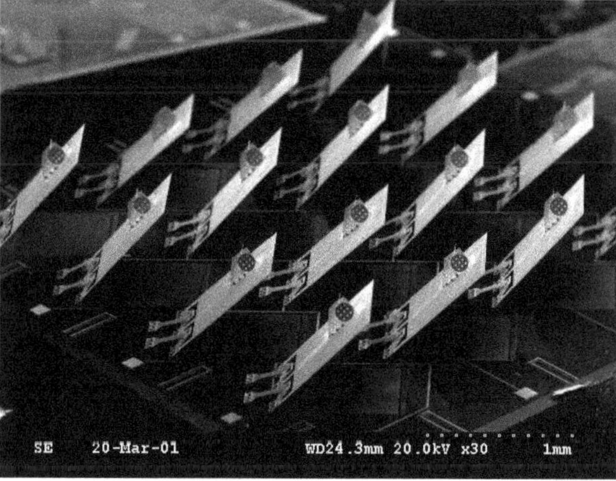

Figure 1. 4 × 4 optical fiber switch. Each of the 16 elements can be addressed separately to provide the needed switching combination [13] (Reproduced with permission from SNCSC. All rights reserved).

These examples belong to the group of quasi loadless actuators as the actuation only has to overcome the restoring counterforce provided by the suspensions of the movable or deformable structures. This is typically the case for microactuator systems used for optical applications, e.g., optical switching. A review about optical switches, which are mostly electrostatically actuated, and not entirely using cooperative architectures, is provided in [15].

3.1.2. Cooperation through Passive Mechanical Coupling Structures

Other actuation systems with weak cooperation are those, where the cooperation of few actuators (2–4) is provided by mechanical coupling structures, which are passive and simple, in most cases spring-like structures or freestanding stiff beams. One purpose for such cooperation is to achieve an actively controlled movement of the actuators back to the non-powered state position by electrostatic actuation, rather than being passively moved back by the elastic counterforce built up in the actuators' beam suspensions, thus, as a consequence, not being entirely dependent on the beam geometry and materials' properties. Another purpose is to extend uniaxial linear actuators to planar movements, i.e., two-degree-of-freedom movements (2-DoF).

A bidirectional mode within a 2-DoF actuation system was presented in [16,17], where two groups of opposing comb drives are cooperatively acting together (one for the x-direction and one for the y-direction) to move a xy-stage, to serve, e.g., as a scanning unit for an integrated atomic force microscope (AFM). A schematic of that set-up is shown in Figure 2. The cooperation between each of the two opposing comb drives is provided by the long freestanding beams connecting the opposing comb drives. These beams are stiff in the direction of the intended electrostatic actuation and allow a (soft) bending in the perpendicular direction. Thus, the length of the beam together with specific stabilizing structures for the movable parts of the comb drives allow a reasonable decoupling of the x and y movements, e.g., 35 dB for a beam length, width and thickness of 600 μm, 2 μm and 4 μm, respectively. Several comb structures are placed in parallel, similar to a muscle fibril (sarcomere). Therefore, the authors in [16] called this design, which reduces the driving voltage, sarcomere actuator.

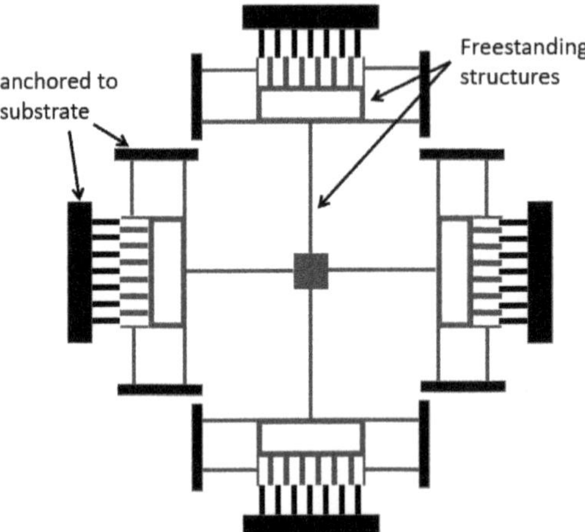

Figure 2. Bidirectional movement is obtained by two opposing comb drives for each axis. By axially loaded long beams, the forces are transmitted to the opposite drive, while the long beams allow bending in respect to forces in a perpendicular direction, thus providing decoupling of the movements in the two axes. Black: structures anchored to the substrate, grey: freestanding structures; adapted from [17].

Recently, a similar sarcomere design was presented as a three-dimensional (3D) polymer interdigitated pillar electrostatic (PIPE) actuator for large forces [18]. With a 3D printed array of parallel as well as stacked comb drives the authors claimed to achieve a work density close to that of a human muscle and a force of around 100 N at 4000 V.

The simple approach of cooperation by connecting beams for bidirectional electrostatic actuation is a quite often used design, e.g., a similar actuation principle was employed in [19]; however, it was combined with self-sensing for accurate nanopositioning.

Furthermore, more sophisticated examples of this type of mechanical coupling of structures will be discussed in Section 3.2.

3.1.3. Cascaded Systems

An early concept of a cascaded system was presented by Minami et al. [20]. The basic driving unit, the so-called distributed electrostatic microactuator (DEMA), consists of two long wave-like insulated electrodes. Due to the wave-form of the electrodes, a large electrostatic force at those locations with a small gap is obtained, which deforms the wave-like structure easily at locations with a wider gap and therefore also provides a large deformation of the driving unit as a whole. By connecting many driving units in series (long DEMA), the total deformation is increased; on the other hand, by connecting many of the wave-formed electrode pairs in parallel, a force amplification is obtained (see Figure 3a). For a miniaturized DEMA, a displacement of 28 µm at 160 V was measured and a force of 6.3 µN at 200 V was simulated by the finite element method (FEM) [20].

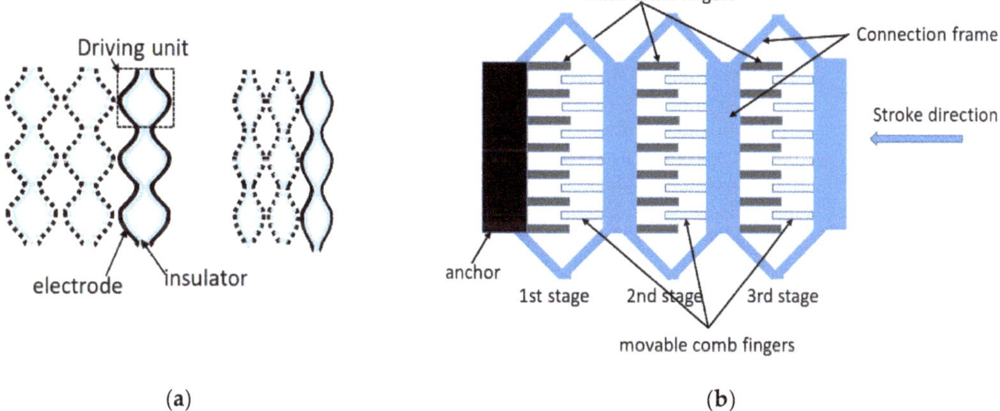

Figure 3. Cascaded electrostatic actuator designs: (**a**) due to a wave-like electrode shape, the structure is easily deformable by an electric potential, by long waves, large deformation is achieved (serial cascading), by parallel wave-like structures, force is increased, adapted from [20]; (**b**) design of a serially cascaded 3-stage comb-drive actuation system, where the 3 stages are connected by a deformable frame structure, adapted from [21].

Another cascaded electrostatic actuation system was proposed by Chiou et al. [21] to overcome the side stability limitation (side pull-in) and extend the stable traveling range of comb-drive actuators by a serially cascaded multi-stage configuration. The three-stage cascaded system architecture is achieved by a connection frame. The first stage is fixed, whereas the second and third stages are movable. Within each stage, one part of the comb drive is fixed, while the other one is movable and can follow the frame deformation. The connection frame was made out of 455 nm thick poly-Si. For the comb structures, a four-metal-layer process with tungsten as vias/contacts was used, thus providing a capacitance height of around 6 µm. The total stroke of the cascaded actuator system is the sum of strokes of the first, second and third stages. Therefore, with the 3× cascaded system shown in Figure 3b, the stable traveling range was extended by 200% compared to a single actuator. However, the side pull-in instability of the cascaded system, relative to its size, is the same as that of a single stage. A total displacement of 10.1 μm was reported at 125 V. However, due to residual stress in the layers, the movable structures showed a slight out-of-plane bending which caused an unbalance between the stroke contributions of the 3 stages. Such cascaded comb drives are often combined with bidirectional electrostatic actuators using opposing comb drives, as the one shown in Figure 2.

Recently, a serially cascaded system with three gap-closing parallel-plate actuators was presented by Schmitt et al. [22,23]. The single actuators are linked by connecting beams. The specific step-like movement of the total system is obtained by a special design of the electrode gaps, which are increasing from the first to the third actuator in the row, and a defined sequence of powering the actuators with the smallest-gap actuator first, such that the gaps of subsequent actuators are decreasing when preceding actuators are activated. In this way a successive pull-in of the actuators in the row is achieved. Guiding sinusoidal-shaped springs guarantee a pure translational movement. The system also allows a bidirectional movement and strokes of up to 35 μm at 50 V, with intermediate steps of 6.9 and 19.4 μm.

Another interesting cascaded design approach was presented in 2015 by Conrad et al. [24]. The basic actuation cell is an electrostatically actuated leverage structure, many of which are placed periodically on a bendable cantilever; thus, we assign this concept under cascaded systems. In this novel approach, the so-called nano electrostatic drive (NED), the actuation by electrostatic forces is combined with a bimorph leverage mechanism to achieve an actuation distance larger than the gap separation between electrodes. The actuator cell has top and bottom electrodes separated by a submicron gap. The asymmetrical compliance of these two mechanical elements, combined with a certain non-planar topography of the gap and electrodes, make up a robust mechanical leverage system capable of producing controllable, repeatable, out-of-plane deflections. Moreover, so-called V- or Λ-shaped electrodes will result in a cantilever bending in the upward or downward direction, respectively. It was demonstrated that a non-optimized cantilever design having a total length of 4 mm and V-shaped actuator cells with a gap of 200 nm achieved a total deflection of 272 nm (136% of the gap) at a driving voltage of 45 V. The authors simulated tens to hundreds of micrometers of beam deflection for a cantilever beam length of 2 mm. The high potential of this novel cooperative system will become more evident when a recent, more elaborate inchworm motor system based on the NED actuator is discussed in Section 3.3.

3.2. Medium Cooperative Architectures

Following the classification presented in Section 2, medium cooperative architectures are characterized by specific and sophisticated coupling, or active joints between actuators, e.g., for achieving controlled movements with complex 3D trajectories. Additionally, cooperative actuator systems and principles providing bistability and stepping mode are presented here.

3.2.1. XYZ Stages and Multi-DoF Systems

A challenging design aspect for xyz-stages in particular, and for multi-degree-of-freedom (multi-DoF) movements in general, is the inherent mechanical coupling across the different movement directions. This is usually addressed by specially designed flexure structures that guide the deflection caused by a certain actuator along its corresponding axis, while limiting its influence in other directions, such that efficient, controllable actuation with a tolerable level of cross-coupling is achieved. Additionally, in different applications, also different degrees of coupling between the actuators are allowed or needed; thus, different approaches of coupling/decoupling the motion into the different directions are discussed in this section.

A very early example of an actuation system with such smart mechanical coupling of the actuators is the xyz-stage presented in [25], which is shown in Figure 4. The basic actuation units are defined by four groups of electrostatic SDAs, an actuation unit that in itself represents a cooperative subsystem, as we will show in Section 3.2.3. Here, each group consists of nine SDAs that are integrated in an actuator plate. The four plates are cooperating to control the position of a central stage, on which four micro-Fresnel lenses have been integrated, through special mechanical couplings that link together each pair of plates that are opposite to one another. The xyz-movement of the central stage is created by appropriate actuations of the opposing plate actuators. To achieve the required xyz-movement, the system needs, in addition to the four in-plane suspended actuator plates, special hinges (so-called polarity hinges) as well as supports (sliding joints) to create the large out-of-plane motion, e.g., an upward movement (case (b) in Figure 4b) is obtained, when the two opposing plates are actuated in opposite directions, and as a result, a critical force is produced at the hinges, which results in a redirection of the planar forces into the z-axis. Once the stage is in the up-position, it can be moved in the x or y directions by an equal actuation of the corresponding pair of opposing plates into the same direction (case (c) in Figure 4b). Therefore, the z-movement as well as the xy-movement are the result of cooperation of the (planar) actuators. A lateral movement up to 120 µm and a vertical movement even up to 250 µm with a resolution of 27 nm was reported with the shown setup.

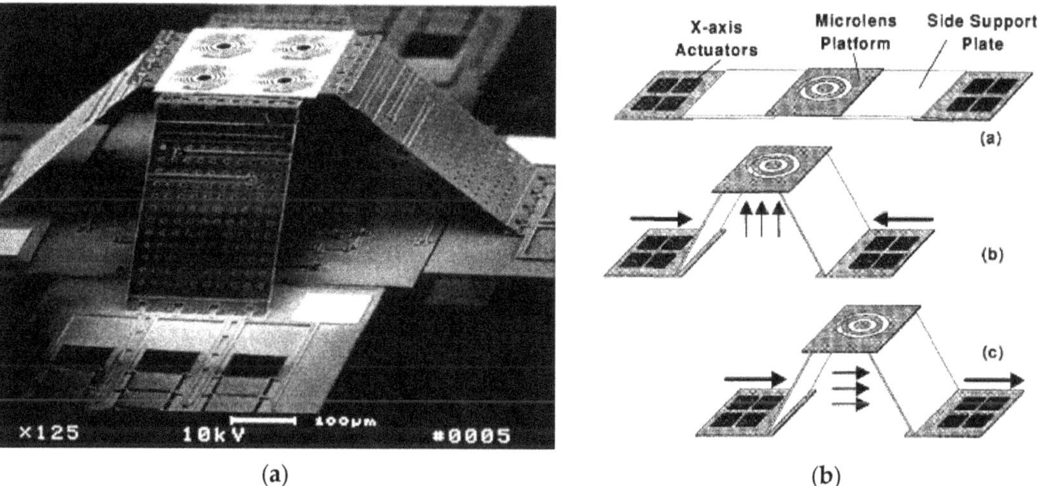

Figure 4. Surface micromachined xyz stage, where the movement in the 3 axes is a result of a cooperative action of 4 actuator plates, each holding a group of electrostatic scratch drive actuators (SDA), in combination with a special hinge design, by which the central movable xyz stage is connected to the actuating plates [25]: (**a**) a SEM image of the fabricated device in the "up position"; (**b**) schematics showing the initial position of the central stage and two opposing actuator plates (top), and the cooperative actions by the actuator plates to achieve the movements indicated by the groups of three arrows (middle and bottom) (Reproduced with permission from IEEE. All rights reserved).

In spite of the title, the "hybrid electrostatic microactuator" presented in [26] is a pure electrostatic system combining vertical comb driving (VCD) and parallel-plate driving (PPD) to realize xyz-movements. The VCD and PPD work together to pull the moving parts of the actuator downwards from an elevated position, which is a result of internal stress in the used structural layer (Ni). The main advantage of the approach presented by Hailu et al. in [26] is that the design can be realized with only one structural layer for both, the VCD and PPD. Even though the VCD and PPD are driven by the same voltage, we count the approach among medium cooperative architectures since the movements of the cooperating actuators are very different, and the system function is relying also on a suitable mechanical stress in the structural layer and special, so-called curve-up beams, which are deformed by the VCD and PPD. For that, the VCD and PPD movements are coupled to the curve-up beams through appropriately designed serpentine springs.

Another approach for realizing a xyz-stage was proposed by Ando [27], where a xyz-movement is obtained by three stages, which are all driven by independent comb-drive actuators and connected by suspensions, some of which are purposefully designed as tilted leaf springs. The out-of-plane movement is a result of the tilt angle of the leaf springs relative to the substrate. To achieve a certain position in the working space, a cooperative action of the three stages (i.e., certain combination of the corresponding electrostatic forces) is required, where the proper relation between the forces depends on the tilt angle of the leaf springs. However, even though most parts of the xyz-stage were fabricated by Silicon-On-Insulator (SOI) technology and surface micromachining, the most critical and important structure for the out-of-plane movement, the tilted leaf springs, was fabricated by focus ion beam scanning electron microscopy (FIB-SEM), which is not suitable for upscaling the fabrication. The stroke is also very limited (1.0 µm for x, 0.13 µm for y, and 0.4 µm for z at 100 V). Additionally, due to the strong coupling of the three movements, an independent control of the x, y and z positions is not possible.

Therefore, to achieve arbitrarily controllable xyz-movements, mechanically less coupled stages can be advantageous. Such a design was presented by Liu et al. [28], where two opposing groups of coupled comb-drive actuators provide the xy-movement and a parallel-plate actuator causes the independent vertical actuation, which is decoupled by long tethering beams. By this straightforward design, movements of about ±12.5 µm at an applied actuation voltage of 30 V in the x and y directions and 3.5 µm at 14.8 V in the z direction were achieved, with a cross-axis coupling of less than 0.3 µm.

Another approach with a direct actuation for the z-movement by a dedicated comb-drive actuator in a 3-DoF micromanipulator with gripping function was presented in [29]. The structures were etched by DRIE in a high-aspect-ratio micromachining (HARM) process, where two DRIE steps were employed to etch trenches of two different depths, especially needed for the z-movement (staggered electrodes to provide electrostatic attraction in the z-direction). The stages are connected by two sets of linear springs allowing movements in the x and y directions and a torsional spring for the z-movement. However, the strokes of the micromanipulator in the xy-plane were only around 1 µm (at 20 V), while the out-of-plane displacement was characterized by an angle of 0.043° at 50 V. The total displacement of the gripper was 5.93 µm at 50 V.

The design of a decoupled dual-axis actuation MEMS microgripper is presented in [30]. By allowing a controlled and independent xy-movement of a so-called actuation arm relative to a sensing arm, which is equipped with electrostatic sensing combs, a gripping mode and a sensing mode are obtained. Therefore, we assign this design to a multi-DoF system. Here, two perpendicularly aligned planar comb-drive actuators were suspended by special folded leaf flexures (FLF). This design allowed a decoupled movement of the microgripper arm towards its (sensing) counterpart from a lateral movement of that arm to conduct defined shear tests of the gripped objects. However, only FEM simulations were presented to prove the decoupling effect of the FLFs in the x and y directions. The simulation showed that an applied input force of 100 µN on the gripping arm in the X-axis (the gripping direction) results in 15.05 µm and 0.1 µm deflections of the arm in the x and y axes, respectively (i.e., around 0.6% of cross movement to the Y-axis); on the other hand, when the same force is applied in the Y-axis, it results in 0.04 µm and 14.65 µm deflections in the x and y axes, respectively (i.e., around 0.2% of cross movement to the X-axis). The Multi-User MEMS Process based on SOI (SOIMUMPs) was suggested for fabrication.

However, a strong coupling between comb drives operating in the x and y directions can also be utilized to provide rotational as well as linear movements. By the cooperative action of two perpendicularly aligned comb drives, which are connected by linking arms, Muthuswamy et al. obtained a rotational motion by appropriate driving (cooperation) of the two comb drives. The rotational motion was then transferred to a linear translation by a rotating gear [31], e.g., with a radius of 36 µm of the rotational motion, a linear translation of 227 µm for every cycle of activation of the x and y drive actuators was reported.

The electrostatic actuation in the presented examples so far is based on attractive electrostatic forces. However, with a suitable design of cooperation, it is also possible to create repulsive electrostatic forces, as demonstrated first in [32], and later utilized by He et al. [33]. The uplifting movement of a central plate in the work of He et al. is provided by a rotational out-of-plane displacement of moving electrodes, as a result of repulsive, upward directed electrostatic forces and the counterforces of anchoring springs. The repulsive forces are created by a special electrode design with four, so-called repulsive-force rotation driving units that cooperatively actuate the central plate from four directions. Each driving unit consists of a moving comb electrode that is placed above an aligned, fixed comb electrode and both are powered with the same potential (V2). Additionally, both of them are interleaved with another unaligned, fixed electrode that is powered with a different potential (V1) (see Figure 5). Due to the resulting equipotential field lines surrounding these electrodes (Figure 5c) a repulsive force acting on the movable electrode is obtained. He et al. achieved a static out-of-plane movement of 86 µm at 200 V by the design principle shown in Figure 5a. By combining the upward movement with a 2D tilt movement of the central plate of +1.5°, points on the surface of that plate can define xyz trajectories, e.g., to realize a vector display. Devices were realized by the surface micromachining foundry process: the three-layer Polysilicon Multi-User MEMS Process (PolyMUMPs).

Besides direct levitation, another advantage of repulsive forces compared to attractive forces is that the actuation is not limited by pull-in. This was utilized by Schaler et al. to realize a 2-DoF micromirror with large tilt angle, which was fabricated in a printed-circuit-board-like process by stacking multiple (4–8) thin-film actuator layers of stainless steel foils that were attached to Kapton foils [34]. With eight layers and at 2000 V, a displacement of around 1.5 mm was reported. Details of the design are discussed in [35]. It is worth noting that this device and the PIPE actuator system mentioned in Section 3.1 [18] are the only devices in this review not to be realized in Si- or SOI-based microtechnology.

3.2.2. Bistable Cooperative Systems

Bistability is an important property to provide reliable and well-defined motion between two stable (secured) positions for an actuator. Examples for which bistability is an essential feature are switching and valve systems. As a result of bistability, bistable actuators do not require a voltage input to maintain either of the two stable positions. Bistability can be actively introduced in microplates or arched microbeams by electrostatically induced deformation. The basic design for such a microbeam or a microplate is shown in Figure 6. Bistability criteria for different designs of microplates were derived in [36,37]. In a recent paper, the effect of in-plane internal stresses on the bistability criteria of electrostatically pressurized, clamped microplates is investigated theoretically using reduced order model and FEM simulation [38]. Bistability occurs in a specific range of transverse pressure created in the beam or plate.

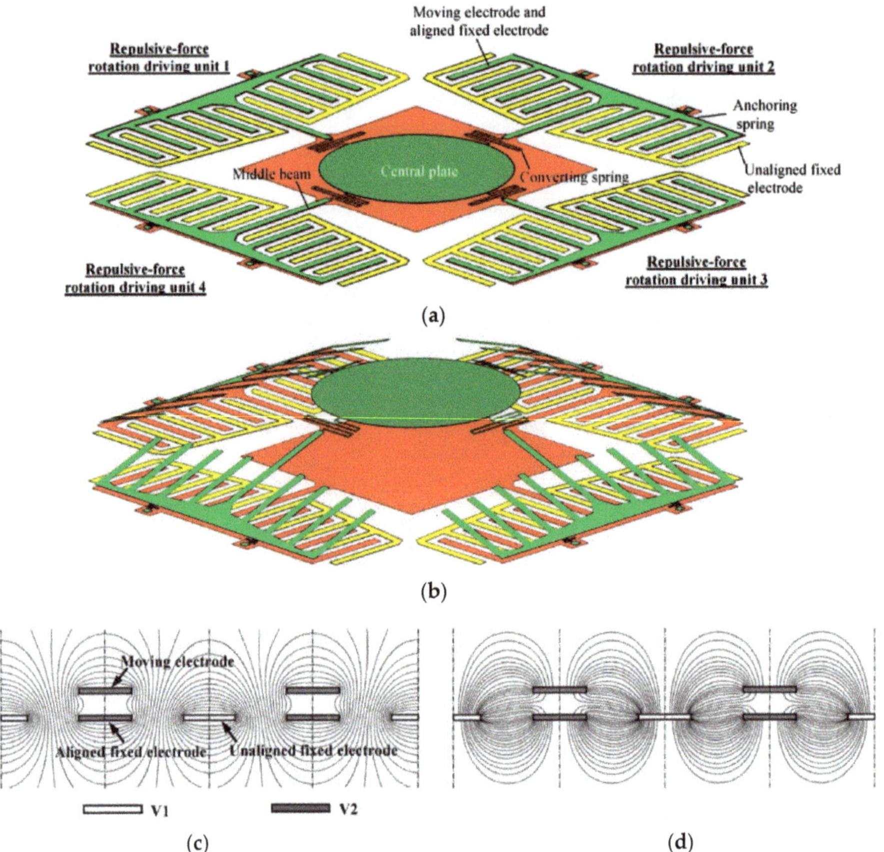

Figure 5. Out-of-plane actuation using repulsive electrostatic forces. Four, so-called repulsive-force rotation driving units are cooperating to lift a central plate: (**a**) without applied voltage; (**b**) with the same voltage applied at the four driving units. The repulsive forces on the moving electrodes are provided by an electrode design, where the moving electrodes are powered with the same potential V2 as opposing aligned, fixed electrodes, while this stack of electrodes is interleaved with a third set of unaligned, fixed electrodes that has a different potential V1; (**c**) equipotential field lines; (**d**) corresponding flux lines [33] (© IOP Publishing. Reproduced with permission. All rights reserved).

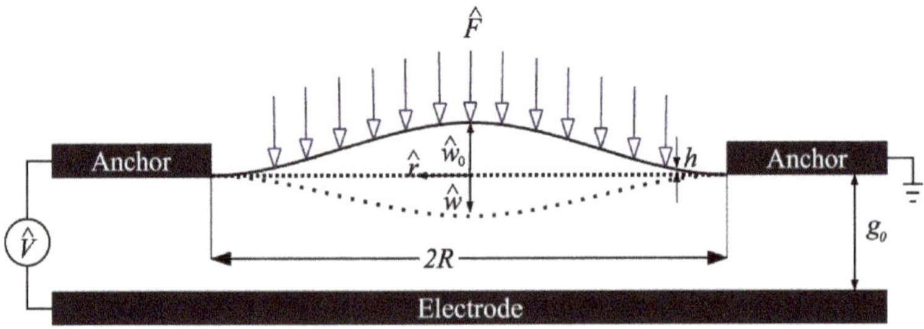

Figure 6. An arched microbeam is driven electrostatically to a 2nd stable position, adapted from [39] (© IOP Publishing. Reproduced with permission. All rights reserved).

However, compared to the theoretical work on bistability in electrostatic actuation systems, only a few realizations have been published. An example of a bistable system using cooperative interaction of two electrostatically activated membranes was presented by Wagner et al. for the realization of a microvalve [40]. Two silicon membranes, which are buckling up due to an intrinsic compressive stress, are defined over two curved cavities. The curved surfaces of the cavities define two separate electrodes, with which the membranes can be pulled down onto the cavities' surfaces. The two cavities are pneumatically coupled by a channel. Therefore, when one membrane is actuated and moves towards its cavity's surface, air is pressed out of this cavity and pressurizes the other cavity, thus pushing up the other membrane, which then closes the inlet valve opening. By the electrostatic activation of this membrane and deactivation of the previously activated one, the valve is opened again. The use of two pneumatically coupled membranes allows the closing and opening of the valve (bidirectional movement) without a need for applying a voltage across the fluid in the microfluidic system.

An example of a bistable microbeam system, where the bistability was obtained by an applied mechanical clamping force, was presented in [41]. Here, employing pre-stressed beams (Euler beams) made the mechanical spring system (folded beam structure) bistable. This bistable structure behaves like a toggle lever, where two opposing electrostatic comb-drive actuators provide the forces to actuate the structure between its bistable positions via the central toggle-point. In this case, two opposing comb drives are activating the toggle lever by driving the so-called toggle point over its instable, central position. Figure 7a shows schematically the mechanically loaded, and thus bistable folded beam structure, while Figure 7b shows a snapshot during the operation of the device, which is realized in SOI-technology. The forces and the stroke are defined by the beam design and the external mechanical load. According to simulation and experimental data, forces of about 50 μN at the two bistable positions and a total stroke of 200 μm between these positions are possible with the shown layout. By combining the bistable toggle lever and the electrostatic actuation, the needed stroke of the two opposing electrostatic comb drives is only slightly larger than half of the distance between the two bistable end positions (i.e., about 100 μm).

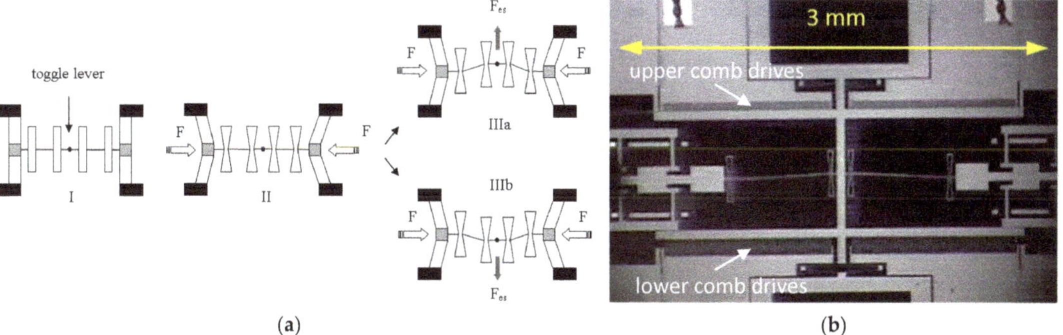

Figure 7. A bistable actuator system, where the bistability is introduced by an external mechanical load acting on a folded beam structure, and the switching between the two bistable positions is performed by electrostatic comb-drive actuation: (**a**) schematic illustration of the so-called toggle lever principle, where the bistability is introduced by a mechanical force F. By two opposing comb drives, electrostatic forces F_{es} are obtained in the "upwards" or "downwards" directions [41]; (**b**) the fabricated bistable switch while in the upper position, which shows the upper electrostatic combs are maximally overlapped, whereas the lower electrostatic combs are almost fully pulled away from one another (this is a snapshot from a video by Markus Freudenreich).

A similar approach for realizing a bistable operation was proposed by Kwon et al. [42], where a bidirectional actuation by means of a single comb drive of a particular electrode design was achieved. In this design, both the fixed and moving comb electrodes had a spade-like form, i.e., the tip part of a comb's finger is broader than the rest of it. As discussed in the introduction, electrostatic forces induce motions that always aim to increase the capacitance between the electrodes. Therefore, depending on the actual position of the broader part of the spade-like, movable electrode relative to that of the fixed counter electrode an "inward" or "outward" directed movement of the movable comb electrode is electrostatically induced to switch the movable electrode to either of the two stable positions defined by a set of chevron-type springs. The inward and outward directed forces are transferred to a shuttle, which is suspended by this set of chevron-type bistable springs. By applying a sufficiently high voltage pulse, either an inward or an outward force is created to overcome the restoring forces of the suspension setup, such that the moveable comb switches from one stable position to the other. Thus, the successful bistable operation is a result of a cooperation between the purposefully designed spade comb drive and chevron-type springs. The bistable microactuator was experimentally tested, where a pulse input of 55 V successfully drove the system between two stable positions with a stroke of around 65 μm and a duty time of 0.32 ms. As such, its performance is similar to that of the bistable switch described in [41]. This device was also fabricated using SOI-technology. Hence, the height of the spade electrodes was defined by a device layer thickness of 80 μm.

3.2.3. Cooperation through Activated Joints Allowing Stepping Mode

Another architecture of cooperation is based on joints, which can themselves be activated by an actuation, e.g., to actively couple different actuators or actuation actions, to obtain a stepping-like movement by a cyclic activation of the joints, combined with "propulsion" actuations within each cycle. Therefore, these actuation systems enable the realization of even more complex cooperative systems, which will be described in Section 3.3.

An early approach towards a stepping motor employing this concept came in the form of the so-called SDA, which was first investigated by Akiyama et al. [10,43]. In this section, we focus on the basic structure of a SDA, which consists of a polysilicon plate with a "scratching" bushing at one of its ends (Figure 8a).

The SDA structure is defined by surface micromachining on a Si-substrate with an insulating film as a passivation coating as well as a suitable "gliding/scratching" surface. The cooperative action that produces a stepping-like movement is provided by different functional elements within the SDA, combined with an appropriate sequence of applied electrical potentials, as shown in Figure 8a. By applying a potential between the plate and the substrate, the deformable plate bends towards the substrate in the first phase of the increasing potential by pull-in, in such a way that the free end of the plate is pulled down onto the surface. Thus, the contact area of the previously free end provides a mechanical grip to the substrate's surface. At the same time, friction forces between the bushing and the substrate's surface prevent a movement of the bushing tip, as a result, the bushing creates a warp in the plate and an elastic strain energy is temporarily stored in the deformed plate, in the so-called priming phase. By further increasing the driving voltage, both the contact area between the plate and the insulated substrate (directly proportional to the "clamping" force of the plate to the surface) and the elastic strain in the rest of the plate increase. Finally, at a certain stepping voltage the bushing is pushed forward (scratches) by the built-up elastic energy, while the previously free, left end of the plate remains still because it is "electrostatically clamped". When turning off the potential, the left end of the plate is de-clamped and the stored elastic energy is released (at least so far as no stiction is occurring). As a result, the now again-free left end of the plate is pulled towards the bushing. Thus, the SDA moves in a step-like movement over the surface. In [43], the SDA is suspended with torsional beams to a frame-like anchored structure (not shown in Figure 8a) in order to transfer the step of the SDA structure to the higher-level system. Step sizes of 10–80 nm depending on the bushing height (1–2 μm) and applied peak voltage (30–200 V) were reported [10].

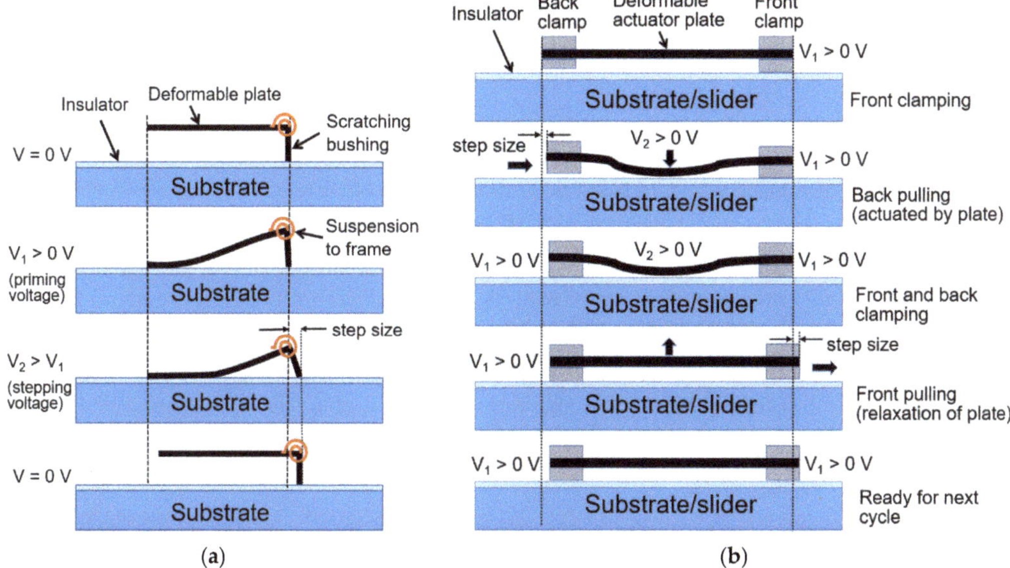

Figure 8. Basic concepts for electrostatically driven stepper motors: (**a**) Cross-sectional schematic of a scratch drive actuator (SDA) (adapted from [44]). The SDA consists of an electrostatically deformable plate and a bushing, and is suspended typically by torsional beams to an anchored frame structure, to which the movement of the SDA is transferred. The frame structure is not shown. (**b**) Cooperative actuation scheme of a shuffle motor showing the driving sequence of the back and front clamps and the deformable actuator plate (adapted from [45]).

In 1997, Akiyama et al. [46] integrated a single SDA into a more complex mechanical system. By connecting the SDA via a link frame to a buckling beam, the movement of the SDA is transmitted to the beam structure, which then changes the deformation of a pre-buckled beam (out-of-plane movement) and thus can be used for the actuation of macrosystems. By the link frame, the steps of the SDA were successfully transferred to a buckling beam, where the relation of SDA displacement and deflection of the buckled beam is depending on the beam geometry and can be therefore adjusted (e.g., different deformation, gear amplification and force transmission are possible). The authors reported a 10 µN force and 100 µm displacement in the z-direction. This is therefore an example of cooperation between one simple actuator (SDA) and a suitable mechanical structure to provide new functions, such as the capability of achieving defined xyz-movements of originally planar actuation architectures or deformation or force amplification.

Li et al. [47] estimated the forces generated by spring-loaded SDAs from the maximum achieved spring deflection. The stiffness coefficient of the employed box springs was determined by FEM analysis. With an applied voltage of 200 V, forces of 250 µN and 850 µN were estimated for one-plate and four-plate SDAs, respectively, thus proving the concept of a force amplification by cooperative action of four SDAs. Additionally, modeling approaches and theoretical grounding attempts concerning the performance characteristics, dynamics and force estimation of SDA implementations can be found in [48–50].

Linderman et al. [44] optimized the driving signal as well as the geometry of SDAs by analytical modeling. Moreover, they implemented SDA actuators in large arrays to form a rotary motor driven by nine SDAs, and a tethered robot made up of 188 SDAs that are driven in parallel, thereby proposing a more advanced cooperative architecture. The robot was capable of pushing a $(2 \times 2 \times 0.5)$ mm³ silicon chip over a distance of 8 mm. Additionally, more robots could be assembled on a ceramic stage to achieve multiple degrees of freedom and bidirectionality. Thereby demonstrating that by integrating a large number of cooperative SDAs, the system performance can be improved considerably. In addition to the xyz-stage introduced in Section 3.2.1 (Figure 4) [25], other applications, in which SDAs were implemented can be found in [51–53].

Even though the basic structure of a SDA and the cooperation therein looks simple, the specific performance is depending on many parameters, which cannot be controlled directly by driving conditions or geometrical design parameters, e.g., the friction between the SDA and the substrate surface is very sensitive to surface roughness and humidity.

These limitations are overcome with another design of electrostatically activated stepping mode, the so-called shuffle motor, which was first presented by the MESA Research Institute, University of Twente. The schematic actuation cycle is shown in Figure 8b. Here, two poly-Si layers were used for an electrostatically driven stepper motor: The first thin layer defines the actuator plate, and the thicker one defines the frame, to which the plate is suspended at two ends. The two ends can be clamped to the substrate (electrostatically activated friction) individually by electrostatic forces between the clamps and the surface. In between the clamping activations, the deformable plate is also electrostatically deformed, creating a strain in the plate, which then pulls the not clamped end towards the clamped one [45,54]. Different from the SDA, the concept provides balanced and well-defined clamping conditions at both ends of the deformed plate. Therefore, the shuffle principle allows the design of more complex inchworm-like motors, which will be discussed in Section 3.3.2.

Another type of stepping mode is obtained by means of an oblique mechanical impact that is transferred from an electrostatically driven resonant structure working as a so-called converter to a circular microrotor, as presented in [55], where two different designs were described, the so-called bi-modal and mono-modal angular vibromotors. Due to the oblique design, by vibration of the converter beams, the rotor, as a passive element, is actuated by tangential friction in combination with a normal impact transfer. In the bi-modal mode, the converter beam can be operated at two primary modes of excitation, which are produced by two separate comb drives that are connected by a bi-modal flexure with the converter beam. Through a pointer tip at the end of the converter beam, the linear oscillation of the converter is transferred to a rotational movement of the microrotor. As the bi-modal converter can be operated at two different primary modes of vibration, it is possible to actively choose the contact sides at which the conversion takes place, and thus the direction of rotation of the rotor. Furthermore, the rotor turning speed can be tuned by adjusting the phase between the driving primary modes, the voltage applied to the combs or the frequency. Thus, a cooperative action of the comb drives and the activated friction is utilized. In the mono-modal vibromotor, two opposing converters are cooperating to actuate the rotor. In this way, the cooperative action is similar to the linear drive shown in Figure 2.

A similar approach of oblique impact transfer was used by Daneman et al. to realize a linear microvibromotor [56]. Here, the impact of four comb resonators was transferred to a slider, which was placed between the impact arms at an angle of 45° to each of them. Two opposing impact arms were used for each direction, thereby accomplishing a bidirectional drive of the slider. A travel range of more than 350 µm was achieved.

3.3. Advanced (Strong) Cooperative Architectures

In this section, cooperative actuator systems consisting of usually many actuators, which have individual excitations for each of them and strong interactions among them, e.g., those featuring hand-over actions, will be reviewed. Inchworm motors are an evident example of such systems; thus, a particular review of their development over the years is provided. This section will be concluded with an overview of recent systems that exhibit even more elaborate cooperative systems, mainly those integrating many inchworm motors for robotic applications.

3.3.1. Microtransportation Systems

The research in the fields of micromanipulation and microtransportation has increased considerably in recent years, partly to cope with the needs of other emerging miniaturization technologies, such as in a micro-assembly application or a micro total analysis system (μ-TAS). Consequently, as mentioned previously, microgrippers [30] and micromanipulators [29] have been proposed. Another noteworthy contribution to this field is the micro transportation system (MTS) by Pham et al. [57]. Their micro conveyor system is an exhibit for cooperative actuation, which comprises many linear and rotational comb-drive actuators that use the ratchet mechanism to drive so-called microcars in straight and curved paths. The system, which consists of several types of modules: a straight module, a turning module, and a separation module (an actuated T-junction), is fabricated using SOI-technology with a device layer of 30 μm and a buried oxide layer of 4 μm. A construction similar to these modules (but of a different actuator design) is shown in Figure 9a. The comb-drive actuators and microcars have sidewalls in the shape of saw teeth, which make up the ratchet mechanism. The comb electrodes are driven with DC-biased sinusoidal signals with $V_{DC} = \frac{1}{2} V_{PP}$, which results in a reciprocating motion of the ratchet racks along the direction of the channel they form. Moreover, a microcar has sliding joints and serpentine springs that allow the microcar's width to vary while being propelled forward. The racks opposite to one another in a channel move with 180° phase shift. As a result, one ratchet rack on one side pushes the microcar forward, while the rack on the other side slides over. If the applied driving voltage is larger than a certain critical value, in the case of the straight module, both ratchet racks will move the car during the actuation cycle, albeit one ratchet actuator moves it by the electrostatic force, while the opposite one uses the restoring elastic force of its deflected beams. The curved-path modules have only one actuated side. Results of successful runs of the straight movement were reported, where a voltage (V_{PP}) of 150 V was applied with frequencies ranging between 5–40 Hz, which corresponded somewhat linearly to microcar velocities ranging between 54–512 μm/s.

Later the same group proposed another design for the MTS, in which the same types of transportation modules were included (Figure 9a); however, in this design, the actuators' ratchet racks had an inward reciprocating movement that "squeezes" the microcar (also referred to as the 'micro container') [58]. Moreover, the microcar exhibited a more elaborate design, which instead of the ratchet racks it had in the previous design, was now fitted with two pairs of 'anti-reverse wings' and 'driving wings' (Figure 9b). The latter pivoted off a spring-supported rotational joint, which when squeezed, rotate backward to generate a reaction force to push the microcar forward, at the same time, the anti-reverse wings serve as a ratchet mechanism and prevent the backward movement of the microcar. A variable travelling velocity of up to 1 mm/s of the microcar was reported, and within a certain applied voltage window (140–160 V) smooth and stable movement through all three types of modules was demonstrated. The system was also fabricated using SOI-technology, with one photolithography mask for a DRIE etch, followed by vapor-phase hydrofluoric acid release. A close-up view of the fabricated microcar can be seen in Figure 9c. Here, a sophisticated cooperation of many actuators and elaborately designed structures, including the microcars, was demonstrated.

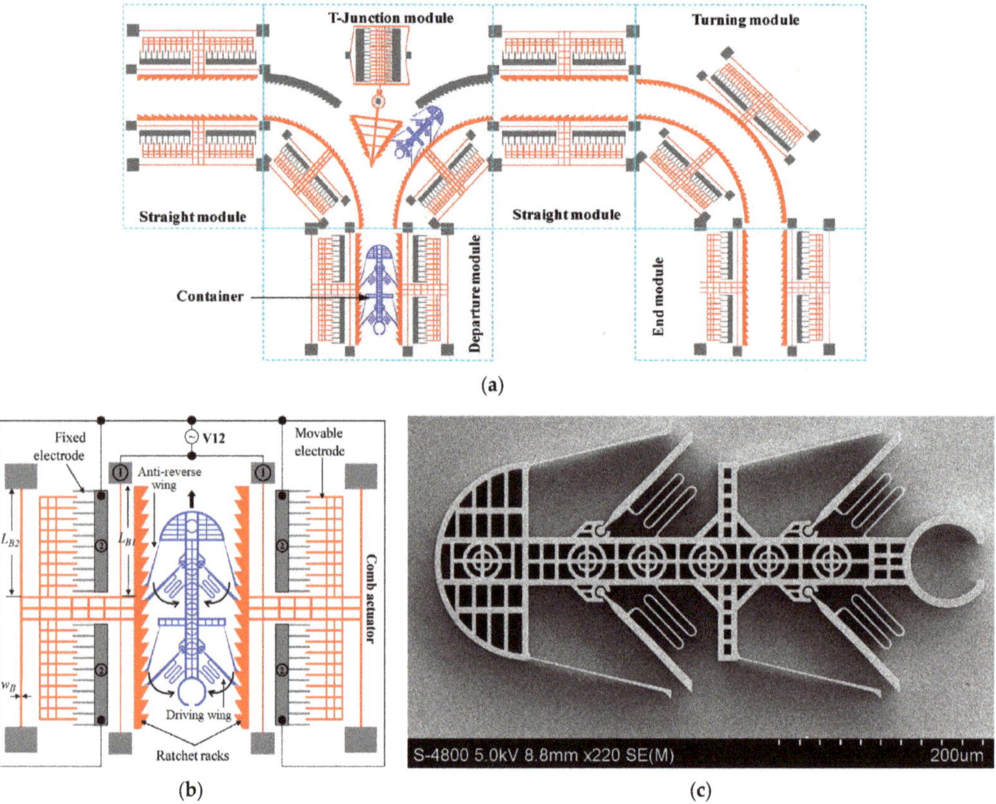

Figure 9. A micro transportation system (MTS): (**a**) a schematic overview of the MTS with all the different types of modules it contains; (**b**) a schematic view of a micro container inside a straight module, which also shows the driving circuit connections; (**c**) a SEM close-up of a fabricated microcar [58] (Reproduced with permission from IEEE. All rights reserved).

3.3.2. Inchworm Motors

The inchworm motor is another cooperative mechanism, the history of which goes back to the 1960s, as the survey by Galante et al. showed [59]. This motor system integrates several actuators to generate large displacements by adding up small steps through a scheme of repeated clutch-drive actuation. The main motivation behind the early propositions of these motors, which typically made use of piezoelectric actuators, was producing high-precision positioning devices, e.g., for precise machining or in scanning electron microscopy. Recently, this linear actuator mechanism has established a strong foothold in the electrostatic realm after increasingly being the subject of research from the 1990s to this day. In this section, many of the contributions made over these years will be viewed, generally in a chronological order; however, as some of the inchworm motors belonged to certain sub-groups, e.g., the shuffle motor (introduced in Section 3.2.3), once introduced, the development of such a group will be viewed as a whole. The electrostatic inchworm motors can be classified, similarly as per the survey by Galante et al., according to their setup into a "walker" type, or a "pusher" type. The former refers to a scheme in which the propulsion actuator is embedded in and moves along with the shaft, whereas the latter indicates that the shaft is driven by a separate, stationary propulsion actuator. From a technological perspective, inchworm motors can also be classified into bulk micromachined/crystalline silicon (c-Si), surface micromachined/poly-Si and SOI/c-Si.

In 1995, several different electrostatic inchworm motors were introduced [60–62]. The one proposed in the pioneering work of Lee and Esashi [60] featured a thick (~200 μm), suspended rotor having the shape of a ladder with tightly spaced steps, which is fabricated by bulk micromachining technology. The rotor is sandwiched with a separation of few micrometers between two fixed plates, made of pyrex glass, which are equipped with strips of thin-film conductors that define the stator electrodes. The electrodes have a certain grouping and a phase shift configuration and use the alignment force component to drive the rotor in one of two directions, hence it is a pusher-type mechanism. Here, the motor had an inchworm-motion type of actuation based on four phase voltages with a step distance of 1.5 μm. The authors claimed to have achieved a force of few mN and a speed of 13 cm/min with 100 V driving voltage at a frequency of 1.4 kHz.

Yeh et al. [61] proposed another stepper motor of the pusher type to drive articulated 3D polysilicon robotic limbs made by surface micromachining. To generate large forces, the motor was designed with comb-drive, gap-closing actuators (GCAs). In one version, the motor had four comb drives, two situated on each side of a shuttle. Every comb drive is fitted with a shoe-like structure that could clamp to the shuttle electrostatically (Figure 10). The two comb drives that are opposite to one another cooperate to propel the shuttle by a specific clamp-drive sequence in one direction. Thus, bidirectional actuation is achieved by having another pair dedicated to the opposite direction (not shown in Figure 10). A comb drive is able to pull the shuttle a distance of 2 μm, which is the gap between the comb electrodes. Figure 10 shows a schematic of the sequence of operation for one half of an inchworm cycle, which is a typical sequence for such a pusher-type inchworm motor. It was estimated that the force generated by the motor was about 6.5 μN while being driven at 35 V. Additionally, the shuttle was moved by a distance of 40 μm by 10 full inchworm cycles before an anchoring spring was broken.

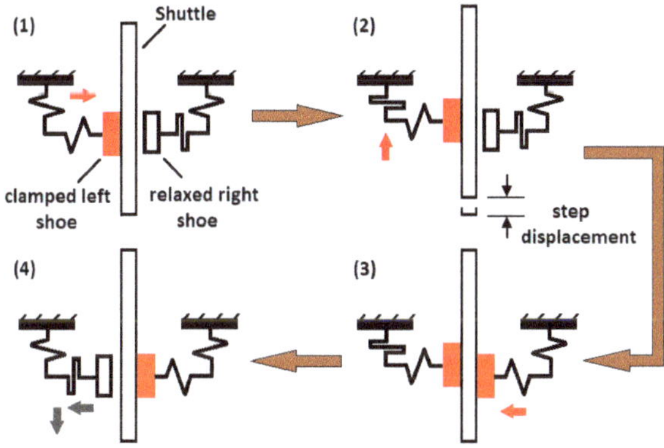

Figure 10. Pusher-type inchworm motor cycle: (**1**) Left shoe clamps to shuttle. (**2**) Left shoe pulls clamped shuttle by a step displacement (the gap distance in this case, 2 μm). (**3**) Right shoe clamps to shuttle, to hold it. (**4**) Left shoe detaches from shuttle to return to its original, idle position. The second half of the cycle then continues with reversed roles of the shoes. The comb-drive, gap-closing actuators that pull the shuttle and a 2nd pair of shoes for the opposite movement direction are no shown. Adapted from [63].

As briefly discussed in Section 3.2.3, the shuffle motor principle was firstly proposed, also in 1995, by a group from the MESA research institute [62], which is another type of polysilicon surface micromachined linear motor that could be considered as a subclass of inchworm motors. The concept of operation shown in Figure 8b indicates that it is of the walker type. After an improved design of the motor, fitted with anti-sticking bumps to overcome the initial sticking problems, and supported with a comprehensive model description, a successful implementation of the motor was realized using three polysilicon layers [45]. The motor provided a high resolution displacement with a step size of about 85 nm and a maximum reach of 43 µm. A force of 43 ± 13 µN was achieved with 25 V and 40 V of driving voltages for the bending plate and clamps, respectively. It is worth noting that the clamps had to be driven by AC voltage to avoid sticking, which in turn increased the slipping and reduced the effective generated force. A displacement speed of 100 µm/s with a driving cycle frequency of 1160 Hz was attained. The attractive features of a shuffle motor, such as inherent bidirectionality, very high precision positioning capability and theoretically unconstrained movement, have garnered a great deal of attention and investigation [64–66]. Furthermore, Sarajlic et al. [67], using trench isolation technology, had further developed the shuffle motor into a 2-DoF planar motor with drastically improved performance characteristics, such as a larger force (0.64 mN) at moderate operating voltages (up to 45 V), an adjustable nanometer-resolution (step size of 41–63 nm), and a cycling frequency of up to 80 kHz.

In 2006, Sarajlic et al. [68] also proposed the "contraction beams micromotor", which is fabricated using surface micromachining and trench isolation technologies as well, and has a built-in mechanical transformation that is similar to the shuffle motor. However, rather than contracting a plate to the substrate, an array of pairs of electrically isolated beams are contracted to one another, such that the pulled-in contraction is transformed in the same plane, contracting the clamping electrodes by a much smaller distance (a step of 10 nm) with a measured large force (0.49 mN) at an actuation voltage of 60 V. This mechanism reduces some drawbacks resulting from contracting the shuffle motor's plate to the substrate, namely, the unwanted induced frictional force at the inactive clamp electrode, and the compromising moment that reduces the clamping force of the activated one.

In 1997, Tas et al. [69] proposed an inchworm motor fabricated with a single-mask surface micromachining process. However, this motor is of the pusher type, and it employs an actuator mechanism similar to Yeh's [61], where a pair of shoe-like structures pull a shuttle sandwiched between them. In this design, two shoes are driven by two cooperative driving units, each consisting of two comb-drive GCAs that have separate actuation beams, which make a right-angle joint at the shoe; one beam clamps the shoe to the shuttle and the other pulls it. The sequence of operation is similar to the one shown in Figure 10, except that the clamping force is frictional rather than electrostatic. The 5 µm high polysilicon structure generated an estimated force of 3 µN in a maximum step size of 2 µm and a total displacement of 15 µm at a driving voltage of 40 V with one cycle of operation per second. However, the asymmetric, parallel spring suspension of the shuttle, combined with a single-sided clamping during propulsion was a limiting factor of the deflection and resulted in a reduced clamping force and an increased slipping as the shuttle moved forward. In the same year, Baltzer et al. [70] proposed a similar inchworm motor with twice the amount of driving units, which provided a double-sided clamping (gripping) of the shuttle during propulsion. The driving units had an in-line design, such that the clamping and driving GCAs were both integrated on a single rigid beam, which had a folded-beam suspension on one end and engaged the shuttle by the other end. Two variants of the motor were presented, one was fabricated using two polysilicon layers and featured a suspended shuttle that also managed a maximum displacement of 15 µm, and the other one had a suspension-free shuttle with tighter clearances that required three polysilicon layers, which achieved a total displacement of 110 µm, being limited only by design. The driving units for both variants were identical. Depending on the applied voltage (15–40 V), the motor could achieve a stepping motion both before and beyond the pull-in point, with the former allowing more exact positioning. The calculated force at an applied voltage of 20 V was >1 µN.

At the beginning of the 21st century, an evolution in the fabrication of such inchworm motors was taking place, marked by the many works that sought the use of crystalline silicon (c-Si) as the building material, in particular with SOI-technology, to fabricate these motors in HAR actuator form. An early example of this development is a pusher-type inchworm motor proposed by Yeh et al. [11], a group that had previous contributions of inchworm motors fabricated with surface micromachining, e.g., [61,63]. The motor had two driving units on one side of a shuttle. The unit consisted of a clutch GCA array, which when activated, brings a frame that encloses another "drive" GCA array in contact with a shuttle to drive it. At the time, their work had pushed the envelope of force generation for such motors by achieving a force of 260 µN with an applied voltage of 33 V. The higher force density was achieved primarily by the higher aspect ratio actuators (up to 25:1). One demonstrator was fabricated out of a 50 µm thick device layer, limited by the elastic restoring forces of a suspended shuttle, it managed a maximum displacement of 80 µm in 2 µm steps. Another, 15 µm thick demonstrator was driven at a frequency of 1 kHz (theoretical limit was 1.4 kHz) and achieved an average speed of 4 mm/s. To guarantee a sufficiently robust transfer of forces to the shuttle and reduce slippage as much as possible, this motor featured interdigitating teeth as a mechanical clutching interface between the driving units and shuttle. Nevertheless, when driven at frequencies higher than 1 kHz, the authors reported some slipping between the clutch and shuttle. It is worth noting that despite using bare silicon teeth for mechanical clutching, motors were operated for more than 20 million cycles, corresponding to 13 h of active operation.

Using driving elements that serve as clamp and propulsion actuators at the same time has been tried before, such as the so-called design A in [71], but unsuccessfully. In 2013, Penskiy et al. [72], proposed an inchworm motor design that merged the traditional clamping (or clutching) and driving mechanisms into one movement. Using an inclined, flexible arm, fitted with a teethed small block at its end, the unidirectional movement of a comb-drive GCA, which is perpendicular to the shuttle, is transformed to a simultaneous clutching and pushing of the shuttle. In this respect, the design is also reminiscent of the linear vibromotor [56], presented previously in Section 3.2.3. To balance the lateral forces, the shuttle is sandwiched between a pair of comb drives that is driven simultaneously, which together with another pair actuate the shuttle consecutively to cause the typical inchworm pushing sequence. Figure 11 shows a SEM image, presenting the layout of the inchworm motor, which achieved a maximum shuttle velocity of 4.8 mm/s at an actuation frequency of 1.2 kHz with a step size of 2 µm, and a force of 1.88 mN at a driving voltage of 110 V.

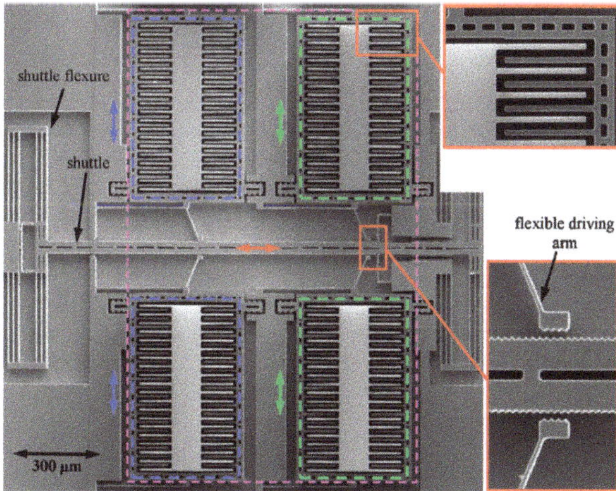

Figure 11. SEM image of the electrostatic inchworm motor that uses an inclined, flexible arm to both clutch and push a suspended shuttle simultaneously, which was proposed by Penskiy et al. [72]. The blue and green dashed lines distinguish the two pairs of GCAs. The GCAs of a single pair are driven in parallel, and the two pairs are driven consecutively to drive the shuttle to the right (© IOP Publishing. Reproduced with permission. All rights reserved).

In 2021, Narimani et al. [73] presented a highly modular, pusher-type inchworm motor based on the NED actuator concept, which has been previously introduced in

Section 3.1.3 as an example of a cascaded actuator system. However, instead of an out-of-plane cantilever deflection, here the S-shaped NED actuators have an in-plane deflection to provide the shuttle propulsion mechanism. A stack of NED actuators makes up a driving block, which can have one of two fixed deflection directions depending on the layout of the NEDs within. Hence, the system also features bidirectional actuation. The shuttle is sandwiched between pairs of NED blocks, which clamp to it electrostatically. The shuttle is suspended by four serpentine flexures. The motor system is realized by c-Si on an SOI wafer using typical HAR silicon micromachining processes, such as DRIE, in addition to atomic layer deposition (ALD) to form insulation islands between the NED electrodes. A maximum shuttle displacement of about 1 mm was achieved with a maximum step deflection of 11.8 µm, by applying 150 V and 130 V for electrostatic clamping and NED driving, respectively. It was estimated that a single NED block with 36 S-beams is capable of producing a maximum force of 0.8 mN.

The authors of this review are also working on a bidirectional, multistable inchworm motor system of the pusher-type, which is based on the cooperation of two actuator subsystems, the mobility transmission electrodes (MTE) and the actuation unit cells (AUC), as shown in Figure 12 [74]. Some details about the system's design concept can also be found in [75]. The purpose of the MTE is to provide multistability to a shaft, whereas the AUC engages the shaft to push it. The modular design of the system aims at providing scalable displacements (several mm) and force capacities (tens of mN). The shaft is composed of two sliding beams that are initially interlocking a central, stationary holder via mechanical suspension springs. The MTEs, placed at both sides of and parallel to the sliding beams, provide electrostatic forces to pull them apart and release them from the stationary holder. When the shaft is freed by the MTEs, the outer shafts' teeth become clutched by those of the AUCs, and a linear movement can be realized by a sequential actuation scheme of the AUCs, which are divided into groups. The AUC is a 2-DoF actuator that is able to both clutch with the main shaft as well as actuate it in either one of the two directions along the shaft. It is worth noting that the multistability feature is a unique capability of this inchworm motor that is rarely found in other inchworm motor realizations, which necessitates the cooperation of another distinct subsystem: the MTE. The multistability in this context is the ability of the shaft to assume multiple positions along its displacement range that are stable in the absence of power. It is of a discrete nature, since it is defined by the inner teeth of the shaft, which in turn defines the stable, interlocked positions the shaft could have with the stationary holder. Furthermore, the shaft is envisioned to be fully untethered by employing proper guiding structures, such as mechanical stoppers and a covering plate, for constraining sideways and out-of-plane deflections, respectively.

3.3.3. More Elaborately Cooperative Systems

An example of an even more sophisticated implementation of inchworm motors, which highlights the potential of such cooperative systems, is the contribution from Hollar et al. in 2003 [76]. This work had integrated essentially the same design of the inchworm motor found in [11] into a 2-DoF robotic leg, which is shown in Figure 13a. Both works emerged from the same research group. In addition to SOI technology, joints and crossover beams were fabricated by polysilicon-based micromachining. With a total displacement of 400 µm and shuttle speeds of up to 6.8 mm/s, improvement of the performance characteristics of the motor was apparent. However, more noteworthy was accomplishing the bidirectional actuation necessary for this application, which was demonstrated in two ways. The simpler approach, that is reversing the control sequence of the inchworm motor and using the elastic restoring force to drive the shuttle, which was enough for driving the robot leg, or by introducing biasing comb drives that determine the actuation direction of the drive GCA, a mechanism that provides equal forces in both directions but complicates the operation with the two additionally required signals. As a result, the latter was only demonstrated by a test structure (Figure 13b) but not implemented in the robot.

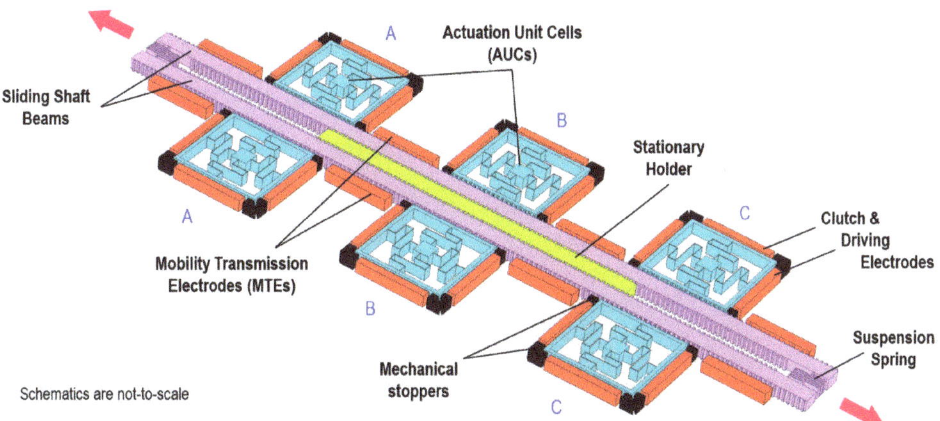

Figure 12. A 3D schematic of a bidirectional, multistable inchworm motor, which has three pairs (A, B and C) of actuation unit cells (AUC) that are distributed at both sides of two connected sliding beams that make the motor shaft. The AUC is a 2-DoF actuator that can both clutch and bidirectionally drive the main shaft. The shaft's beams are pulled together by suspension springs that interlock a central stationary holder at idle states. When activated, the mobility transmission electrodes (MTE) release the shaft from the holder and allow it to slide in either of the two indicated directions (magenta arrows) by the cooperative actuation of the pairs of AUCs.

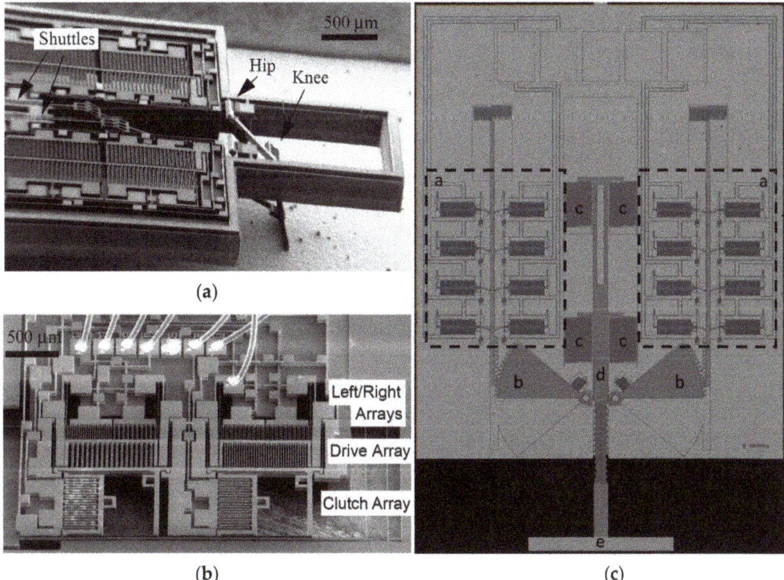

Figure 13. Examples of more elaborately cooperative systems for robotic applications that are based on inchworm motors: (**a**) a SEM image of a 2-DoF robotic leg, where each DoF (hip or knee) is driven by an inchworm motor (not fully shown). The leg can be seen folded down and touching the bottom of the die package [76]; (**b**) a SEM image for an implementation of a bidirectional inchworm motor using biasing actuators. The so-called left/right arrays are comb-drive GCAs that bias the frame of

the drive GCA one way or the other, to allow either forward or reverse drive actuation [76] (Reproduced with permission from IEEE. All rights reserved); (c) a so-called inchworm-of-inchworms motor topology implemented in a 5.0×6.4 mm^2 jumping microrobot, which primarily consists of: (a) electrostatic inchworm motors, (b) pinions with 10:1 force amplification factor, (c) energy storing serpentine springs, (d) a main shuttle, and (e) a foot. (adapted from Figure 1 in [77], reprinted with the permission of the author and the Transducer Research Foundation).

Following the novel implementation of the flexible, inclined arm to drive the inchworm motor's shuttle that was proposed by Penskiy et al. in 2012 [72], several more complex cooperative systems have used this optimized design and made it somewhat of a blueprint for more advanced inchworm motor implementations, such as a so-called 'inchworm-of-inchworms' motor topology with force amplification for a jumping microrobot by Greenspun et al. (Figure 13c) [77]; a MEMS airfoil actuator by Kilberg et al. [78]; another jumping microrobot by Schindler et al. [79]; an aerodynamic control of miniature rockets by Rauf et al. [80]; a microgripper with 15 mN of force and 1 mm of displacement by Schindler et al. [81]; and a remarkable, intricate realization of a cooperative electrostatic actuation in the form of a multichip terrestrial robot developed by Contreras [82]. In the latter, each leg of the six-legged robot had a 2-DoF planar silicon linkage, featuring rotary joints, that is operated by two separate inchworm motors. The construction of the robot legs is similar to the one found in [83]. Therefore, the multichip robot has 12 motors, 6 linkages and 12 degrees of freedom in total. The robots walking was demonstrated after externally supplying it with power and control signals via connected wires [82].

In Appendix A, a table listing some of the electrostatic linear motors reviewed in this paper, with comparison of their operational and performance characteristics as well as the technology used for their fabrication, is given.

3.4. Hybrid System Architectures for Cooperative Actuation Systems

The cooperative microactuator systems presented so far are driven purely electrostatically and the three subsections were addressing the different level of cooperation of these systems. As a last type of electrostatic cooperative actuation systems, hybrid system architectures that combine electrostatic actuation with other types of actuation principles are reviewed in this section. The examples cover different levels of cooperation, including even inchworm implementations. This Section is also a kind of link to other review papers of the Special Issue on "Cooperative Microactuator Devices and Systems", which describe cooperative microactuators using purely other actuation principles. In this section, we will begin by introducing the motivation for and benefits of such hybrid cooperative actuation systems. The specific examples present architectures in which electrostatic actuators are cooperating with thermal (electrothermal, allotropic phase transformation and thermal pneumatic), piezoelectric and electromagnetic actuators.

One purpose for such hybrid cooperative systems is to introduce a bidirectional mode, where for one direction, the common electrostatic actuation is used, while the movement in the opposite direction is triggered or performed by a separate principle. Another motivation for hybrid cooperation actuator systems is to bring bistability into the system, which is in many cases a precondition for a reliable function of such systems. Additionally, such hybrid cooperative systems can overcome the inherent constraints of pure electrostatic actuators, which are in most architectures, as previously discussed, the limited stroke and force. Furthermore, hybrid systems are also used for closing the gap between micromachined, electrostatic MEMS actuators and the "macro-world" tolerances, e.g., those of the standard machining techniques.

In many hybrid actuator systems, the additional non-electrostatic actuation principle is just supporting the action of the primary electrostatic actuator, especially by bringing the electrodes close together, such that the electrostatic actuation can take over or be used for a hold function with very little power consumption. Early examples were shown for radio frequency (RF) MEMS switches based on a variable capacity, where additional actuation principles work in series with the electrostatic actuation to allow a lower driving voltage for electrostatic switching, by reducing the gap between the movable and fixed electrodes with the extra actuator [84]. In this way, hybrid systems help to overcome the limitations due to the scaling law discussed in chapter 1, which requires even with very sophisticated, deformable electrodes, such as those shown in Figure 3a, at least at some locations, electrode distances in the range of μm and below. With the support of the additional non-electrostatic actuators, the typical problems of pure electrostatic actuation, such as stiction upon pull-in and dielectric charging can be overcome, while keeping its advantage of low power consumption, e.g., for high frequency switching.

For example, in [84], an additional thermal actuator was integrated to support the primary electrostatic actuation. This approach was analyzed further with suitable models in [85], where the concept of a bidirectional actuator that combines electrostatic actuation with thermal actuation was presented as a so-called electrothermomechanical (ETM) actuation. By an appropriate beam design, it was possible to achieve a bidirectional movement, in this case provided by the cooperative action of electrostatic and electrothermal forces. The layout allowed active control of the different states by changing the potential boundary conditions. However, the reported stroke of the bidirectional actuator was only about ± 1.5 μm.

A similar approach of combining electrothermal and electrostatic actuations is described in [86]. Here, by a 2nd metalization layer, a bidirectional switching is already obtained with the electrothermal actuation alone, whereas the electrostatic actuation provides the holding state with a low-power consumption. The layout is a standard cantilever supported by four beams, which are realized with the two separate metalization layers. Two beams are dedicated for downward movement and the other two for upward movement. The electrostatic "holding electrode" (membrane) is formed by the 2nd metalization layer. A thermal switching time of 47 μs was reported; however, it required an electric current of 0.23 A, whereas the measured electrostatic switching time was 4.5 μs at an applied voltage of 15.4 V; therefore, the electrostatic switching time was $10\times$ shorter than its thermal counterpart. The total power consumption was 3.24 μJ per switching cycle.

Another concept of cooperative actuation using thermal principles was proposed in [87], where many SDAs, similar to those described in Section 3.2.3, are connected to a shape-memory alloy (SMA) wire to achieve a 3D movement of a catheter by combining the rotational movement provided by the highly integrated electrostatic SDAs (array of 1430 SDAs to create a reasonable torque) with a macro flexure movement by the SMA wire. SMAs make use of the allotropic phase transformation within certain materials, such as Ti-Ni, which was used in [87].

A cooperation of a current-driven thermopneumatic actuation with an electrostatic actuation for a microvalve is described in [88]. Here, the fluid on the inlet side of the microvalve is heated up by a suspended resistor grid. The rising pressure pushes the valve diaphragm towards the valve seat. However, the final closure of the outlet valve is provided by an electrostatic pull-in movement. An electrostatic latch at the outlet opening is also used to detect the pull-in state of the diaphragm, such that the electric power of the pneumatic actuator can then be reduced or even turned off. The motivation for the cooperative design of the valve was achieving a large throw (i.e., large travel distance for closing the outlet valve to provide large flow in the open state), a low leak rate (especially critical with particles in the fluid larger than, e.g., 30 μm, if only electrostatic actuation is provided), a power reduction (compared to an exclusively thermopneumatic-driven actuator) and a fast response time. However, in the presented layout, complete electrostatic latching was not achieved. The authors suggested several design changes to reach electrostatic latching; however, no functional device was presented to the best of our knowledge.

In [89], a piezoelectric support of an electrostatic RF MEMS variable capacitor for RF-switching was proposed. Here, a thin film of lead zirconate titanate (PZT) was embedded in a sacrificial layer process. The piezoelectric actuator supports the pull-in-type electrostatic actuation to increase the capacitance and allows a lower switching voltage (pull-in voltage is considerably reduced), while keeping the pull-out voltage almost constant, which is important to maintain high restoring forces. Thus, by lowering the operation voltage, stiction and dielectric charging are reduced in the hybrid design.

Hybrid concepts with piezoelectric actuation have also been employed for realizing inchworm motors. For example, in 2004, Toda et al. [90] presented a proof-of-concept, hybrid inchworm motor for extended out-of-plane actuation. It utilized in-plane electrostatic forces to alternately clamp a moveable platform to either a driver plate or stationary structures (holders). When clamped to the driver plate, which is driven in the Z-axis by an external piezoelectric actuator, the platform gains a small elevation, then the platform is clamped to the holders to maintain the gained elevation, while the driver plate, i.e., the PZT actuator, is reset to the initial position to start a new cycle. Although the system was realized by a 50 μm thick device layer of an SOI wafer, the motor could only achieve about 0.5 μm of elevation in 10 actuation cycles. Later in 2007, Toda et al. [91] proposed another hybrid inchworm motor combining electrostatic actuation for clutching a slider and piezoelectric actuation for pushing it. The operation and the realized system are shown in Figure 14. The system consists of two electrostatic clutching units (A and B) and a PZT-stack actuator. Unit A is stationary, while unit B is fixed to one side of the PZT actuator, whose other side is fixed to the substrate; hence, unit B can be translated back and forth. Therefore, the PZT actuator can drive the slider in a bidirectional manner, while it is clamped by unit B (see Figure 14a). This cooperative inchworm motor features a power-free clutching by the actuation units A and B, which is achieved by mechanically prestressing the tether beams that suspend the clutching structures in these units. This prestressing results from inserting a slider of a width that is slightly larger than the space available between the clutching structures, which are driven by the movable electrodes of four opposing comb drives on both sides of the slider. This provides a zero-power latching and thus a secure operation. The PZT actuator created a defined step size of 59 nm/cycle and a large force to overcome the residual friction when powering the electrostatic clamps. By FEM, a clutching force of 4×4 mN = 16 mN was simulated, where the basic clutching force of 4 mN is a result of the attractive electrostatic forces (40 mN), the repelling bending forces of the tether beams (20 mN), and a friction coefficient of 0.2 ($(40 - 20)$ mN $\times 0.2 = 4$ mN). The MEMS part consists of c-Si, which is fabricated using a SOI-process.

A hybrid actuation system with cooperation of electromagnetic and electrostatic actuators for a miniaturized contactless levitation system was presented by Poletkin et al. in 2015 [92]. Two coils provide induction based forces, one for levitation, the other for stabilization of the levitated proof mass during levitation. The electrostatic forces provided by two different electrode arrangements take over additional functions. Firstly, under pullin, the electrostatic actuation works oppositely to the electromagnetic levitation force. This means that by keeping a certain levitation height even with increased induction current, a larger voltage can be applied. Therefore, the electrostatic force will change the effective stiffness of the contactless electromagnetic suspension. Additionally, electrostatic forces are used for vertical and angular actuation of the proof mass. 3D microcoils obtained by means of a wire bonder were assembled on a Pyrex substrate, whereas the electrostatic actuators were realized on a Si wafer with DRIE etching of both sides of the wafer. Currents for electromagnetic actuation of 100–130 mA provide levitation heights between 30–189 μm. The measured stiffness of the angular stiffness changed by electrostatic actuation by more than a factor of 2 (from 0.9×10^{-8} Nm/rad to 0.4×10^{-8} Nm/rad). Model descriptions of that system considering both, inductive and electrostatic forces were later provided in [93,94].

A similar concept was also presented in 2015 by Sari et al. [95]. Electromagnetic levitation and stabilization are cooperating with an electrostatic propulsion drive of the levitated object. However, here Lorentz forces through an alternating current were used to generate electromagnetic forces. Therefore, electromagnetic and electrostatic actuators could be realized on one silicon wafer.

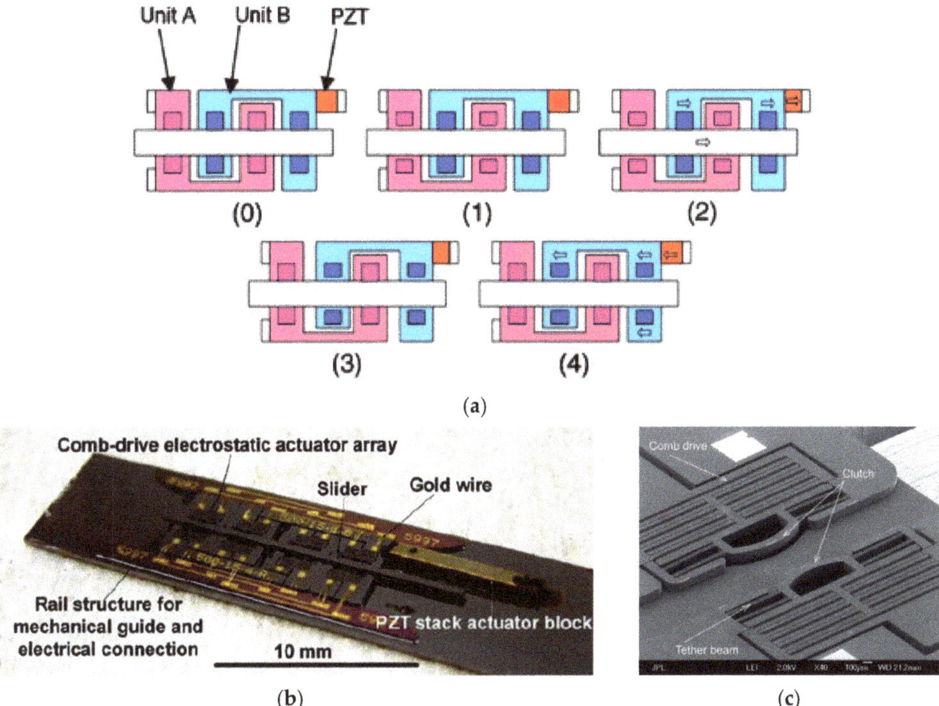

Figure 14. An inchworm microactuator realized by the cooperation of 8 electrostatic comb drives and a PZT-stack for a bidirectional stepping mode [91]: (**a**) one cycle of an inchworm motor operation, where at (**0**) all the clutches (red and blue squares) are pressing on the central long slider, in (**1**) unit A is released from the slider, in (**2**) the PZT actuator pulls unit B and the slider to the right (one step), in (**3**) with still engaged PZT actuator, unit A is clamped to the slider, while unit b is released and in (**4**) the PZT actuator is turned off so that unit B is moved back, while the slider is latched by unit A; (**b**) the fabricated system (MEMS part out of c-Si); (**c**) tethered clutches connected to a group of comb drives (© IOP Publishing. Reproduced with permission. All rights reserved).

4. Discussion

While the possibility of even 3D-like motion by planar, surface micromachined structures was already shown by researchers more than 20 years ago [13,25], none of these have been commercialized due to their sophisticated mechanical function, especially the needed hinges to provide the out of plane motion. Therefore, we conclude that a key point to cooperative electrostatic microactuators is how to create bistability and even multistability in such systems. However, only a very limited numbers of papers were found, where functional, electrostatically driven bistable microactuator systems were presented. Ongoing research on the theory and models of electrostatic bistable systems, which is documented, e.g., by several actual papers describing the bistability conditions in arched microbeams or microplates, shows that this is still a topic of fundamental research. We conclude that especially snap-in and snap-out is a critical aspect and hard to control by electrostatic actuation under the pull-in regime.

Due to these limitations, which hold especially true for pure electrostatic cooperative actuator systems, hybrid systems could in theory be expected to play an important role for large force applications and for providing bistability. However, such hybrid systems, which combine different types of actuation principles, have their own technological challenges, and it is much more difficult to create a system model of them compared to pure electrostatic actuators systems. Additionally, with the integration of another actuation principle, which needs specific materials (piezoelectric, SMA, magnetostrictive), the big advantages of electrostatic actuation systems, namely, the CMOS compatibility and scalability for high volume production, are lost. This is not the case for electrothermal actuation. If there are only a few papers dealing with hybrid systems in general, there are, of course, even fewer combining electrostatic and electrothermal actuation. We conclude that a disadvantage of the cooperation of electrostatic and electrothermal actuators is the very different driving principle (voltage versus current driven) and the very large difference in power and reaction time, which make this concept practically not feasible. This conclusion is supported by the fact that most work on hybrid actuation systems was published already more than 10 years ago. Additionally, we observe a lack of continuity in the proposed concepts as mostly only one paper has been published by the according groups. We conclude that despite of interesting features, hybrid systems are sophisticated to realize and to scale up for commercialization and are also not reliable. Even the rather simple—and due to integration in a standard surface micromachining process, also scalable—approach of the piezoelectric actuation cooperating with the electrostatic actuation presented in [89], by the research team of Toshiba, did not pave the way to a commercial product, as today, Toshiba uses a pure electrostatic actuation principle for RF-MEMS variable capacitors [96].

Although electrostatic actuation in principle has large force capability when the gap between the electrodes is reduced, numerous problems and challenges arise when using this gap reduction effect. Besides stiction and viscous damping within small gaps, such systems are very sensitive to environmental conditions, especially humidity and pollution. Additionally, on a submicrometer scale of distances or gaps, also severe technological challenges occur, such as surface roughness of side walls (defining the electrode surface and thus the homogeneity of electric field for driving), which have a strong influence on the devices' performances.

Other challenges for the realization of cooperative electrostatic actuator systems are related to technological aspects, e.g., in DRIE, which is typically used for the realization of HAR microstructures, the limited anisotropy and several other limitations such as notching [97] have to be overcome for improving the reliability and for further miniaturization. Additionally, as a result of miniaturization, the surface to volume relation is also increasing. Therefore, the overall systems' properties are mostly defined by the surfaces within such systems, which are hard to control and exposed to major external influences. This is a severe challenge in respect to system reliability, especially when such systems should provide large forces.

Electrostatic motors are examples of strong cooperative architectures, which can overcome the limitations of a stand-alone electrostatic actuator. From their inception, the primary challenged constraint was, especially at first, the inherently limited travel range. Later, the generation of larger forces was also aimed for by many researchers. The high resolution displacement provided by small defined steps can also be a major advantage for high precision positioning applications, which is especially characteristic of some inchworm motor types. Here, we will discuss some of the pros and cons of the different realizations of linear electrostatic motors, with a special focus on the inchworm motors.

The early works on inchworm motors in the mid-1990s were mainly based on poly-Si and typically generated forces in the range of few to some tens of μN. The force density was small, especially in the absence of any unique mechanical leverage. With the establishment of SOI wafer technology, which provides the easily releasable thicker c-Si structures, accompanied by the more optimized HAR silicon micromachining processes, e.g., the DRIE, better performance characteristics could be achieved. For instance, upgrading from a thin poly-Si structural layer to a thicker c-Si device layer of a SOI wafer enlarged the force capacity substantially and solved an adhesion problem between contacting surfaces in [11], which could have been the result of having larger restoring forces by the now stiffer springs, or the reduced adhesion by virtue of the larger DRIE sidewall roughness, as reasoned by the authors.

It is worth noting, however, that some electrostatic motor types that have appealing characteristics have almost exclusively been fabricated in thin poly-Si layers, e.g., SDA and shuffle motors. These motors, in addition to the contraction beams motor, which are based on orthogonal mechanical conversion of force or deflection, have exceptionally high capacity for precision positioning with typically reported step resolutions between 10–80 nm [47,67,68], and could be driven at high frequencies (up to 80 kHz) [67,68]. Additionally, some of these walker-type inchworm motors can be built without a need for guiding rails or mechanical suspensions [44,67], which imposes less constraints on the maximum travel range and direction when compared to pusher-types that usually include suspended shuttles. This also makes the former more suited for realizing multi-DoF and planar motors [53,67]. Furthermore, surface-micromachined motors out of poly-Si could more readily accommodate untethered shuttle configurations in their monolithic fabrication, which lends itself to achieving longer travel distances [70].

Nevertheless, they also have some inherent drawbacks and have shown some typical shortcomings. SDA and shuffle motors rely on a certain frictional component with the substrate in order to function properly. They also require insulated electrodes, and as a result, the dielectric insulation is prone to trap charges (dielectric charging), which influences the friction profile and consequently the device behavior during operation [65]. It could also cause stiction leading to device malfunction [54], or even permanent sticking, which led to destruction upon repositioning in [44]. Furthermore, the typically thin bending plate in these motors imposes an inherent limitation on the maximum generated force due to the induced plate stretching that occurs beyond a certain force level [65]. Additionally, the efficiency of these devices, by design, is subject to question. For instance, large friction forces caused by adhesion forces between inactive clamps and the substrate have been reported [45]; moreover, the bending plate's deflection towards the substrate in a shuffle motor, causes unwanted friction at the released clamp site, which opposes the motor's shuffling motion, and introduces a moment at the activated clamp site, which could significantly reduce its clamping force [68]. In addition, in standard surface micromachined motors, the thickness of the functional layer (poly-Si or thin c-Si device layer) is typically an order of magnitude smaller than HAR motors made out of such SOI wafers that have thicker device layers (e.g., >10 µm), this is a substantial drawback for applications that engages mesoscale loads, where the fabrication and force transfer become much more challenging. This conclusion is inferred from the prevailing trend witnessed in recent years towards HAR inchworm motors for applications involving a micro-macro transfer of force and displacement [77–82,98].

The use of interlocking structures such as teeth, jaws, pawls, etc., is another easily observable trend in recent years, especially in contributions that aimed for a large force capacity [11,74,77–82,98]. This trend was needed in HAR devices as a means for a reliable transfer of forces between the different parts of an inchworm motor, which solved a clutch slipping problem as reported in [11], for instance. These structures produce an impact-induced mechanical wear resulting from the colliding surfaces, which generates less uniformly distributed mechanical loads that lead to higher loading surface stresses, at teeth edges, for example. This is a kind of wear that is additional to other kinds that are associated with the rubbing of sliding planar surfaces found in all types of electrostatic motors, of which the adhesive wear was reported to dominate (as opposed to the abrasive and corrosive kinds of wear) [99]. Although some authors have reported relatively long life spans of their systems with reliable outcomes [11], it is well-understood that the wear in these systems, and the microparticles contamination and debris produced by their operation, have led to rapid loss of efficiency even after a short operation time [82], and sometimes complete device failure [100].

Many of the proposed inchworm motors rely on the passive restoring elastic forces of suspension springs to reset the actuated body, e.g., a shuttle, to its initial position, thereby lacking the capacity of bidirectional positioning. This has been overcome in some other inchworm motors that inherently provide a bidirectional movement mechanism, such as the shuffle motors [62], the contraction beams motor [68], the multi-phase, alignment-based motors [60] and the PZT-based hybrid inchworm motor in [91]. Others have a built-in bidirectionality in the design of the propulsion actuator by having double the amount of driving electrodes [70,74], or introducing a biasing mechanism for presetting the direction of propulsion, with a reported minor influence on the force density [76]. On the other hand, other motors achieved bidirectionality by duplicating the unidirectional cooperative actuators for the opposite direction [61]. An approach that could be viewed as the least efficient utilization of space, should this perspective be critical, e.g., for achieving high force density motors. The NED-based inchworm motor in [73], has two oppositely oriented pairs of unidirectional NED-blocks; however, when one pair of NED-blocks is used to drive the shuttle in a certain direction, the other is used momentarily to hold the shuttle while the former recovers from deflection; therefore, there is an overlap of inherent and duplicate-based bidirectional designs. It is worth noting that the reported successful bidirectional actuation through reversing the control sequence of the inchworm motor in [76] could be a viable option for many other inchworm motors that have suspended shuttles, albeit the generated driving force will not be equal in the two directions.

Though electrostatic linear motors largely perform the same task, depending on their type and configuration, their cooperation-based operation and control entail different levels of complexity, e.g., the number of individual actuators in the system and the number of required control signals. When we compare some of the motors presented in this review from this perspective, we find that SDAs are among the simplest motors to control, since aside from the ground electrode, they are driven in parallel using one signal per direction, regardless of how many SDA plates are integrated in the motor. To remedy the dielectric charging issue, SDAs have been driven with AC signals of alternating sign, whether in pulse [10], square [46], triangular or sinusoidal form [50]. Sometimes to keep them in a primed position for optimized performance, SDAs have also been driven with a DC-bias [44]. On the other hand, inchworm motors, especially the walker-type, conventionally need three control signals to perform a complete cycle, two for the clamping electrodes and one for the driving electrode. This configuration is true for shuffle and contraction beams motors, which typically also uses bipolar, modulated AC potentials to avoid dielectric charging [45]. By reversing the sequence of these signals these motors can move in the opposite direction. The multi-phase, alignment-based motor in [60] used four AC signals to drive the four-phase motor, which are also used for both directions of actuation. Moreover, pusher-type inchworm motors typically require four signals (two for each driving unit: one for clutching/clamping and one for driving) to actuate the shuttle in one direction [11,69]. However, concerning the required added complexity to achieve bidirectionality, the motor configuration discussed in the previous paragraph plays a decisive role. Naturally, the duplicate-based bidirectionality will require double the amount of control signals, i.e., eight. While motors that have integrated bidirectional electrodes in their basic driving units, require one additional signal for each driving unit; thus, some require six different signals for two driving units [70], while others require nine signals for three groups of actuation unit cells, with one more signal to achieve multistability, making the total ten signals [74]. Similarly, the biasing mechanism proposed by Hollar et al. [76] required two additional signals for predetermining the direction of actuation, bringing the total to six signals for each inchworm motor. However, due to this added complexity, the authors preferred the reverse control approach to accomplish a bidirectional drive of their robotic leg, which required a total of eight signals to drive both, the hip and knee inchworm motors. On the other hand, by virtue of the double-purposed driving units, as discussed previously, the NED-based motor requires only four signals to accomplish a bidirectional drive [73]. Another noteworthy novelty in this regard is the control simplification that accompanied the combination of the clutch and driving mechanisms into one movement by an inclined arm, which resulted in a successful unidirectional inchworm motor movement that requires only two input signals [72].

5. Conclusions and Outlook

Despite the ubiquitous electrostatic microactuators that are found and used today in billions of commercial products and devices, systems and commercial examples of highly cooperating electrostatic microactuators are rare, e.g., the micromirror device of TI has a relative simple design architecture with isolated microactuators (electrostatically tilted micromirrors), where the cooperation is basically provided through the driving electronic circuit (CMOS drive logic) and the individual excitation capability within the chip. As such, it is a somewhat "weak" cooperation facilitated on the system level by the capability to drive each of the more than 1 million micromirrors individually to produce an image pattern. Additionally, all micromirrors are actuated in the same way (tilted electrostatically by around $\pm 10°$).

We explain the lack of commercial products using strong cooperative electrostatic actuator systems by the planar character of surface micromachining, which, on the other hand, is a prerequisite for the success story of electrostatic actuators in consumer applications.

To overcome the critical wear and microparticles phenomena, when using mechanically interlocking structures, well-grounded research into the durability of these force-transfer mechanisms in HAR actuators with experimentally based solutions, e.g., using functional anti-wear coatings deposited by atomic layer deposition, would be a highly beneficial endeavor for improving the efficacy of these systems in the future.

Just recently, W. Fang reported that thin piezoelectric films, in particular aluminum nitride (AlN) and PZT, have been established in foundries and equipment suppliers in Taiwan's strong semiconductor industry, thus allowing the integration of these materials in CMOS processes [101]. This might be a game changer, especially for hybrid cooperative actuation systems, as piezoelectric and electrostatic actuators can be monolithically integrated in a large scale production.

The progress of miniaturization towards the nanotechnology will open the path towards controlled gaps in the range of some 10–100 nm, and thus enable much more powerful devices than reported so far. In any case, due to the different challenges, further research is needed to provide reliable cooperative systems, e.g., for large force applications.

Author Contributions: Conceptualization, A.A. and U.M.; investigation, A.A. and U.M.; methodology, U.M.; formal analysis, A.A. and U.M.; resources, U.M.; writing—original draft preparation, A.A. and U.M.; writing—review and editing, A.A. and U.M.; visualization, A.A. and U.M.; supervision, U.M.; project administration, U.M.; funding acquisition, U.M. Moreover, A.A. and U.M. have contributed equally to this publication. However, concerning the inchworm motors in particular, A.A. was responsible for their review, critical analysis and discussion. All authors have read and agreed to the published version of the manuscript.

Funding: This research is funded by the German Research Foundation (Deutsche Forschungsgemeinschaft—DFG) under the umbrella of the priority program SPP2206—Cooperative Multistage Multistable Micro Actuator Systems (KOMMMA), project number ME 2093/5-1.

Conflicts of Interest: The authors declare no conflict of interest. Moreover, the funders had no role in the design of the study; in the collection, analyses, or interpretation of data; in the writing of the manuscript; or in the decision to publish the results.

Appendix A

Table A1. Comparison of the performance and operational characteristics for some of the electrostatic linear motors reviewed in this paper.

Year	Ref.	Actuators	Walker v. Pusher	Stroke (Step)	Max Disp.	Force	Voltage [V]	Speed	Freq. of Operation	Uni-/Bidirectional	Fabrication Technique	Clutch Mechanism	Propulsion Mechanism	Comment
1993	[10]	Scratch drive actuator (SDA)	Walker	(10–80) nm	-	-	40–150	(10–80) μm/s	1 kHz	Uni	SM poly-Si	-	Plate bending	
1995	[43]	SDA	Walker	-	~120 μm	63 μN	$\pm$112	-	50 Hz	Uni	SM poly-Si	-	Plate bending	
2001	[44]	SDA robot (Scratchuator)	Walker	30 nm [1]	8 mm [1]	85 μN [2]	200 V_{AC} [2]	-	1 kHz [2]	Bi	SM poly-Si	-	Plate bending	(1) Robot made of 188 SDAs, (2) array of 4 SDAs
2002	[47]	SDA	Walker	25 nm [2]	60 μm *	(250 [1], 850 [3]) μN	200 [1,3], 290 [2]	250 μm/s [2]	10 kHz [2]	Uni	SM poly-Si	-	Plate bending	(1) one SDA, (2) two SDAs, (3) four SDAs, * limited by design
1995	[61]	Stepper (inchworm) motor	Pusher	2 μm	40 μm	6.5 μN	35	-	-	Bi	SM poly-Si	Electrostatic	Comb-drive pull	
1995	[60]	Attachment/ detachment motor	Pusher	1.5 μm	-	few mN	100	-	1.4 kHz	Bi	BM c-Si	Electrostatic	Electrostatic alignment	
1998	[45]	Shuffle motor	Walker	85 nm	43 μm	43 μN	Clamp: 40, Drive: 25	100 μm/s	1.16 kHz **	Bi	SM poly-Si	Electrostatic	Plate bending	** limited by driving electronics
2005	[67]	Shuffle motor with 2 DoF	Walker	(41–63) nm	60 μm *	0.64 mN	Clamp: 36, Drive: 45	$\leq$3.6 mm/s	$\leq$80 kHz **	Bi	SM poly-Si with TI tech.	Electrostatic	Plate bending	2-DoF (planar motion), * limited by design, ** limited by driving electronics
2006	[68]	Contraction beams motor	Walker	10 nm	140 μm	0.49 mN	Clamp: 50, Drive: 60	$\leq$0.78 mm/s	$\leq$80 kHz **	Bi	SM poly-Si with TI tech.	Electrostatic	Beams bending	** limited by driving electronics
1997	[69]	Stepper (inchworm) motor	Pusher	2 μm	15 μm	3 μN	40	4 μm/s	1 Hz	Uni	SM poly-Si	Frictional	Comb-drive pull	
1997	[70]	Stepper (inchworm) motor	Pusher	(0.5–3) μm	(15 [1], 110 [2],*) μm	>1 μN	15–40	-	-	Bi	SM poly-Si	Frictional	Comb-drive pull	(1) design A: suspended slider, (2) design B: free slider, * limited by design
2002	[11]	Inchworm motor	Pusher	2 μm	80 μm	260 μN	33	4 mm/s	1 kHz	Uni	HARSM, SOI	Teeth	Comb-drive pull	
2013	[72]	Inchworm motor	Pusher	2 μm	124 μm	1.88 mN	110	4.8 mm/s	1.2 kHz	Uni	HARSM, SOI	Teeth	Comb-drive pull (inclined arm)	

Table A1. *Cont.*

Year	Ref.	Actuators	Walker v. Pusher	Stroke (Step)	Max Disp.	Force	Voltage [V]	Speed	Freq. of Operation	Uni-/Bidirectional	Fabrication Technique	Clutch Mechanism	Propulsion Mechanism	Comment
2021	[73]	NED-based inchworm motor	Pusher	≤11.8 µm *	997 µm	1.4 mN	Clamp: 150, Drive: 130	-	500 Hz	Bi	HARSM, SOI	Electrostatic	Nanoscopic elect. Drive	* function of driving voltage
2021	[98]	Inchworm motor	Pusher	4 µm	80 mm	15 mN	100	5 mm/s	-	Uni	HARSM, SOI	Teeth	Comb-drive pull (inclined arm)	

SM poly-Si: Surface micromachining (poly-Si), **BM c-Si**: Bulk micromachining (c-Si), **HARSM, SOI**: HAR silicon micromachining (c-Si), SOI, **SM poly-Si with TI tech.**: Surface micromachining (poly-Si) with trench isolation technology.

References

1. Xie, H.; Fedder, G.K. Fabrication, characterization, and analysis of a DRIE CMOS-MEMS gyroscope. *IEEE Sens. J.* **2003**, *3*, 622–631. [CrossRef]
2. Maenaka, K. MEMS inertial sensors and their applications. In Proceedings of the 5th International Conference on Networked Sensing Systems, (INSS 2008), Kanazawa, Japan, 17–19 June 2008; IEEE: New York, NY, USA, 2008; pp. 71–73, ISBN 978-4-907764-31-9.
3. Xie, H.; Fedder, G.K. Vertical comb-finger capacitive actuation and sensing for CMOS-MEMS. *Sens. Actuators A Phys.* **2002**, *95*, 212–221. [CrossRef]
4. Howe, R.T. Surface micromachining for microsensors and microactuators. *J. Vac. Sci. Technol. B* **1988**, *6*, 1809. [CrossRef]
5. Breguet, J.-M.; Johansson, S.; Driesen, W.; Simu, U. A review on actuation principls for few cubic millimeter sized mobile micro-robots. In Proceedings of the 10th International Conference on New Actuators, (ACTUATOR 2006), Bremen, Germany, 14–16 June 2006; pp. 374–381.
6. Judy, M.W. Evolution of integrated inertial MEMS technology. In Proceedings of the Solid-State, Actuators, and Microsystems Workshop Technical Digest, Hilton Head, SC, USA, 6–10 June 2004; Sulouff, R., Kenny, T.W., Eds.; Transducer Research Foundation, Inc.: San Diego, CA, USA, 2004; pp. 27–32.
7. GSMArena.com. Samsung I8530 Galaxy Beam: Complete Phone Specifications. Available online: https://www.gsmarena.com/samsung_i8530_galaxy_beam-4566.php (accessed on 25 January 2023).
8. Judy, J.W. Microelectromechanical systems (MEMS): Fabrication, design and applications. *Smart Mater. Struct.* **2001**, *10*, 1115–1134. [CrossRef]
9. Chen, C.-H.; Yeh, J.A.; Wang, P.-J. Electrical breakdown phenomena for devices with micron separations. *J. Micromech. Microeng.* **2006**, *16*, 1366–1373. [CrossRef]
10. Akiyama, T.; Shono, K. Controlled stepwise motion in polysilicon microstructures. *J. Microelectromech. Syst.* **1993**, *2*, 106–110. [CrossRef]
11. Yeh, R.; Hollar, S.; Pister, K. Single mask, large force, and large displacement electrostatic linear inchworm motors. *J. Microelectromech. Syst.* **2002**, *11*, 330–336. [CrossRef]
12. Sampsell, J.B. Digital micromirror device and its application to projection displays. *J. Vac. Sci. Technol. B* **1994**, *12*, 3242. [CrossRef]
13. Liu, A.Q.; Zhang, X.M.; Murukeshan, V.M.; Zhang, Q.X.; Zou, Q.B.; Uppili, S. Optical Switch Using Draw-Bridge Micromirror for Large Array Crossonnects. In *Transducers '01 Eurosensors XV*; Obermeier, E., Ed.; Springer: Berlin/Heidelberg, Germany, 2001; pp. 1296–1299. ISBN 978-3-540-42150-4.
14. de Dobbelaere, P.; Falta, K.; Gloeckner, S.; Patra, S. Digital MEMS for optical switching. *IEEE Commun. Mag.* **2002**, *40*, 88–95. [CrossRef]
15. Plander, I.; Stepanovsky, M. MEMS technology in optical switching. In Proceedings of the 14th IEEE International Scientific Conference on Informatics, Poprad, Slovakia, 14–16 November 2017; IEEE: New York, NY, USA, 2017; pp. 299–305; ISBN 978-1-5386-0888-3.
16. Jaecklin, V.P.; Linder, C.; de Rooij, N.F.; Moret, J.M.; Bischof, R.; Rudolf, F. Novel polysilicon comb actuators for xy-stages. In Proceedings of the IEEE Micro Electro Mechanical Systems, Travemunde, Germany, 4–7 February 1992; IEEE: New York, NY, USA, 1992; pp. 147–149, ISBN 0-7803-0497-7.
17. Indermuehle, P.-F.; Linder, C.; Brugger, J.; Jaecklin, V.P.; de Rooij, N.F. Design and fabrication of an overhanging xy-microactuator with integrated tip for scanning surface profiling. *Sens. Actuators A Phys.* **1994**, *43*, 346–350. [CrossRef]
18. Ni, D.; Heisser, R.; Davaji, B.; Ivy, L.; Shepherd, R.; Lal, A. Polymer interdigitated pillar electrostatic (PIPE) actuators. *Microsyst. Nanoeng.* **2022**, *8*, 18. [CrossRef] [PubMed]
19. Del Corro, P.G.; Imboden, M.; Pérez, D.J.; Bishop, D.J.; Pastoriza, H. Single ended capacitive self-sensing system for comb drives driven XY nanopositioners. *Sens. Actuators A Phys.* **2018**, *271*, 409–417. [CrossRef]
20. Minami, K.; Kawamura, S.; Esashi, M. Fabrication of distributed electrostatic micro actuator (DEMA). *J. Microelectromech. Syst.* **1993**, *2*, 121–127. [CrossRef]
21. Chiou, J.-C.; Lin, Y.-J.; Kuo, C.-F. Extending the traveling range with a cascade electrostatic comb-drive actuator. *J. Micromech. Microeng.* **2008**, *18*, 15018. [CrossRef]
22. Schmitt, L.; Schmitt, P.; Barowski, J.; Hoffmann, M. Stepwise Electrostatic Actuator System for THz Reflect Arrays. In Proceedings of the International Conference and Exhibition on New Actuator Systems and Applications, (ACTUATOR 2021), Online Event, 17–19 February 2021; VDE VERLAG: Berlin, Germany, 2021; pp. 1–4, ISBN 978-3-8007-5454-0.
23. Schmitt, L.; Hoffmann, M. Large Stepwise Discrete Microsystem Displacements Based on Electrostatic Bending Plate Actuation. *Actuators* **2021**, *10*, 272. [CrossRef]
24. Conrad, H.; Schenk, H.; Kaiser, B.; Langa, S.; Gaudet, M.; Schimmanz, K.; Stolz, M.; Lenz, M. A small-gap electrostatic micro-actuator for large deflections. *Nat. Commun.* **2015**, *6*, 10078. [CrossRef]
25. Fan, L.; Wu, M.C.; Choquette, K.D.; Crawford, M.H. Self-assembled microactuated XYZ stages for optical scanning and alignment. In Proceedings of the International Conference on Solid State Sensors and Actuators, (TRANSDUCERS '97), Chicago, IL, USA, 16–19 June 1997; pp. 319–322, ISBN 0-7803-3829-4.
26. Hailu, Z.; He, S.; Ben Mrad, R. Hybrid micro electrostatic actuator. *Microsyst. Technol.* **2016**, *22*, 319–327. [CrossRef]
27. Ando, Y. Development of three-dimensional electrostatic stages for scanning probe microscope. *Sens. Actuators A Phys.* **2004**, *114*, 285–291. [CrossRef]

28. Liu, X.; Kim, K.; Sun, Y. A MEMS Stage for 3-Axis Nanopositioning. In Proceedings of the IEEE International Conference on Automation Science and Engineering, (CASE 2007), Scottsdale, AZ, USA, 22–25 September 2007; IEEE: New York, NY, USA, 2007; pp. 1087–1092, ISBN 978-1-4244-1153-5.
29. Chang, H.-C.; Tsai, J.M.-L.; Tsai, H.-C.; Fang, W. Design, fabrication, and testing of a 3-DOF HARM micromanipulator on (111) silicon substrate. *Sens. Actuators A Phys.* **2006**, *125*, 438–445. [CrossRef]
30. Yang, S.; Xu, Q. Design and Analysis of a Decoupled XY MEMS Microgripper with Integrated Dual-Axis Actuation and Force Sensing. *IFAC-PapersOnLine* **2017**, *50*, 808–813. [CrossRef]
31. Muthuswamy, J.; Okandan, M.; Jain, T.; Gilletti, A. Electrostatic microactuators for precise positioning of neural microelectrodes. *IEEE Trans. Biomed. Eng.* **2005**, *52*, 1748–1755. [CrossRef] [PubMed]
32. Tang, W.C.; Lim, M.G.; Howe, R.T. Electrostatic Comb Drive Levitation And Control Method. *J. Microelectromech. Syst.* **1992**, *1*, 170–178. [CrossRef]
33. He, S.; Ben Mrad, R.; Chong, J. Repulsive-force out-of-plane large stroke translation micro electrostatic actuator. *J. Micromech. Microeng.* **2011**, *21*, 75002. [CrossRef]
34. Schaler, E.W.; Jiang, L.; Fearing, R.S. Multi-layer, Thin-film Repulsive-force Electrostatic Actuators for a 2-DoF Micro-mirror. In *Actuator 18: International Conference and Exhibition on New Actuators and Drive Systems, Bremen, Germany, 25–27 June 2018: Interactive Conference Proceedings*; Borgmann, H., Ed.; VDE VERLAG: Berlin, Germany, 2018; pp. 299–303. ISBN 978-3-8007-4675-0.
35. Schaler, E.W.; Zohdi, T.I.; Fearing, R.S. Thin-film repulsive-force electrostatic actuators. *Sens. Actuators A Phys.* **2018**, *270*, 252–261. [CrossRef]
36. Gorthi, S.; Mohanty, A.; Chatterjee, A. Cantilever beam electrostatic MEMS actuators beyond pull-in. *J. Micromech. Microeng.* **2006**, *16*, 1800–1810. [CrossRef]
37. Medina, L.; Gilat, R.; Krylov, S. Bistability criterion for electrostatically actuated initially curved micro plates. *Int. J. Eng. Sci.* **2018**, *130*, 75–92. [CrossRef]
38. Kumar, S.; Bhushan, A. Interaction of transverse pressure and in-plane internal stresses on bi-stability of electrostatically rectangular microplates. *Eng. Res. Express* **2022**, *4*, 45042. [CrossRef]
39. Das, M.; Bhushan, A. Investigation of an electrostatically actuated imperfect circular microplate under transverse pressure for pressure sensor applications. *Eng. Res. Express* **2021**, *3*, 045023. [CrossRef]
40. Wagner, B.; Quenzer, H.J.; Hoerschelmann, S.; Lisec, T.; Juerss, M. Bistable microvalve with pneumatically coupled membranes. In Proceedings of the 9th IEEE Annual International Workshop on Micro Electro Mechanical Systems, San Diego, CA, USA, 11–15 February 1996; IEEE: New York, NY, USA, 1996; pp. 384–388, ISBN 0-7803-2985-6.
41. Freudenreich, M.; Mescheder, U.; Somogyi, G. Simulation and realization of a novel micromechanical bi-stable switch. *Sens. Actuators A Phys.* **2004**, *114*, 451–459. [CrossRef]
42. Kwon, H.N.; Hwang, I. H.; Lee, J. H. A pulse operating electrostatic microactuator for bi-stable latching. *J. Micromech. Microeng.* **2005**, *15*, 1511–1516. [CrossRef]
43. Akiyama, T.; Fujita, H. A quantitative analysis of Scratch Drive Actuator using buckling motion. In Proceedings of the IEEE Micro Electro Mechanical Systems, (MEMS 1995), Amsterdam, The Netherlands, 29 January–2 February 1995; IEEE: Piscataway, NJ, USA; pp. 310–315, ISBN 0-7803-2503-6.
44. Linderman, R.J.; Bright, V.M. Nanometer precision positioning robots utilizing optimized scratch drive actuators. *Sens. Actuators A Phys.* **2001**, *91*, 292–300. [CrossRef]
45. Tas, N.; Wissink, J.; Sander, L.; Lammerink, T.; Elwenspoek, M. Modeling, design and testing of the electrostatic shuffle motor. *Sens. Actuators A Phys.* **1998**, *70*, 171–178. [CrossRef]
46. Akiyama, T.; Collard, D.; Fujita, H. Scratch drive actuator with mechanical links for self-assembly of three-dimensional MEMS. *J. Microelectromech. Syst.* **1997**, *6*, 10–17. [CrossRef]
47. Li, L.; Brown, J.G.; Uttamchandani, D. Study of scratch drive actuator force characteristics. *J. Micromech. Microeng.* **2002**, *12*, 736–741. [CrossRef]
48. Honarmandi, P.; Zu, J.W.; Behdinan, K. Analytical study and design characteristics of scratch drive actuators. *Sens. Actuators A Phys.* **2010**, *160*, 116–124. [CrossRef]
49. Chen, S. Improved model of rectangular scratch drive actuator. *J. Micro/Nanolith. MEMS MOEMS* **2011**, *10*, 13016. [CrossRef]
50. Abtahi, M.; Vossoughi, G.; Meghdari, A. Dynamic Modeling of Scratch Drive Actuators. *J. Microelectromech. Syst.* **2015**, *24*, 1370–1383. [CrossRef]
51. Kanamori, Y.; Aoki, Y.; Sasaki, M.; Hosoya, H.; Wada, A.; Hane, K. Fiber-optical switch using cam-micromotor driven by scratch drive actuators. *J. Micromech. Microeng.* **2005**, *15*, 118–123. [CrossRef]
52. Kanamori, Y.; Yahagi, H.; Hane, K. A microtranslation table with scratch-drive actuators fabricated from silicon-on-insulator wafer. *Sens. Actuators A Phys.* **2006**, *125*, 451–457. [CrossRef]
53. Donald, B.R.; Levey, C.G.; McGray, C.D.; Paprotny, I.; Rus, D. An Untethered, Electrostatic, Globally Controllable MEMS Micro-Robot. *J. Microelectromech. Syst.* **2006**, *15*, 1–15. [CrossRef]
54. Tas, N.; Wissink, J.; Sander, L.; Lammerink, T.; Elwenspoek, M. The shuffle motor: A high force, high precision linear electrostatic stepper motor. In Proceedings of the International Conference on Solid State Sensors and Actuators, (TRANSDUCERS '97), Chicago, IL, USA, 16–19 June 1997; pp. 777–780, ISBN 0-7803-3829-4.
55. Lee, A.P.; Pisano, A.P. Polysilicon angular microvibromotors. *J. Microelectromech. Syst.* **1992**, *1*, 70–76. [CrossRef]

56. Daneman, M.J.; Tien, N.C.; Solgaard, O.; Pisano, A.P.; Lau, K.Y.; Muller, R.S. Linear microvibromotor for positioning optical components. *J. Microelectromech. Syst.* **1996**, *5*, 159–165. [CrossRef]
57. Pham, P.H.; Dao, D.V.; Amaya, S.; Kitada, R.; Sugiyama, S. Novel Micro Transportation Systems Based on Ratchetmechanism and Electrostatic Actuators. In Proceedings of the International Conference on Solid-State Sensors, Actuators and Microsystems & Eurosensors, (TRANSDUCERS '07 & EUROSENSORS XXI), Lyon, France, 10–14 June 2007; IEEE: New York, NY, USA, 2007; pp. 451–454; ISBN 1-4244-0841-5.
58. Dao, D.V.; Pham, P.H.; Sugiyama, S. Multimodule Micro Transportation System Based on Electrostatic Comb-Drive Actuator and Ratchet Mechanism. *J. Microelectromech. Syst.* **2011**, *20*, 140–149. [CrossRef]
59. Galante, T.P.; Frank, J.E.; Bernard, J.; Chen, W.; Lesieutre, G.A.; Koopmann, G.H. Design, modeling, and performance of a high-force piezoelectric inchworm motor. In Proceedings of the 5th Annual International Symposium on Smart Structures and Materials, San Diego, CA, USA, 1 March 1998; Regelbrugge, M.E., Ed.; SPIE: Washington, DC, USA, 1998; pp. 756–767.
60. Lee, S.-K.; Esashi, M. Design of the electrostatic linear microactuator based on the inchworm motion. *Mechatronics* **1995**, *5*, 963–972. [CrossRef]
61. Yeh, R.; Kruglick, E.; Pister, K. Microelectromechanical Components For Articulated Microrobots. In Proceedings of the International Conference on Solid State Sensors and Actuators, (TRANSDUCERS '95), Stockholm, Sweden, 25–29 June 1995; IEEE: New York, NY, USA, 1995; pp. 346–349.
62. Tas, N.R.; Legtenberg, R.; Berenschot, J.W.; Elwenspoek, M.C.; Fluitman, J.H.J. The Electrostatic Shuffle Motor. In Proceedings of the Micromechanics Europe '95 Workshop, Copenhagen, Denmark, 3–5 September 1995; pp. 128–131.
63. Yeh, R.; Kruglick, E.; Pister, K. Surface-micromachined components for articulated microrobots. *J. Microelectromech. Syst.* **1996**, *5*, 10–17. [CrossRef]
64. Zhou, Y.-H.; Yang, X. Numerical analysis on snapping induced by electromechanical interaction of shuffling actuator with nonlinear plate. *Comput. Struct.* **2003**, *81*, 255–264. [CrossRef]
65. deBoer, M.P.; Luck, D.L.; Ashurst, W.R.; Maboudian, R.; Corwin, A.D.; Walraven, J.A.; Redmond, J.M. High-Performance Surface-Micromachined Inchworm Actuator. *J. Microelectromech. Syst.* **2004**, *13*, 63–74. [CrossRef]
66. Patrascu, M.; Stramigioli, S. Physical Modelling Of The μWalker, A MEMS Linear Stepper Actuator. *IFAC Proc. Vol.* **2006**, *39*, 743–748. [CrossRef]
67. Sarajlic, E.; Berenschot, E.; Fujita, H.; Krijnen, G.; Elwenspoek, M. Bidirectional electrostatic linear shuffle motor with two degrees of freedom. In Proceedings of the 18th IEEE International Conference on Micro Electro Mechanical Systems, (MEMS 2005), Miami Beach, FL, USA, 30 January–3 February 2005; IEEE: New York, NY, USA, 2005; pp. 391–394, ISBN 0-7803-8732-5.
68. Sarajlic, E.; Berenschot, E.; Tas, N.; Fujita, H.; Krijnen, G.; Elwenspoek, M. Fabrication and Characterization of an Electrostatic Contraction Beams Micromotor. In Proceedings of the 19th IEEE International Conference on Micro Electro Mechanical Systems, (MEMS 2006), Istanbul, Turkey, 22–26 January 2006; IEEE: New York, NY, USA, 2006; pp. 814–817, ISBN 0-7803-9475-5.
69. Tas, N.R.; Sonnenberg, A.H.; Sander, A.; Elwenspoek, M.C. Surface micromachined linear electrostatic stepper motor. In Proceedings of the 10th IEEE Annual International Workshop on Micro Electro Mechanical Systems, Nagoya, Japan, 26–30 January 1997; IEEE: New York, NY, USA, 1997; pp. 215–220, ISBN 0-7803-3744-1.
70. Baltzer, M.; Kraus, T.; Obermeier, E. A linear stepping actuator in surface micromachining technology for low voltages and large displacements. In Proceedings of the International Conference on Solid State Sensors and Actuators, (TRANSDUCERS '97), Chicago, IL, USA, 16–19 June 1997; pp. 781–784, ISBN 0-7803-3829-4.
71. Tas, N.R.; Sonnenberg, T.; Molenaar, R.; Elwenspoek, M. Design, fabrication and testing of laterally driven electrostatic motors employing walking motion and mechanical leverage. *J. Micromech. Microeng.* **2003**, *13*, N6–N15. [CrossRef]
72. Penskiy, I.; Bergbreiter, S. Optimized electrostatic inchworm motors using a flexible driving arm. *J. Micromech. Microeng.* **2013**, *23*, 15018. [CrossRef]
73. Narimani, K.; Shashank, S.; Langa, S.; Gomez, R.P.; Ruffert, C.; Scholles, M.; Schenk, H. Highly Modular Microsystem Inchworm Motor Based on a Nanoscopic Electrostatic Drive. In Proceedings of the MikroSystemTechnik Congress 2021, (MST 2021), Stuttgart-Ludwigsburg, Germany, 8–10 November 2021; VDE VERLAG: Berlin, Germany; pp. 1–4, ISBN 978-3-8007-5656-8.
74. Albukhari, A.; Mescheder, U. Investigation of the Dynamics of a 2-DoF Actuation Unit Cell for a Cooperative Electrostatic Actuation System. *Actuators* **2021**, *10*, 276. [CrossRef]
75. Kloub, H. Design Concepts of Multistage Multistable Cooperative Electrostatic Actuation System with Scalable Stroke and Large Force Capability. In Proceedings of the International Conference and Exhibition on New Actuator Systems and Applications, (ACTUATOR 2021), Online Event, 17–19 February 2021; VDE VERLAG: Berlin, Germany; pp. 1–4, ISBN 978-3-8007-5454-0.
76. Hollar, S.; Bergbreiter, S.; Pister, K. Bidirectional inchworm motors and two-DOF robot leg operation. In Proceedings of the 12th International Conference on Solid State Sensors, Actuators and Microsystems: Digest of Technical Papers, (TRANSDUCERS '03), Boston, MA, USA, 8–12 June 2003; IEEE: New York, NY, USA, 2003; pp. 262–267, ISBN 0-7803-7731-1.
77. Greenspun, J.; Pister, K. First leaps of an electrostatic inchworm motor-driven jumping microrobot. In Proceedings of the Solid-State, Actuators, and Microsystems Workshop: Technical Digest, Hilton Head, SC, USA, 3–7 June 2018; Lamers, T., Rais-Zadeh, M., Eds.; Transducer Research Foundation, Inc.: San Diego, CA, USA, 2018; pp. 159–162, ISBN 978-1-940470-03-0. [CrossRef]

78. Kilberg, B.; Contreras, D.S.; Greenspun, J.; Gomez, H.; Liu, E.; Pister, K. MEMS airfoil with integrated inchworm motor and force sensor. In Proceedings of the Solid-State, Actuators, and Microsystems Workshop: Technical Digest, Hilton Head, SC, USA, 3–7 June 2018; Lamers, T., Rais-Zadeh, M., Eds.; Transducer Research Foundation, Inc.: San Diego, CA, USA, 2018; pp. 306–309; ISBN 978-1-940470-03-0. [CrossRef]

79. Schindler, C.B.; Greenspun, J.T.; Gomez, H.C.; Pister, K.S.J. A Jumping Silicon Microrobot with Electrostatic Inchworm Motors and Energy Storing Substrate Springs. In Proceedings of the 20th International Conference on Solid-State Sensors, Actuators and Microsystems & Eurosensors, (TRANSDUCERS & EUROSENSORS XXXIII), Berlin, Germany, 23–27 June 2019; IEEE: New York, NY, USA, 2019; pp. 88–91, ISBN 978-1-5386-8104-6.

80. Rauf, A.M.; Kilberg, B.G.; Schindler, C.B.; Park, S.A.; Pister, K.S.J. Towards Aerodynamic Control of Miniature Rockets with MEMS Control Surfaces. In Proceedings of the 33rd IEEE International Conference on Micro Electro Mechanical Systems, (MEMS 2020), Vancouver, BC, Canada, 18–22 January 2020; IEEE: New York, NY, USA, 2020; pp. 523–526, ISBN 978-1-7281-3581-6.

81. Schindler, C.B.; Gomez, H.C.; Acker-James, D.; Teal, D.; Li, W.; Pister, K.S.J. 15 Millinewton Force, 1 Millimeter Displacement, Low-Power MEMS Gripper. In Proceedings of the 33rd IEEE International Conference on Micro Electro Mechanical Systems, (MEMS 2020), Vancouver, BC, Canada, 18–22 January 2020; IEEE: New York, NY, USA, 2020; pp. 485–488, ISBN 978-1-7281-3581-6.

82. Contreras, D.S. Walking Silicon: Actuators and Legs for Small-Scale Terrestrial Robots. Ph.D. Thesis, University of California, Berkeley, CA, USA, 1 May 2019.

83. Contreras, D.S.; Drew, D.S.; Pister, K.S.J. First steps of a millimeter-scale walking silicon robot. In Proceedings of the 19th International Conference on Solid-State Sensors, Actuators and Microsystems, (TRANSDUCERS '17), Kaohsiung, Taiwan, 18–22 June 2017; Fang, W., Ed.; IEEE: Piscataway, NJ, USA, 2017; pp. 910–913, ISBN 978-1-5386-2732-7.

84. Robert, P.; Saias, D.; Billard, C.; Boret, S.; Sillon, N.; Maeder-Pachurka, C.; Charvet, P.L.; Bouche, G.; Ancey, P.; Berruyer, P. Integrated RF-MEMS switch based on a combination of thermal and electrostatic actuation. In Proceedings of the 12th International Conference on Solid State Sensors, Actuators and Microsystems: Digest of Technical Papers, (TRANSDUCERS '03), Boston, MA, USA, 8–12 June 2003; IEEE: New York, NY, USA, 2003; pp. 1714–1717, ISBN 0-7803-7731-1.

85. Alwan, A.; Aluru, N.R. Analysis of Hybrid Electrothermomechanical Microactuators With Integrated Electrothermal and Electrostatic Actuation. *J. Microelectromech. Syst.* **2009**, *18*, 1126–1136. [CrossRef]

86. Chae, U.; Yu, H.-Y.; Lee, C.; Cho, I.-J. A Hybrid RF MEMS Switch Actuated by the Combination of Bidirectional Thermal Actuations and Electrostatic Holding. *IEEE Trans. Microw. Theory Tech.* **2020**, *68*, 3461–3470. [CrossRef]

87. Bourbon, G.; Minotti, P.; Langlet, P.; Masuzawa, T.; Fujita, H. Three-dimensional active microcatheter combining shape memory alloy actuators and direct-drive tubular electrostatic micromotors. In Proceedings of the Micromachining and Microfabrication Conference, (Micromachined Devices and Components IV), Santa Clara, CA, USA, 20 September 1998; French, P.J., Chau, K.H., Eds.; SPIE: Washington, DC, USA, 1998; pp. 147–158.

88. Potkay, J.A.; Wise, K.D. A Hybrid Thermopneumatic and Electrostatic Microvalve with Integrated Position Sensing. *Micromachines* **2012**, *3*, 379–395. [CrossRef]

89. Ikehashi, T.; Ohguro, T.; Ogawa, E.; Yamazaki, H.; Kojima, K.; Matsuo, M.; Ishimaru, K.; Ishiuchi, H. A Robust RF MEMS Variable Capacitor with Piezoelectric and Electrostatic Actuation. In Proceedings of the IEEE MTT-S International Microwave Symposium Digest, San Francisco, CA, USA, 11–16 June 2006; IEEE: New York, NY, USA, 2006; pp. 39–42, ISBN 0-7803-9541-7.

90. Toda, R.; Yang, E.-H. Fabrication and Characterization of Vertical Inchworm Microactuator. In Proceedings of the ASME International Mechanical Engineering Congress and Exposition, Anaheim, CA, USA, 13–10 November 2004; ASME: New York, NY, USA, 2004.

91. Toda, R.; Yang, E.-H. A normally latched, large-stroke, inchworm microactuator. *J. Micromech. Microeng.* **2007**, *17*, 1715–1720. [CrossRef]

92. Poletkin, K.; Lu, Z.; Wallrabe, U.; Badilita, V. A New Hybrid Micromachined Contactless Suspension With Linear and Angular Positioning and Adjustable Dynamics. *J. Microelectromech. Syst.* **2015**, *24*, 1248–1250. [CrossRef]

93. Poletkin, K.; Korvink, J. Modeling a Pull-In Instability in Micro-Machined Hybrid Contactless Suspension. *Actuators* **2018**, *7*, 11. [CrossRef]

94. Poletkin, K. On the Static Pull-In of Tilting Actuation in Electromagnetically Levitating Hybrid Micro-Actuator: Theory and Experiment. *Actuators* **2021**, *10*, 256. [CrossRef]

95. Sari, I.; Kraft, M. A MEMS linear accelerator for levitated micro-objects. *Sens. Actuators A Phys.* **2015**, *222*, 15–23. [CrossRef]

96. Corporate Research & Development Center, Toshiba. Low-Cost and Reliable Package for RF-MEMS Tunable Capacitor. Available online: https://www.global.toshiba/ww/technology/corporate/rdc/rd/fields/10-e13.html (accessed on 26 January 2023).

97. Laermer, F.; Franssila, S.; Sainiemi, L.; Kolari, K. Deep Reactive Ion Etching. In *Handbook of Silicon Based MEMS Materials and Technologies*; Elsevier: Amsterdam, The Netherlands, 2015; pp. 444–469. ISBN 9780323299657.

98. Teal, D.; Gomez, H.C.; Schindler, C.B.; Pister, K.S.J. Robust Electrostatic Inchworm Motors for Macroscopic Manipulation and Movement. In Proceedings of the 21st International Conference on Solid-State Sensors, Actuators and Microsystems, (TRANSDUCERS '21), Online Event, 20–25 June 2021; IEEE: New York, NY, USA, 2021; pp. 635–638.

99. van Merlijn Spengen, W. MEMS reliability from a failure mechanisms perspective. *Microelectron. Reliab.* **2003**, *43*, 1049–1060. [CrossRef]

100. Subhash, G.; Corwin, A.D.; de Boer, M.P. Operational Wear and Friction in MEMS Devices. In Proceedings of the ASME International Mechanical Engineering Congress and Exposition, Microelectromechanical Systems, Anaheim, CA, USA, 13–19 November 2004; ASMEDC: Washington, DC, USA, 2004; pp. 207–209, ISBN 0-7918-4714-4.
101. Fang, W.; Li, S.-S.; Li, M.-H. Leveraging semiconductor ecosystems to MEMS. In Proceedings of the 36th IEEE International Conference on Micro Electro Mechanical Systems, (MEMS 2023), Munich, Germany, 15–19 January 2023.

 *actuators*

Article

Non-Inchworm Electrostatic Cooperative Micro-Stepper-Actuator Systems with Long Stroke

Lisa Schmitt [1], Peter Conrad [1], Alexander Kopp [2], Christoph Ament [2] and Martin Hoffmann [1,*]

1   Microsystems Technology (MST), Ruhr-University Bochum, Universitätsstraße 150, DE-44801 Bochum, Germany; lisa.schmitt-mst@rub.de (L.S.)
2   Control Engineering, University Augsburg, Eichleitnerstraße 30, DE-86159 Augsburg, Germany
*   Correspondence: martin.hoffmann-mst@rub.de; Tel.: +49-(0)234-32-27700

Abstract: In this paper, we present different microelectromechanical systems based on electrostatic actuators, and demonstrate their capacity to achieve large and stepwise displacements using a cooperative function of the actuators themselves. To explore this, we introduced micro-stepper actuators to our experimental systems, both with and without a guiding spring mechanism; mechanisms with such guiding springs can be applied to comb-drive and parallel-plate actuators. Our focus was on comparing various guiding spring designs, so as to increase the actuator displacement. In addition, we present systems based on cascaded actuators; these are converted to micromechanical digital-to-analog converters (DAC). With DACs, the number of actuators (and thus the complexity of the digital control) are significantly reduced in comparison to analog stepper-actuators. We also discuss systems that can achieve even larger displacements by using droplet-based bearings placed on an array of aluminum electrodes, rather than guiding springs. By commutating the voltages within these electrode arrays, the droplets follow the activated electrodes, carrying platforms atop themselves as they do so. This process thus introduces new applications for springless large displacement stepper-actuators.

Keywords: cooperative electrostatic actuators; long stroke; large displacement; stepper actuators; spring-less actuators; EWOD; digital microfluidics; liquid bearings; MEMS; SOI

 check for updates

Citation: Schmitt, L.; Conrad, P.; Kopp, A.; Ament, C.; Hoffmann, M. Non-Inchworm Electrostatic Cooperative Micro-Stepper-Actuator Systems with Long Stroke. *Actuators* **2023**, *12*, 150. https://doi.org/10.3390/act12040150

Academic Editor: Qingan Huang

Received: 16 February 2023
Revised: 16 March 2023
Accepted: 27 March 2023
Published: 30 March 2023

1. Introduction

To start, a definition of "long-stroke microactuation" is needed, as its definition varies depending on the point of view. In some applications, the required actuation range is comparable to the spatial dimensions of springs or other key components; in others, it can be up to about 10 times higher. The travel range is usually limited from a few micrometers up to a few tens of micrometers. Consequently, we define "long stroke" as actuations of a few hundred micrometers or more. This violates the common assumption that micromechanical springs or cantilevers can exhibit only a "small" deflection and stay within the proportional range. Consequently, there are very few examples in the literature of long-stroke microactuators being based on silicon-based MEMS. However, in most long-stroke systems the moving payload is limited to, e.g., a mirror or a platform that has to be moved.

Concerning the use of electrostatic comb-drive actuators to achieve large displacements, Grade et al. [1] varied the length of their electrodes and used springs in an initially bent configuration, thus achieving displacements of up to 150 μm at less than 150 V; however, electrostatic lateral instability (see below) appears with any larger displacement [2]. Zhou et al. [3] used tilted folded-beams, increasing the lateral stability; compared to a conventional serpentine spring, the achieved displacement of their microactuator was almost double, increasing from 33 μm to 61 μm. Lateral stability also improved with the placement of symmetrical and tilted serpentine springs on both sides of the actuators [4]. The combination of multiple comb-drive actuators allowed not only bi-directional deflection, but also deflection in preferred (as well as orthogonal) directions [5]. Xue et al. [5] built an XYZ microstage for 3D atomic force microscopy (AFM), wherein the comb-drive actuators were guided by folded flexure springs, achieving a maximum deflection of 50.5 μm at 80.9 V. A deflection of 245 μm at 120 V was achieved by Olfatnia et al. [6] using a bending mechanism based on a clamped paired double parallelogram (C-DP-DP).

Parallel-plate actuators were our second basic type of electrostatic actuators, which can achieve a continuous displacement that is usually only stable up to 1/3 of the electrode distance, although numerous investigations are aimed at controlling deflection beyond this pull-in effect [7]. Nonetheless, this "digital" switching also offers opportunities for stepwise actuation at relatively low voltages. Additionally, coating the electrodes with an isolating layer (e.g., SiN or SiO_2) can prevent current flow, should the electrodes touch. Legtenberg et al. [8] insulated curved electrodes with silicon nitride and reported an experimental electrode tip deflection of 30 μm at 40 V. Preetham et al. [9,10] used both bent and insulated electrodes, achieving peak-to-peak deflections of 19.5 μm and forces of 43 μN at a voltage of ±8 V [9]. In another study [11], long, flexible, and electrically insulated electrodes (in combination with cascaded cantilevers of varying stiffness) achieved deflections of more than 62.5 μm at a voltage of less than 30 V. In another [12], a cascaded mechanism was introduced, comprising similar actuators with long and flexible electrodes; these actuators were arranged in a chain and connected using mechanical springs. The electrode distance of each actuator increases, such that their 16-step actuator displaced up to 230.7 ± 0.9 μm at 54 V.

For MEMS applications, electrostatic forces have a significant advantage compared to, e.g., thermal or magnetic actuation; this is because their lack of a permanent current flow results in a higher energy efficiency, charge transfers being only required for a change in their position, or to compensate for the small leak of currents that may discharge the electrodes. For this reason, long-stroke electrostatic actuation is widely used in MEMS; e.g., in THz systems [13], for optical switching [11], in capacitors [14], micromirrors [15], relays [16], and biomedical systems [17]. The electrostatic actuators benefit from their fast response to electrical signals [1,18] and their suitability for harsh environments [9]. Their use in underwater [9] and close-to-body systems for haptic sensing [19] requires electrostatic actuators that can achieve large displacements, short switching times, and a high reaction force at low voltages. Depending on its application, a mechanical spring can define an initial position, and allow a reset of the microsystem back into this position. In contrast, springless actuators are adequate in those applications that require the preservation of a stable position after actuation.

Long-stroke electrostatic actuators face the principal challenge that electrostatic forces fade with increasing stroke. In microsystems, the moving part (rotor) is usually held in place with solid springs that exhibit an almost linear increase in restoring forces. This can be overcome by using a freely moving rotor that has been clamped with additional actuators, a so-called inchworm-like actuation, which has been successfully demonstrated in studies [20–22]. For this inchworm-like actuation, the challenge lies in the precise measurement of its rotor position.

In this study, we present our investigation into non-inchworm cooperative electrostatic actuators with long stroke, stepwise displacement. Two main concepts are discussed: firstly, different combinations of spring-guided electrodes, and secondly, concepts based on liquid bearings, using virtual springs caused by surface tension effects. Section 2 is devoted to spring-based actuators, wherein we have primarily focused on the main challenges to achieving high and step deflections with the use of electrostatic spring-based actuators, in addition to presenting systems using long-stroke parallel-plate and comb-drive actuations. Section 3 is about spring-less long stroke/infinite stroke actuators with electrostatic actuation, wherein liquid microdroplets act as virtual springs. Section 4 shows a summary and an outlook.

2. Cooperative Spring-Based Long-Stroke Electrostatic Actuators

2.1. Electrostatic Vertical Attracting and Tangential Forces

With the rise in the use of microactuators, the importance of electrostatic forces has become apparent. The "small" gaps in microsystems, in combination with the *Paschen effect*, allows for the use of electrical field strengths far above the values achievable in macroscopic systems before causing an electrical breakdown.

A limiting factor for long-stroke electrostatic actuators is the small level of force, which decreases rapidly as the gap between the actuated electrodes increases. Electrostatic forces are proportional to the (already quite small) constant of vacuum permittivity; $\varepsilon_0 = 8.854 \times 10^{-12} \frac{As}{Vm}$. The material-related relative permittivity ε_r is almost 1 for gaps filled with gases, and typically <100 for common polar liquids (about 80 for water). Therefore, electrostatic actuators play no role in macroscopic actuation, and their being considered "long-stroke" depends on the point of view. For this paper, we have defined "Long-stroke" as meaning approximately 100 µm.

The electrostatic force is usually calculated from the directional derivative $F = -\frac{\partial W}{\partial C}$ of the energy stored in a capacitor:

$$W = \frac{1}{2}CU^2 \tag{1}$$

Generally, common electrostatic actuators use either the vertical attracting force or the tangential force, both being always present between the plates of a capacitor. With the vertical attracting force, the parallel-plate actuators achieve stable displacements up to 1/3 of the electrode distance, as shown in Figure 1a. The displacement of the comb-drive actuators is based on the tangential force caused between two not fully overlapping plates (Figure 1b). Unfortunately, both forces do not exhibit a linear force–deflection graph that pairs well with linear guiding springs.

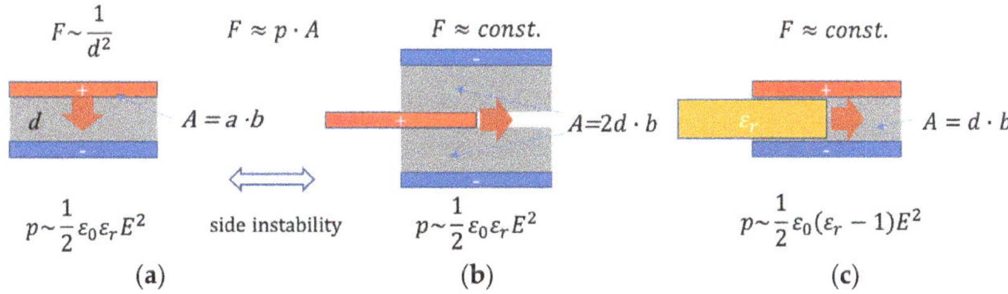

Figure 1. Overview of electrostatic actuators with tangential and vertical attracting forces. The electrostatic pressure and the relevant areas A for integration are given, respectively.

A normalized expression with respect to the electrode area is introduced here, comparable to an "electrostatic pressure" p_e that is also commonly used in dielectric elastomer actuators [23]. Additionally, the voltage is replaced by the local electrostatic field $E = \frac{U}{d}$, where d is the shortest relevant distance between the electrodes, as fringing fields are initially neglected here.

With b being the out-of-plane width, this results in:

- Vertical parallel-plate actuation:

$$F_v = \frac{1}{2}\varepsilon_0\varepsilon_r\frac{A}{d^2}U^2, normalized : p_{e,v} \sim \frac{1}{2}\varepsilon_0\varepsilon_r E^2 \tag{2}$$

- Tangential parallel-plate comb actuation:

$$F_t = \varepsilon_0\varepsilon_r\frac{b}{d}U^2, normalized : p_{e,t} \sim \frac{1}{2}\varepsilon_0\varepsilon_r E^2 \tag{3}$$

- The relevant normalization area is given using the gap area in the direction of movement: $A = b \cdot 2d$, and for standard "dry" actuators with gas-filled gap d, the dielectric constant ε_r is almost 1. As the area to which the pressure applies does not change with actuation, the resulting force is constant as long as both plates do not fully overlap. Additionally, a tangential actuation of dielectrics, particularly liquids, can be approximated in a similar way. In this case, any dielectric (dielectric constant $\varepsilon_r > 1$) is drawn into an air-filled ($\varepsilon_r \approx 1$) capacitor (see Figure 1c). Here, the calculation of the electrostatic pressure results in a quite similar formula: $p_{e,liquid} \approx \frac{1}{2}\varepsilon_0(\varepsilon_r - 1)E^2$; note that this electrostatic pressure is usually much larger due to the factor $(\varepsilon_r - 1)$ and will be used for the springless long-stroke actuation.
- The electrostatic pressure is proportional to E^2.

2.2. Guiding Mechanisms for Long-Stroke Actuators

Two important parameters for the design of cooperative long-stroke microactuators are the required load and the force at the maximum stroke. Whereas solutions for large forces at small stroke scales are well known, a long stroke is much more difficult, as most of the microactuators make use of rigid springs and the force of the spring increases with the size of the stroke. Furthermore, to guarantee a translational displacement, the springs have to exhibit anisotropic spring constants. However, the stiffness' in all other directions directly influences the achievable stroke of the actuator system. Guiding springs with optimized designs should provide a low spring constant in the actuated direction, but a strong guidance both in the other spatial directions and with respect to parasitic rotational movements. Therefore, the reaction force F_x and stiffness c_x in the direction of deflection must be low, but the stiffnesses in all other directions need to be much higher. The selectivity S is a suitable parameter for this demand:

$$S = \frac{c_{i,i \neq x}}{c_x} \tag{4}$$

A high selectivity is naturally provided with clamped beams [24], as shown in Figure 2a. However, actuators guided by clamped beams achieve only small displacements due to the strongly non-linear force–displacement characteristic (FDC) increasing sharply with displacement [25]. Figure 2b shows serpentine springs (also called folded or meander springs), which exhibit a constant low spring stiffness in deflection direction, and are therefore often used for guiding electrostatic actuators. Serpentine springs allow deflection in one direction, only. Due to their sharply decreasing selectivity in line with deflection, the serpentine springs tend to exhibit significant lateral instability when guiding long-stroke actuators [26]. Even the smallest unbalances during fabrication can cause a rotation due to uncompensated torque in the cantilevers.

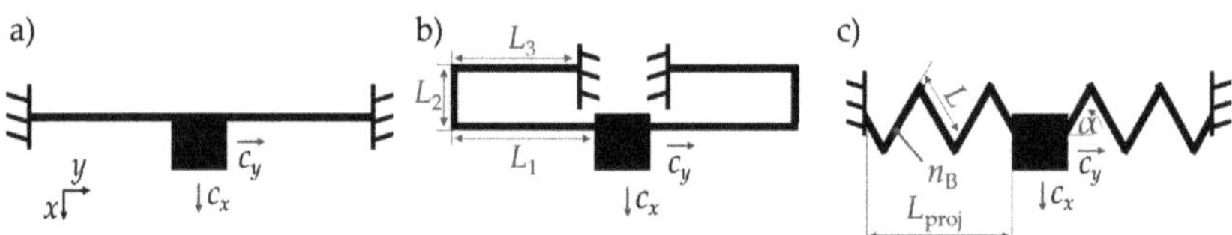

Figure 2. (**a**) a clamped-clamped beam, (**b**) a serpentine spring, (**c**) a triangular spring.

Further common spring geometries for in-plane deflection include the M-shaped springs with non-linear, asymmetric force–displacement characteristics [27]. An advanced geometry is provided for both triangular and sinusoidal springs [28]. The triangular springs (Figure 2c) have an almost linear force–displacement characteristic with a high selectivity, achieving large deflections in both positive and negative x-directions. These springs are highly variable in appearance, as the number of spring segments (n_B), the length L of each beam, and the inclination angle α can be varied. The triangular springs can be transferred to sinusoidal springs. The sinusoidal shape reduces the local stress appearing at the kink from one beam to the next and thus reduces the risk of rupture [28].

Figure 3 compares the c_x- and c_y- stiffness of a triangular and a serpentine spring. Both springs exhibit the same stiffness and selectivity in idle mode. With increasing stroke, the stiffness c_x quadratically increases for the triangular spring, while it remains constant for the serpentine spring. The stiffness c_y shows a sharp drop for the serpentine spring, while it remains almost constant for the triangular spring. The selectivity S resulting from (4) decreases much faster for the serpentine spring than for the triangular spring.

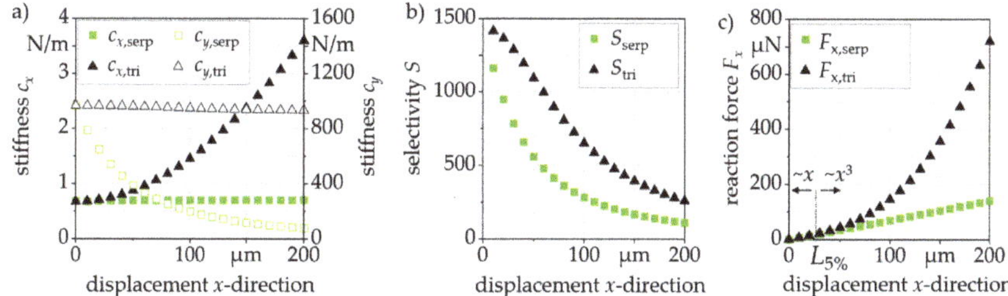

Figure 3. Comparison of the triangular and serpentine springs: (**a**) stiffness, (**b**) selectivity (COMSOL *Multiphysics* solid-state simulation results), (**c**) spring reaction force $F_{x,\mathrm{serp}}$ and $F_{x,\mathrm{tri}}$.

Figure 3c shows the force reaction of both springs. The spring force F_x of a triangular spring in the x-direction has been approximated for an even number of beam elements n_B [28]:

$$
F_{x,\mathrm{tri}} = \frac{12AEI_Z}{L(-1+n_B)\left(AL^2(-1+n_B)^2\cos(\alpha)^2+12I_Z\sin(\alpha)^2\right)}\cdot x \\
+ \frac{144AEI_Z\left(1+n_B\right)}{5LL_{\mathrm{proj}}n_B{}^2\left(48I_Z(-1+n_B)^2\cos(\alpha)^2+AL^2\left(1-8n_B+4n_B^2\right)\sin(\alpha)^2\right)}\cdot x^3, n_B \in \{2,4,\ldots\} \tag{5}
$$

Here, L_{proj} is the projected length of the triangular spring, I_Z the second moment of area, E the Young's modulus and the area A is the product of the beam length L and the substrate depth. From (5), the force $F_{x,\mathrm{tri}}$ depends on the displacement via the factors x and x^3. Consequently, the force–displacement characteristic has both a linear and a non-linear component. According to [28], the linear part of the force $F_{x,\mathrm{tri,linear}}$ is described as

$$
F_{x,\mathrm{tri,linear}} = x\cdot\left(\frac{(n_B-1)^3L^3\cos^2(\alpha)}{12EI_z} + \frac{(n_B-1)L\sin^2(\alpha)}{EA}\right)^{-1}, n_B\in\{2,4,\ldots\} \tag{6}
$$

The linearity of the triangular spring increases with increasing angle of inclination α. With an increasing number of beams, the linearity of the spring decreases; however, the selectivity increases in the ratio $S \sim n_B^2$ [28]. $L_{5\%}$ describes the region in which the force–displacement characteristic deviates by a maximum of 5% from a linear curve. $L_{5\%}$ is presented in Figure 3c and approximately described as follows [28]:

$$
L_{5\%} \approx \sqrt{5\%\cdot\frac{5L^2\cos^2(\alpha)}{12}}\cdot\sqrt{\left(\frac{48I_z(n_B-1)^2+AL^2\left(1-8n_B+4n_B^2\right)\tan^2(\alpha)}{AL^2(n_B-1)^2+12I_z\tan^2(\alpha)}\right)}, n_B\in\{2,4,\ldots\} \tag{7}
$$

In addition to the above springs, bi-stable curved springs (with their negative spring rate [29,30], as shown in Figure 4) should be highlighted for use in electrostatic actuation. These springs can achieve high deflections, but their force–displacement characteristic is nonlinear. These guiding mechanisms can even achieve constant force responses [31,32]. Springs that directly apply a constant force are space saving, and a simple alternative to the complex force–feedback systems that are often used to control the actuator force [33,34].

Boudaoud et al. [35] used a comb-drive actuator guided by bending beams clamped on both sides, connected to a complex capacitive force sensor. Such constant force mechanisms (CFM) achieve a constant output force at specific deflection levels by applying an external force, such as a voltage [36,37]. In the constant force region, the stiffness of such springs is ideally 0 N/m. A constant force can also be achieved by means of, e.g., buckling a spring mechanism [37–39], which results in increased deflections and reduced switching voltages [40].

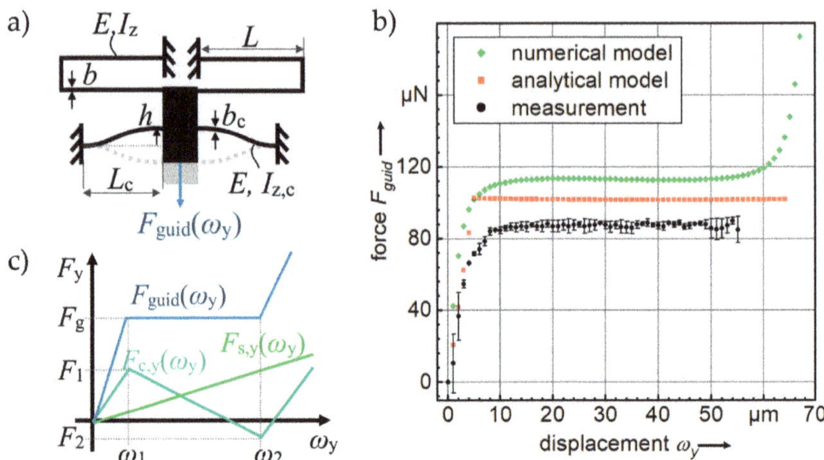

Figure 4. Constant-force mechanisms: (**a**) Setup, (**b**) the constant force characteristics resulting from a mechanical anti-spring with the negative spring rate $F_{c,y}$, and (**c**) a spring with the linear spring rate $F_{s,y}$ combined with a spring with bi-stable spring rate $F_{c,y}$ resulting in a constant force F_{guid}, as presented in [39].

The electrode shape can also be used to generate constant forces or to realize large displacements. There are several approaches for special comb drive finger designs, such as those in [25,41]. Engelen et al. presented a finger shape, providing a constant available force that is up to 1.8 times larger than using a straight finger design, and reducing the actuator size [42]. By varying the shape and length of the comb-drive fingers, the comb-drive actuators can achieve larger displacements, as demonstrated by Grade et al. in [1]. In [43] it is shown that comb drives with variable-gap profiles can be designed to deliver specific driving force profiles.

2.3. Electrostatic Comb Drives for Long-Stroke Actuation

Tangential electrodes appear to be ideal for long-stroke actuators; their force does not depend on the geometry in direction of movement as long as the two electrodes do not overlap completely. The total force can easily be increased using comb-drive actuators with a reasonable number of fingers, while the parallel-plate forces simply cancel out due to their opposed direction. A decrease in the lateral distance d increases the force, allowing for a dense comb array and, thus, higher forces. Theoretically, the stroke seems to be almost "infinite" as long as the two electrodes completely overlap. The most critical limiting effect results from the increasing overlap and thus the resulting parasitic vertical forces.

However, as a closer view shows, this is an incomplete picture of events. The use of comb-drive actuators is in fact limited by two main properties: On one hand, the electrostatic force stays constant over the complete travel range while the retracting force of the springs increases with the stroke, which results in an increased driving voltage. On the other hand, the area at which the parallel-plate forces are acting increases with a larger overlap. For comb-drives, where the tangential derivative is important, this seems to be less critical. Unfortunately, the well-investigated side instability depends on $\frac{1}{d^2}$ and any unavoidable small asymmetries from fabrication cause a deflection to one of the two electrodes, which in turn leads to the moving part sticking to their current position [7,26]. The side-instability is significantly influenced by the guiding spring. Legtenberg et al. [26] have demonstrated that serpentine springs work well for long strokes thanks to their low stiffness c_x. The characteristic displacement of the voltage-dependent displacement is shown in Figure 5. Lateral pull-in due to lateral instability occurs in Figure 5c. However, the stiffness of serpentine springs in off-axis direction decreases with increasing deflection, as shown in a long-stroke system (Figure 6a), causing lateral instability in the comb-drive actuator.

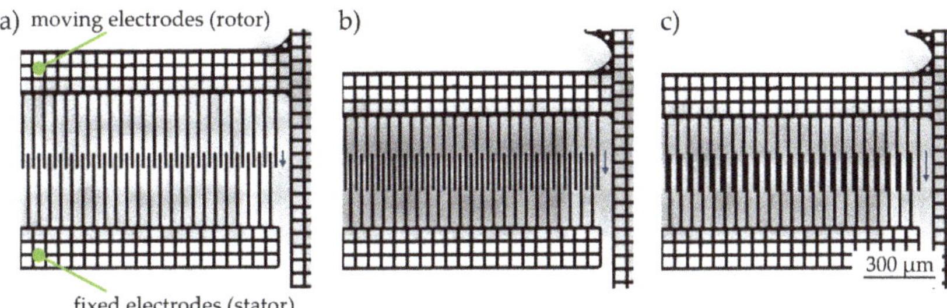

Figure 5. Experimental voltage-dependent displacement of a comb-drive actuator (**a**) after applying a voltage, (**b**) further increase in the voltage and displacement, (**c**) sudden pull-in due to lateral instability.

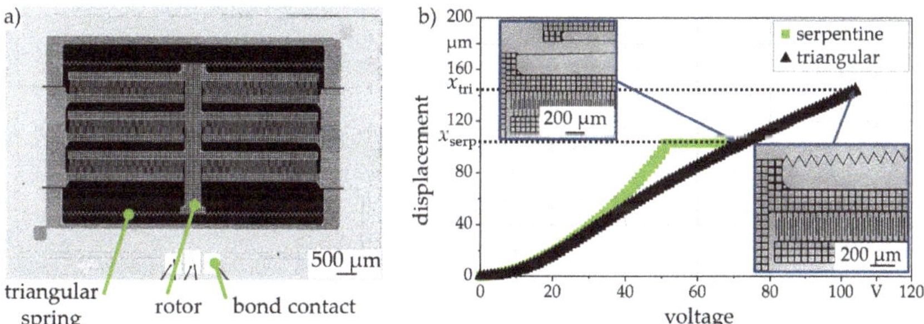

Figure 6. (**a**) Microscopic image of a characterized comb-drive actuator with a triangular spring, (**b**) characterization result comparing the displacement of comb-drive actuators.

Within the stable region of the comb-drive actuator, the vertical forces on the left and right of the electrodes cancel out:

$$F_{\mathrm{el}} = \frac{n\varepsilon_0 t_F(x+x_0)}{2(d+y)^2}U^2 - \frac{n\varepsilon_0 t_F(x+x_0)}{2(d+y)^2}U^2 \tag{8}$$

The unstable electrode deflection starts as soon as the first derivative of the electrostatic force orthogonal to the deflection direction exceeds the stiffness of the spring c_y. The opposing forces of the electrodes no longer cancel out, meaning that the rotor approaches the stator, blocking further deflection in the x-direction:

$$c_y > \frac{2n\varepsilon_0 t_F(x+x_0)}{d^3}U^2 \tag{9}$$

The voltage U_{SI} at which the lateral instability occurs depends on the stiffness c_x and c_y and thus the selectivity S, as well as on the actuator geometry:

$$U_{SI}^2 = \frac{d^2 c_x}{2\varepsilon_0 t_F n}\left(\sqrt{2\frac{c_y}{c_x} + \frac{x_0^2}{d^2}} - \frac{x_0}{d}\right) = \frac{d^2 c_x}{2\varepsilon_0 t_F n}\left(\sqrt{2S + \frac{x_0^2}{d^2}} - \frac{x_0}{d}\right) \tag{10}$$

The maximum stable displacement x_{SI} depends on the electrode spacing, the initial overlap, and the selectivity of the spring. For a spring with very high selectivity ($c_y \gg c_x$), the maximum deflection reaches:

$$x_{SI} = d\sqrt{\frac{c_y}{2c_x}} - \frac{x_0}{2} = d\sqrt{\frac{1}{2}S} - \frac{x_0}{2} \tag{11}$$

According to (11), the deflection increases in line with increasing spring selectivity S [26]. Therefore, the springs presented in Figure 3 are suitable to fulfil this requirement. Figure 6a shows a microscopic image of a fabricated actuator with triangular springs. Figure 6b shows the voltage-dependent displacement of two comb-drives: The deflection of the actuator with serpentine springs was approximately quadratic to the applied voltage ($x \sim U^2$), the maximum deflection was limited to 103 µm and 70 V whereas the actuator with triangular springs achieved a maximum deflection of 145 µm at 104 V, an increase of more than 40%. Furthermore, the voltage-displacement characteristic remained almost linear.

2.4. Large Displacements Based on Parallel-Plate Actuators

2.4.1. Electrostatic Parallel-Plate Actuators

Comb-drive actuators usually exhibit a low force, and their displacement requires an active position control. In contrast, parallel-plate actuators allow for a high reaction force. However, the displacement of parallel-plate actuators is limited by the electrode distance d_0, as shown in Figure 7a. With respect to long-stroke actuation, the attractive force between two parallel-plate electrodes seems to be less suitable, as d^2 usually needs to be much smaller than the area of the electrodes. This would result in an increased size of $\frac{A}{d^2}$ for a reasonable force at $d = d_0$, but approaching an infinite force for a vanishing gap $d \to 0$.

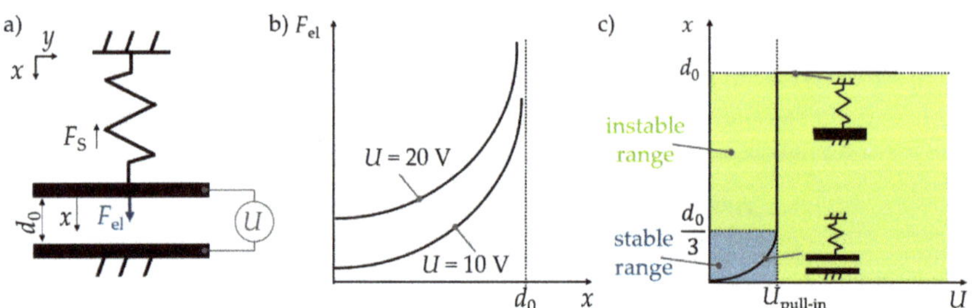

Figure 7. (**a**) Setup of an electrostatic parallel-plate actuator, (**b**) exemplary force–displacement characteristic curve, (**c**) exemplary electrode displacement depending on the applied voltage.

The characteristic force–displacement curve of a parallel-plate actuator operating against a linear spring is outlined in Figure 7b. It does not allow control of the position across the whole range of the gap d_0. As the electrodes move closer together (with $x = d_0 - d$), the electrostatic force increases with $F \sim \frac{1}{(d_0-d)^2}$ with $\lim_{d \to d_0} \frac{1}{(d_0-d)^2} = \infty$, theoretically approaching an infinitely high electrostatic force.

Assuming a rotor guided by a simple linear spring would result in a non-linear behavior, the restoring mechanical force F_S counteracts the electrostatic force F_{el} between the electrodes. At a displacement of $d = \frac{1}{3}d_0$, the mechanical spring force can no longer balance the steeply increasing electrostatic force; at this point, $F_{el} = F_S$, $\frac{\partial F_{el}}{\partial d} = \frac{\partial F_S}{\partial d}$ applies. Pull-in occurs, the movable electrode is abruptly pulled towards the counter electrode, $(d_0 - d) \to 0$, as depicted in Figure 7c. With A being the surface and c the linear spring constant, the pull-in voltage $U_{\text{pull-in}}$ is simply identified as follows:

$$U_{\text{pull}-\text{in}} \geq \sqrt{\frac{8cd_0^3}{27\varepsilon_0\varepsilon_r A}} \tag{12}$$

Despite this, the pull-in condition allows for a very efficient low-voltage high-force actuator by means of the cascaded stiffness of the springs, as demonstrated in [11]; it can thus achieve a stable "digital" behavior (undeflected/fully deflected at $U = 0/U = U_{\text{pull-in}}$). For this purpose, the electrodes are electrically insulated (e.g., using SiO_2) to prevent a short when the electrodes contact. Here, the bending-plate actuators are combined with triangular/serpentine springs (see Section 2.2) as shown in Figure 8. As known from [44], the stator features electrodes which connect to a rigid cantilever, and the rotor features thin electrodes which connect to the actuated platform, guided by flexible springs. The cascaded stiffness is part of the specific design, as the thinnest cantilevers bend at the start of the process, then becoming more and more attracted, and finally end up fixed to the stiffer cantilevers.

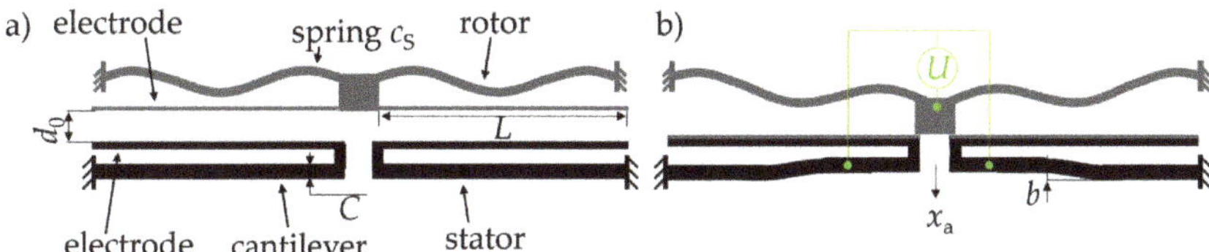

Figure 8. Schematic of the bending-plate actuator: (**a**) setup, (**b**) completely pulled-in actuator.

The switching speed is limited by squeeze film damping, as the air within the gap between the electrodes has to be squeezed out; a process that is accelerated by higher forces at higher voltages [45]. During pull-in, the stiff cantilever also slightly bends the distance b towards the electrodes, which reduces the displacement of the bending-plate actuator x_a (Figure 8b). Therefore, it is not only the spring stiffness c_s, but also the cascaded cantilever thicknesses that strongly influences the pull-in voltage [44]. For the actuator shown in Figure 8, the pull-in voltage is described as follows [11]:

$$U_{\text{pull}-\text{in}} \geq \sqrt{\frac{2E_{Si}C^3 d_0^3}{27\varepsilon_0\varepsilon_r L^4}} \tag{13}$$

Using insulating layers prevents an electrical short during actuation. The high field strengths across thin electrical insulation bear the risk of self-charging of the dielectric, which can result in a permanent static charge. The worst-case scenario here would be that the actuator is unable to release after switching off the voltage. Replacing the DC voltage with an AC voltage reduces the risk of permanent charging but increases the power consumption as a frequency-dependent AC current is required to charge and discharge the electrodes. It has to be noted that the AC frequency needs to be significantly higher than the mechanical resonance of the system to prevent a parasitic resonant excitation of the mechanical system. For DC actuation, the required permanent electrical power is almost close to zero after charging only once. Alternatively, an electrical short can be prevented using local spacers or pull-down resistors [12]

2.4.2. Electrostatic Parallel-Plate Actuators with Multiple Stable Steps

To achieve long-stroke actuation, systems based on cascaded actuators are our objective. An effective way to achieve such systems is to stack bending-plate actuators with stepwise increasing gaps and connecting springs c_i [12,44,46]. From the initial position (Figure 9a), the actuation starts by applying a voltage to the first actuator a_1, after which the system moves for $x_{actuator,1}$ and the gap at actuator a_2 decreases. By additionally applying a voltage to the second actuator a_2, the system displaces again, now for the distance $x_{actuator,2}$. When all three actuators a_1, a_2 and a_3 are activated, the system reaches $x_{actuator,3}$. The required voltage does not increase; a_1–a_3 operate at the same pull-in voltage.

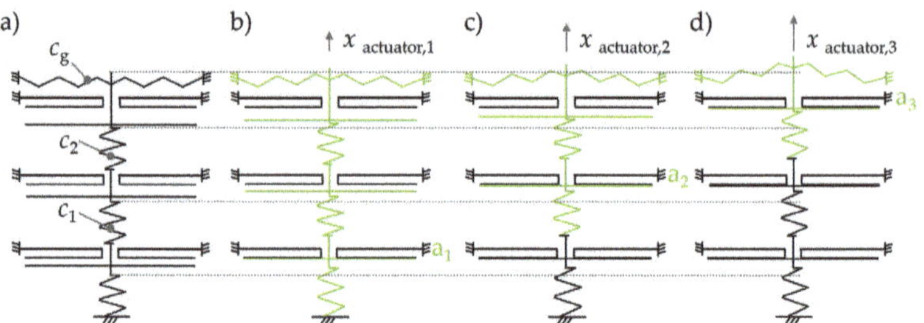

Figure 9. An actuator with three steps: (**a**) initial position, (**b**) the first actuator a_1 is activated and the resulting system displacement is $x_{actuator,1}$, (**c**) the actuators a_1 and a_2 are activated, the system displacement being $x_{actuator,2}$, (**d**) all actuators a_1, a_2 and a_3 are activated, resulting in the system displacement $x_{actuator,3}$.

For large displacements at low voltages, the stiffnesses of the guiding c_g and connecting springs c_i are crucial parameters. The influence of the spring stiffness on the total system displacement $x_{actuator,i}$ depending on the activated actuators a_i [12] is described as follows:

$$x_{actuator,i} = \frac{x_i}{c_g \cdot \left(\frac{n-i}{c_i} + \frac{1}{c_g} \right)}, \text{für } i \leq n \tag{14}$$

For this study, an actuator in line with Figure 9 was realized and evaluated. The actuators were sequentially activated. Figure 10 shows the resulting displacement due to the step-by-step activation of the actuators. The red dots mark the actuators with an applied voltage. In [12], we present the evaluated experimental displacements as well as the dynamic switching behavior, showing eigenfrequencies between 90 Hz (avg. c_g = 2.90 N/m) and 139 Hz (avg. c_g = 0.17 N/m), depending on the stiffness of the guiding spring. A video of the stepwise displacing actuator is available at https://etit.ruhr-uni-bochum.de/mst/forschung/projekte/marie/ accessed on 28 March 2023.

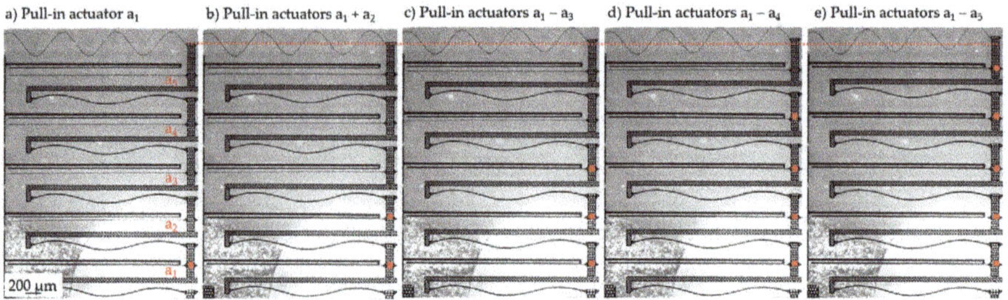

Figure 10. Displacement of the stepwise actuator; the voltage is applied to (**a**) actuator a_1, (**b**) actuator a_1 and a_2, (**c**) actuators a_1 to a_3, (**d**) actuators a_1 to a_4, (**e**) all actuators.

2.4.3. Digital-to-Analog Converting Actuators (DAC)

For the step-like long-stroke actuators presented in Section 2.4.2, the step count was equal to the number of actuators, resulting in a complex system wiring. In this section, we present digital-to-analog converters (DACs) where the number of addressable positions increases exponentially with the number of [47]. A true mechanical DAC converts a binary input code into an analog mechanical output displacement proportional to the analog value of the input code. Thus, the control and wiring of true DACs is significantly less complex than for conventional actuators (for N bit N actuators, as compared to 2^N actuators). In-plane microelectromechanical DACs based on electrostatic actuators have been demonstrated; Toshiyoshi et al. [48] and Sarajlic et al. [47] presented DACs based on comb-drive actuators, and achieved maximum binary encoded displacements of up to 5.8 µm at 150 V and 8.6 µm at 45 V, respectively.

The DAC for long-stroke operation, shown in Figure 11a, consists of three 'bits' [49–51]. Each 'bit' features guiding springs c_{gi}, connecting springs c_{ci}, and two actuators ($a_{i,\text{down}}$ and $a_{i,\text{up}}$). Here, $a_{i,\text{up/down}}$ represents a Boolean variable associated to the ith slider of the DAC, which can be either set to a value of 1, if a voltage is applied, or 0 for a non-active actuator. Therefore, the slider is either displaced upwards, when activating the upper actuator ($a_{i,\text{up}} = 1$ and $a_{i,\text{down}} = 0$, Figure 11d), or downwards, when activating the lower actuator ($a_{i,\text{up}} = 0$ and $a_{i,\text{down}} = 1$, Figure 11b). In our first attempt, the two actuators of each bit were linked by a rigid slider, so each actuator can be individually set in an up or down position, resulting in a mere $2^3 = 8$ positions. In our second approach, each 'bit' could address four different states: 0–0, 1–0, 0–1 and 1–1. In Figure 11d, the upwards actuator $a_{i,\text{up}}$ has been activated and position 1–0 is thus shown. In Figure 11b, position 0–1 is depicted by activating the downwards actuator $a_{i,\text{down}}$. When activating both actuators simultaneously ($a_{i,\text{up}} = a_{i,\text{down}} = 1$), the intermediate position 1–1 is achieved, as shown in Figure 11c. In position 1–0, the bit achieves its largest displacement, in position 1–1 an intermediate displacement and in position 0–1 the smallest displacement [52]. Position 0–0 is an undefined state not suitable for actuation.

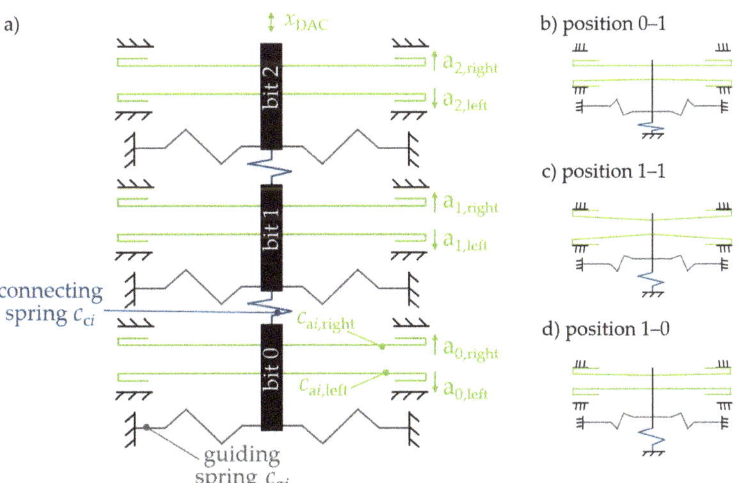

Figure 11. (**a**) Setup of a 3-bit DAC; a single bit in (**b**) position 0–1, (**c**) position 1–1, (**d**) position 1–0, adapted from © [2023] IEEE. Reprinted, with permission, from [L. Schmitt, X. Liu, A. Czylwik and M. Hoffmann, "Large Displacement Actuators with Multi-Point Stability for a MEMS-Driven THz Beam Steering Concept," *Journal of Microelectromechanical Systems*, 2023], [52].

With $N = 3$ connected bits, the resolution of the DAC subsequently increased from $2^3 = 8$ to $3^3 \triangleq 27$, as presented in Figure 12a. Bit 0 proved to be the least significant bit (LSB).

The displacement of the LSB resulted in a partial deformation of the connecting springs c_{c1} and c_{c2}. As the output of the most significant bit (MSB) is not connected to a further spring, the entire MSB displacement resulted in the DAC displacement. Therefore, the impact of each bit on the final system displacement appears to add up along the chain of the connecting springs. The DAC resolution can be enlarged by adding more sliders, such that a combination of, e.g., four or five sliders would result in a resolution of $3^4 \triangleq 81$ or $3^5 \triangleq 243$, respectively. By adding more bits to the DAC, the overall DAC displacement x_{DAC} increased, too. As the DAC is a mechanically connected system, the stiffness of the springs and actuators is important for the linearity of the actuation and a large stroke, which has also been investigated by Sarajlic et al. [47]. The DAC stiffness is important for large displacements when connecting the DAC to a mechanical leverage, which results in an increased DAC displacement (experimentally, up to 234.5 µm, as shown in [53]).

Pos.	Bit 2	Bit 1	Bit 0	Pos.	Bit 2	Bit 1	Bit 0
1	0-1	0-1	0-1	15	1-1	1-1	1-0
2	0-1	0-1	1-1	16	1-1	1-0	0-1
3	0-1	0-1	1-0	17	1-1	1-0	1-1
4	0-1	1-1	0-1	18	1-1	1-0	1-0
5	0-1	1-1	1-1	19	1-0	0-1	0-1
6	0-1	1-1	1-0	20	1-0	0-1	1-1
7	0-1	1-0	0-1	21	1-0	0-1	1-0
8	0-1	1-0	1-1	22	1-0	1-1	0-1
9	0-1	1-0	1-0	23	1-0	1-1	1-1
10	1-1	0-1	0-1	24	1-0	1-1	1-0
11	1-1	0-1	1-1	25	1-0	1-0	0-1
12	1-1	0-1	1-0	26	1-0	1-0	1-1
13	1-1	1-1	0-1	27	1-0	1-0	1-0
14	1-1	1-1	1-1				

(a)

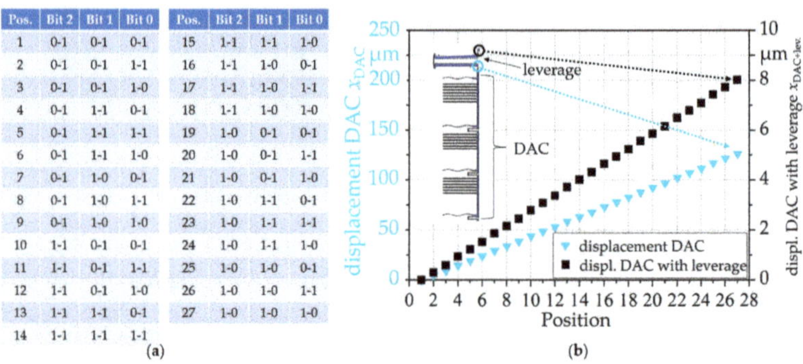

(b)

Figure 12. (**a**) Presentation of the 27 positions in logical order addressed by the 3^3 DAC, (**b**) Simulated displacement of a 3-bit DAC with a 3^3 resolution: the DAC performs a binary encoded stepwise displacement that is finally enlarged using a mechanical leverage (an amplification ratio of ~40) to a stepwise and large mechanical displacement. Experimental results are shown in [51–53].

A video of the mechanical DAC is available at https://etit.ruhr-uni-bochum.de/mst/forschung/projekte/marie/, accessed on 28 March 2023.

3. Springless Long Stroke/Infinite Stroke Actuators

3.1. Actuator Concept

As demonstrated before, a limiting factor for long stroke lengths is the solid spring. With optimized designs, the range can be extended, but for displacements in the mm or even cm range, solutions with non-rigid bearings are better suited. A promising alternative is the use of liquid bearings: surface tension in droplets in the micro-range provides a roller bearing and a virtual spring between two surfaces at the same time, although several constraints have to be fulfilled. The calculation of the spring constant is different from rigid cantilever systems, and the properties of the surfaces as well as of the liquid and their mutual interaction all have an important impact. Nonetheless, the actuation of droplets by applying electrostatic fields is a well-known approach (e.g., in digital microfluidics [53,54], where the movement and manipulation of single droplets is a common application). Especially in medicine and biotechnological applications, microfluidics continues to attract more and more attention. Digital microfluidics (DMF) in particular enables the realization of lab-on-a-chip applications [55]. Unlike in conventional microfluidics, individual droplets are manipulated in DMF. This technology allows for transport, separation, as well as improved mixing and merging processes of the smallest liquid volumes [53]. In combination with other analytical tools, for example, new types of point-of-car platforms can be developed [56]. In addition to the very small volumes that can be processed, the low power consumption and the lack of need for valves and pumps are further advantages of DMF [57].

For cooperative long-stroke actuators, the droplets become part of a liquid roller bearing or "wheel", as well as of the actuator itself: three or more droplets are tethered to a moving platform, e.g., a small substrate, while the actuation is induced using the on-chip travel path.

Electrowetting on Dielectrics (EWOD) is probably the best-known method of actuation in DMF systems [56,57]. EWOD describes the behavior of a conductive or polar liquid on an electrode isolated using a dielectric when a potential difference is applied between them; basically, the fluid is electrically connected. The applied voltage results in a change in the contact angle between the substrate and the drop, which can be described using the Young–Lippmann equation:

$$\cos(\theta) = \cos(\theta_0) + \frac{\varepsilon_0 \varepsilon_d}{2d\gamma_{lv}} U^2 \tag{15}$$

Here, θ_0 is the contact angle between liquid and solid surface at $U = 0$; ε_0 is the permittivity of vacuum, and ε_d is the permittivity of the dielectric layer; d is the thickness of the dielectric, γ_{lv} is the surface tension of the liquid, and U is the applied voltage.

The electrostatic force acting on the droplet can be derived from the change in the energy within the capacitors, or more generally, using the Maxwell stress tensor, see [58,59].

At microscale level, surface effects such as surface tension dominate, such that drops can also be used as liquid bearings in microsystems. In [60], a MEMS micromotor was realized based on such a liquid bearing. In [61], a liquid motor was presented using EWOD, with a droplet serving as a liquid bearing. A further application of EWOD with liquid bearings is shown in [62], wherein four drops carried a micromirror. Changing the contact angle between the drop and the support tilted the mirror, but the mirror was not moved 'in plane'.

The system for long-stroke actuation presented in this study used a combination of lateral droplet transport and the properties of microdroplets as liquid bearings. It consisted of a movable platform resting on four droplets. There were no solid springs connecting the platform with the substrate. The platform was instead moved laterally on a long scale by applying electric fields between planar electrode arrays on the substrate, whereby the drop itself was not in direct contact with an electrode.

Two different shapes of electrode arrays consisting of interdigital electrodes have been investigated so far; one had trapezoidal boarders, and the other one had sine shaped boarders, as shown in Figure 13. The interdigital structures ensured an increased contact line between the droplet and the neighboring electrode. The gap between the two electrodes was set to be about 20 µm. This system was designed for a drop with a contact area radius of 600 µm, corresponding to a drop of, e.g., DI water with a volume of 1.2 µL, and a contact angle of 120° in air. This demonstrator used DI water, but other polar fluids are also suitable, e.g., low-vapor pressure polar fluids such as ethylene glycol ($\varepsilon_r = 37$) and glycerol carbonate ($\varepsilon_r = 110$).

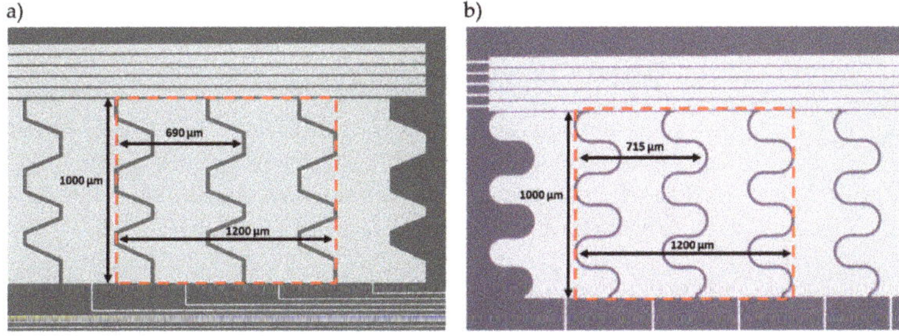

Figure 13. Size and dimensions of (**a**) trapezoidal shaped electrodes and (**b**) sine shaped electrodes. The five finger electrodes on the side of the structure are not relevant for the long stroke actuation

The electrodes were realized using lithographically structured aluminum on silicon, with a 2 μm thermal SiO_2 layer to insulate the electrodes. The dielectric layer consisted of 110 nm Al_2O_3, deposited using atomic layer deposition (ALD). To achieve a reliable hydrophobic surface with a contact angle >90°, a 1% fluoropolymer solution (FluoroPel PFC1601V-FS) was applied by means of spin coating (Figure 14).

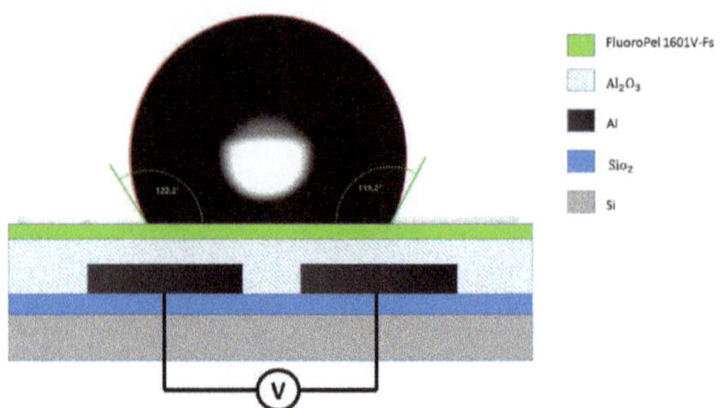

Figure 14. Schematic structure of the actuator system.

The moving platform was a 220 μm thick silicon chip, likewise coated with fluoropolymer. The hydrophobic coating was locally removed from the contact areas where the droplets were to be attached as bearings.

3.2. System Modelling

For an approximated model describing the dynamic behavior of the long-stroke actuator for precise motion control, the system is decomposed into the four droplet bearings suspending the actuator platform. The elastic behavior of the droplets is represented by the spring constant c. A single droplet can be described using the mechanical analogy depicted in Figure 15.

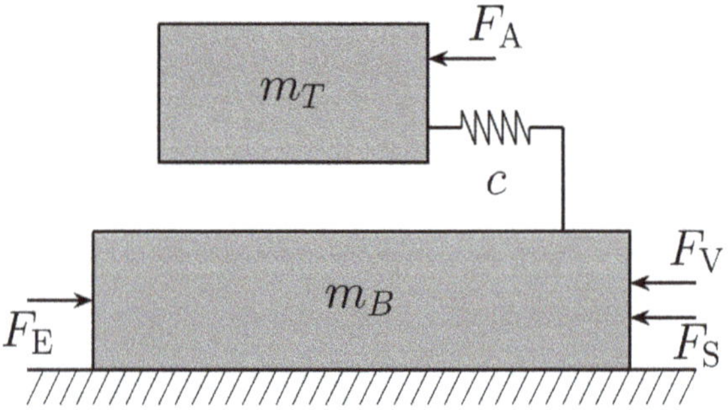

Figure 15. Mechanical analogy for droplet modelling.

The platform mass m_T was connected to the bottom mass m_B, corresponding to the droplet mass, by a linear spring with the stiffness c. In combination with the quadratic air friction force F_A, this enabled the emulation of the droplets ringing behavior. The air friction within the model is given as follows:

$$F_A = \mathrm{sign}(\dot{\rho})d_T\dot{\rho}^2 \tag{16}$$

where $\dot{\rho}$ denotes the velocity of the top mass and d_T denotes the friction coefficient. The combination of the nonlinear friction force F_S and the linear friction force F_V reflects the interaction at the interface between droplet and chip. This allowed us to incorporate the adhesion of the droplet, inhibiting the movement below a certain actuation voltage in the model. These friction forces are modelled as

$$F_s = H_0 e^{-\frac{\dot{p}^2}{v_T^2}} \operatorname{atan}(\alpha \dot{p}) \tag{17}$$

$$F_V = d_2 \dot{p} \tag{18}$$

where $\dot{p}$ denotes the velocity of the bottom mass and H_0 and v_T are calibration coefficients to the behavior of the maximum (force) of the nonlinear friction force F_S. The calibration coefficient α allowed us to adjust the gradient of F_S for a non-moving droplet. The electrostatic force F_E can be derived by means of differentiating the electrostatic energy of the droplet when applying a voltage U, with respect to the position of the droplets base, modelled as the position of the bottom mass m_B, as discussed before. From the FEM-simulation, we derived electrostatic forces up to 9.85 µN for a single droplet.

Using the aforementioned forces, a model for the droplet behavior can be deduced, including unknown parameters such as the spring stiffness c. Such parameters can only be identified using experimental data. The simulation result with these identified parameters for a single droplet (alongside of the used experimental data, as well as the error e between simulation and measurement) are depicted in Figure 16.

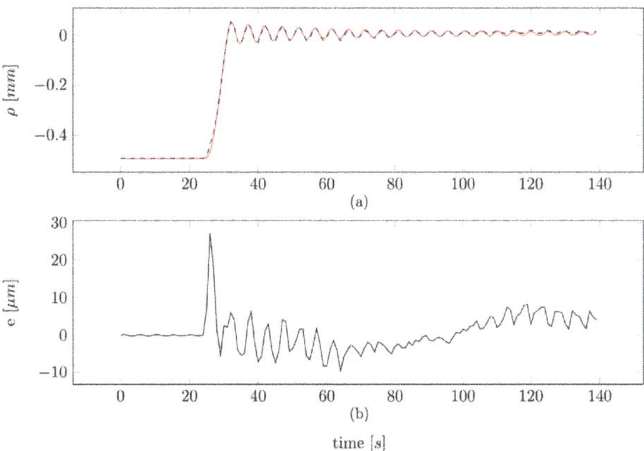

Figure 16. Simulation results (red, solid) and measurement data (black, dashed) for a single droplet (**a**) when combined with the simulation error e (**b**); when the error is in the single-digit percentage range.

The whole actuator was considered as a platform, inextricably connected to the four suspending droplets. To derive a dynamic model, the platform with side length $2a$ was attached at each corner to the four corresponding top masses of the mechanical analogy depicted in Figure 15. Taking only the translational degree of freedom along the electrode array into account, the governing differential equations can be derived using the Lagrangian formalism. Considering the positions p_i of the respective droplet base and ρ the position of the platform as generalized coordinates, and the friction and electrostatic forces as non-conservative forces of a Lagrangian formalism, a set of governing differential equations can be derived. Reformulating these equations using the state vector

$$x - \begin{bmatrix} p_1 & \dot{p}_1 & p_2 & \dot{p}_2 & p_3 & \dot{p}_3 & p_4 & \dot{p}_4 & \rho & \dot{\rho} \end{bmatrix}^T \tag{19}$$

results in a system of first-order differential equations typically used for simulation or control applications:

$$
\dot{x} = \begin{bmatrix}
x_2 \\
\frac{1}{m_{B,1}}\left(F_{E,1}(x_1, V_1) - F_{S,1}(x_2) - d_1 x_2 - k_1(x_1 + a - x_9)\right) \\
x_4 \\
\frac{1}{m_{B,2}}\left(F_{E,2}(x_3, V_2) - F_{S,2}(x_4) - d_2 x_4 - k_2(x_3 + a - x_9)\right) \\
x_6 \\
\frac{1}{m_{B,3}}\left(F_{E,3}(x_5, V_3) - F_{S,3}(x_6) - d_3 x_6 - k_3(x_5 + a - x_9)\right) \\
x_8 \\
\frac{1}{m_{B,4}}\left(F_{E,4}(x_7, V_4) - F_{S,4}(x_8) - d_4 x_8 - k_4(x_7 + a - x_9)\right) \\
x_{10} \\
\frac{k_1(x_1+a-x_9)+k_2(x_3+a-x_9)+k_3(x_5-a-x_9)+k_4(x_7-a-x_9)-\text{sign}(x_{10})d_\rho x_{10}^2}{m_P+m_{T,1}+m_{T,2}+m_{T,3}+m_{T,4}}
\end{bmatrix}
\tag{20}
$$

where the friction coefficient d_ρ incorporates the friction of the platform and all $d_{T,i}$. To reflect the behavior of an existing actuator, the parameters of the system model (20) had to be determined. This was achieved using a system identification approach, as described for a single droplet in [63].

3.3. Experimental Results

3.3.1. Coordinated Droplet Actuation for Moving Platforms

The droplets were positioned on the electrodes by means of applying a voltage on the required electrodes. Figure 17a shows the four liquid bearings in their initial position. To ensure a smooth movement of the platform, all drops needed to be actuated synchronously. For this purpose, the neighboring electrode in front of the drop was switched on, so that the electrostatic force pulled the drop into the field. In this configuration, reproduceable droplet movements can be achieved at voltages as low as $U = 20$ V [64].

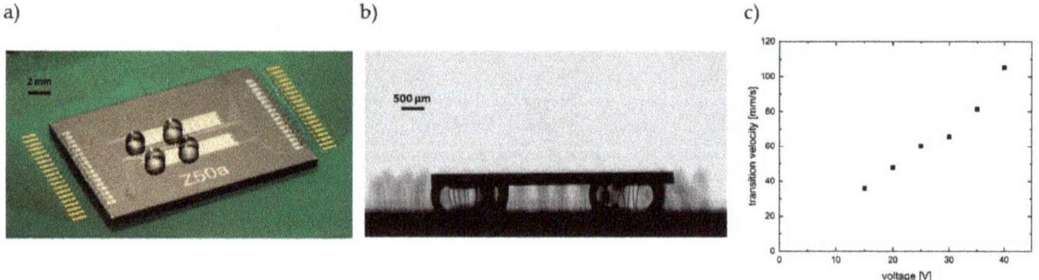

Figure 17. (**a**) four droplets in their initial positions on the two electrode arrays; (**b**) four droplets on the electrode arrays serving as liquid bearings, carrying a 220 µm thick silicon platform; (**c**) Transition velocity of one single liquid bearing without a payload between two electrodes, as a function of the applied voltage.

Figure 17b shows a cross-sectional view of the platform on four droplets. The drops were slightly deformed by the weight of the platform, resulting in a modified contact surface.

A stable movement with the payload of the platform could be achieved at 40 V. A video of the moving platform is available at https://etit.ruhr-uni-bochum.de/mst/forschung/projekte/kommma-raeumliche-aktorik/, accessed on 28 March 2023. The speed of the moving platform depended strongly on the switching period of the electrodes. For a switching time of 85 ms at 42 V, the speed v of the platform was approximately 5.5 mm/s. The demonstrator platform allowed a maximum distance of 1500 µm (Figure 18). Theoretically, the stroke of this actuator system is infinite and determined only by the number of control electrodes.

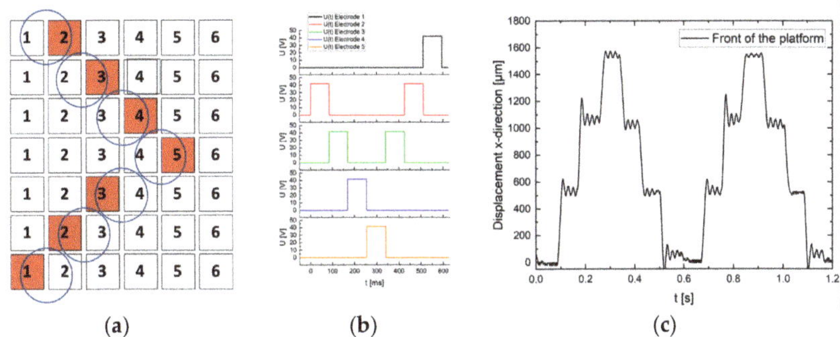

(a) (b) (c)

Figure 18. (a) Actuation scheme for every liquid bearing. Five control electrodes were switched on (red) sequentially for 85 ms at 42 V;**(b)** voltage input timing sequence.. First period guaranteed a stable initial position of the moving platform. **(c)** Platform displacement in *x*-direction. Displacement was measured at the front side of the platform (cf. Figure 19).

Figure 19. Three positions of the moving platform at 40 V with a switching period of *t* = 100 ms.

3.3.2. Multistability

Unfortunately, the described scheme exhibited a disadvantage: if no voltages are applied, the droplets on the non-wetting surface tend to roll off the electrode area if there is any kind of tilt of the electrode substrate. Therefore, it is necessary to "fix" the droplets in their positions. This also allows us to achieve a multi-stable behavior. Such a multi-stable step-like behavior of the platform system was achieved using mechanical barriers introduced into the system in places where the droplet was intended to stick. In a first attempt, photoresist piles were created on the electrode structures (Figure 20). Four of these trapezoidal photoresist structures featured a barrier to hold the droplet in a stable position even when no voltage is applied, or if the chip is tilted.

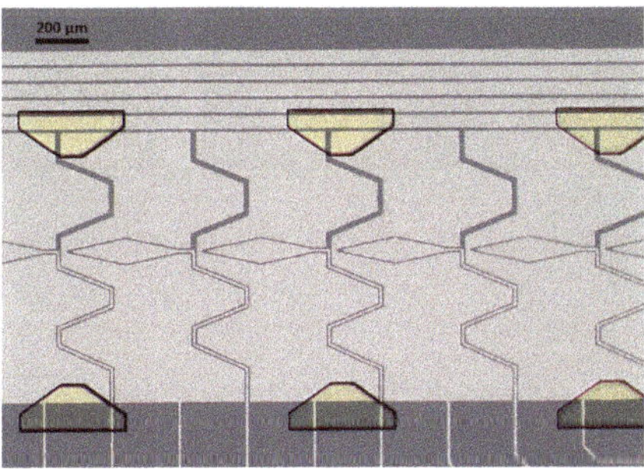

Figure 20. Physical photoresist barriers with a height of approximately 9 µm. Neither the five segmented finger electrodes to the side nor the electrodes in the center of the structure are relevant for the long stroke actuation.

Here, the dielectric Al_2O_3-layer had a reduced thickness of 60 nm. The photoresist is applied on this dielectric, featuring a height of approx. 9 μm. Subsequently, the completed structure is hydrophobically coated with FluoroPel PFC1601V-FS by means of spin coating. The hydrophobic state is shown in Figure 21.

a)   b)

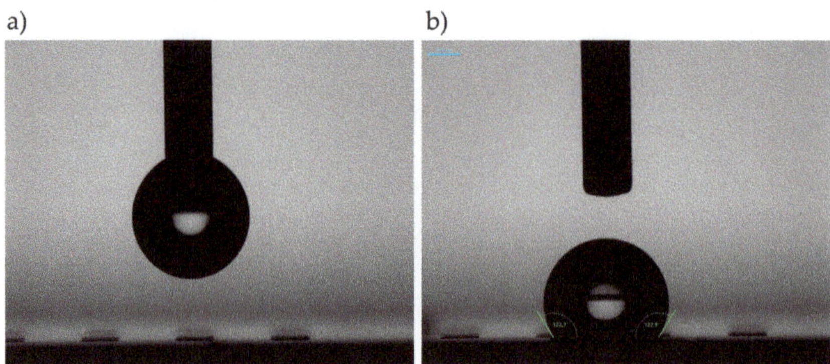

Figure 21. (**a**) Droplet above the photoresist barriers; (**b**) droplet after placement on the hydrophobic structures.

Figure 22 shows four droplets without load that are moved synchronously by means of applying 25 V to the control electrodes. The increased voltage (compared to 20 V in [64]) is needed to overcome the additional barriers. Figure 23 shows the setup with barriers, four liquid bearings and the platform on top. In this case, the platform consisted of a 100 μm thick glass substrate, likewise coated with FluoroPel, and the droplets stuck locally to the platform. Again, two electrodes had to be activated simultaneously to move the platform. A voltage of 35 V was applied to the activated electrodes for 500 ms. The platform was able to move step-by-step. The forward movement is shown in Figure 23. The use of mechanical barriers to create multi-stable states for the drop-based long-stroke actuator was thus successfully demonstrated. These energy states can easily be adjusted using the height and shape of the barriers.

a)   b)   c)

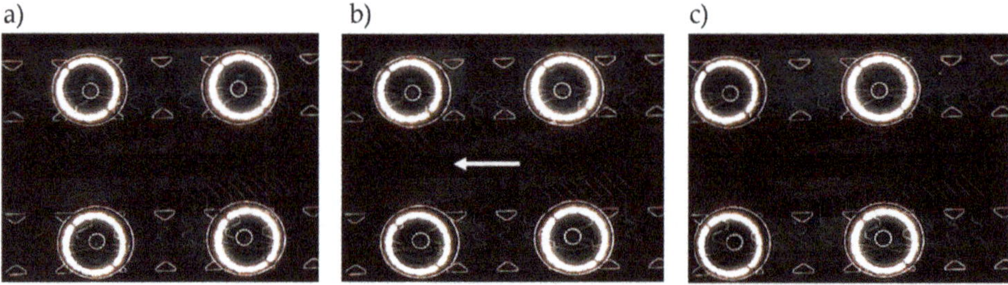

Figure 22. Four droplets moved synchronously from one stable position to the next. (**a**) Four droplets positioned between the photoresist barriers; (**b**) initial movement of the droplets; (**c**) four droplets reaching the next stable position, as defined by the photoresist structures.

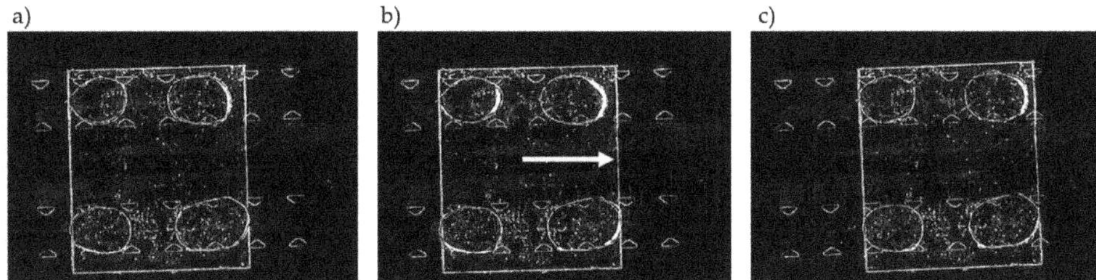

Figure 23. (**a**) Starting position of the platform on four liquid bearings, which are positioned between the photoresist barriers (top view); (**b**) initial moving of the platform; (**c**) liquid bearings of the platform reaching the next stable position, as defined by the photoresist structures.

4. Summary and Outlook

This contribution provided an overview of non-inchworm long-stroke actuators and presented some recent improvements based on evolved spring concepts for common electrostatic actuator systems and a spring-free system with liquid bearings. Such systems have allowed us to integrate step-like actuators with an theoretically infinite stroke length. It should be noted that all systems require actuation voltages <100 V. This voltage range can easily be supplied using surplus up-conversion from low voltages. As almost no current is needed, the power consumptions of these systems is low, and high-impedance voltage sources create no safety risk. Our own unpublished work has shown that such actuators can be powered using RFID or NFC systems.

Electrostatic actuation for long-stroke application can facilitate yet further improvements. Such solutions strongly depend on the constraints of the system environment, and requires both careful design and a selection of the spring-actuator system that is far more meticulous than the comparatively simple basic principles of electrostatic force effects.

Author Contributions: Conceptualization, M.H.; methodology, L.S., P.C. and A.K.; software, L.S., P.C. and A.K.; validation, M.H., C.A., L.S., P.C. and A.K.; formal analysis, L.S., P.C. and A.K.; investigation, L.S., P.C. and A.K.; resources, M.H. and C.A.; data curation, M.H., C.A., L.S., P.C. and A.K.; writing—original draft preparation, M.H., L.S., P.C. and A.K.; writing—review and editing, M.H. and L.S.; visualization, L.S., P.C. and A.K.; supervision, M.H. and C.A.; project administration, M.H. and C.A.; funding acquisition, M.H. and C.A. All authors have read and agreed to the published version of the manuscript.

Funding: Funded by the Deutsche Forschungsgemeinschaft (DFG, German Research Foundation) –Project-ID 287022738—Collaborative Research Centre TRR 196—Mobile Material Characterization and Localization by Electromagnetic Sensing, C12 and Priority Program SPP 2206—KOMMMA—Cooperative Multistage Multistable Microactuator Systems (HO 2284/14-1 and AM 120/9-1).

Data Availability Statement: The data can be provided upon reasonable request.

Acknowledgments: The authors thank the Center for Interface Dominated High Performance Materials (ZGH), Ruhr University Bochum (RUB), for the opportunity to use deep etching equipment. The ALD layers have been fabricated within the BMBF-ForLab-Bochum-PICT2DES-grant number: 16ES0941 and in cooperation with David Zanders, Inorganic Materials Chemistry Group, RUB.

Conflicts of Interest: The authors declare no conflict of interest.

References

1. Grade, J.; Jerman, H.; Kenny, T. Design of large deflection electrostatic actuators. *J. Microelectromechanical Syst.* **2003**, *12*, 335–343. [CrossRef]
2. Jerman, J.H.; Grade, J.D.; Drak, J.D. Electrostatic Microactuator and Method for Use Thereof. U.S. Patent 5,998,906, 7 December 1999.
3. Zhou, G.; Dowd, P. Tilted folded-beam suspension for extending the stable travel range of comb-drive actuators. *J. Micromechanics Microengineering* **2002**, *13*, 178 183. [CrossRef]

4. Chen, C.; Lee, C. Design and modeling for comb drive actuator with enlarged static displacement. *Sens. Actuators A Phys.* **2004**, *115*, 530–539. [CrossRef]

5. Xue, G.; Toda, M.; Li, X.; Wang, X.; Ono, T. Assembled Comb-Drive XYZ-Microstage With Large Displacements and Low Crosstalk for Scanning Force Microscopy. *J. Microelectromechanical Syst.* **2021**, *31*, 54–62. [CrossRef]

6. Olfatnia, M.; Sood, S.; Gorman, J.J.; Awtar, S. Large Stroke Electrostatic Comb-Drive Actuators Enabled by a Novel Flexure Mechanism. *J. Microelectromechanical Syst.* **2012**, *22*, 483–494. [CrossRef]

7. Zhang, W.-M.; Yan, H.; Peng, Z.-K.; Meng, G. Electrostatic pull-in instability in MEMS/NEMS: A review. *Sens. Actuators A Phys.* **2014**, *214*, 187–218. [CrossRef]

8. Legtenberg, R.; Gilbert, J.; Senturia, S.D.; Elwenspoek, M. Electrostatic curved electrode actuators. *J. Microelectromechanical Syst.* **1997**, *6*, 257–265. [CrossRef]

9. Preetham, B.S.; Lake, M.A.; Hoelzle, D.J. A curved electrode electrostatic actuator designed for large displacement and force in an underwater environment. *J. Micromechanics Microengineering* **2017**, *27*, 095009. [CrossRef]

10. Preetham, B.S.; Mangels, J.A. Performance evaluation of a curved electrode actuator fabricated without gold/chromium conductive layers. *Microsyst Technol* **2018**, *24*, 3479–3485.

11. Hoffmann, M.; Nüsse, D.; Voges, E. Electrostatic parallel-plate actuators with large deflections for use in optical moving-fibre switches. *J. Micromechanics Microengineering* **2001**, *11*, 323–328. [CrossRef]

12. Schmitt, L.; Hoffmann, M. Large Stepwise Discrete Microsystem Displacements Based on Electrostatic Bending Plate Actuation. *Actuators* **2021**, *10*, 272. [CrossRef]

13. Sun, Q.; He, Y.; Liu, K.; Fan, S.; Parrott, E.P.J.; Pickwell-MacPherson, E. Recent advances in terahertz technology for biomedical applications. *Quant. Imaging Med. Surg.* **2017**, *7*, 345–355. [CrossRef]

14. Bakri-Kassem, M.; Mansour, R.R. High Tuning Range Parallel Plate MEMS Variable Capacitors with Arrays of Supporting Beams. In Proceedings of the 19th IEEE International Conference on Micro Electro Mechanical Systems, Istanbul, Turkey, 22–26 January 2006; pp. 666–669. [CrossRef]

15. Zhang, J.; Zhang, Z.; Lee, Y.C.; Bright, V.M.; Neff, J. Design and invention of multi-level digitally positioned micromirror for open-loop controlled applications. *Sens. Actuators A Phys.* **2003**, *103*, 271–283. [CrossRef]

16. Spencer, M.; Chen, F.; Wang, C.C.; Nathanael, R.; Fariborzi, H.; Gupta, A.; Kam, H.; Pott, V.; Jeon, J.; Liu, T.-J.K.; et al. Demonstration of Integrated Micro-Electro-Mechanical Relay Circuits for VLSI Applications. *IEEE J. Solid-State Circuits* **2010**, *46*, 308–320. [CrossRef]

17. Velosa-Moncada, L.A.; Aguilera-Cortés, L.A.; González-Palacios, M.A.; Raskin, J.-P.; Herrera-May, A.L. Design of a Novel MEMS Microgripper with Rotatory Electrostatic Comb-Drive Actuators for Biomedical Applications. *Sensors* **2018**, *18*, 1664. [CrossRef]

18. Kostsov, E.; Alexei, S. Fast-response electrostatic actuator based on nano-gap. *Micromachines* **2017**, *8*, 78. [CrossRef]

19. Leroy, E.; Hinchet, R.; Shea, H. Multimode Hydraulically Amplified Electrostatic Actuators for Wearable Haptics. *Adv. Mater.* **2020**, *32*, 2002564. [CrossRef]

20. Toda, R.; Yang, E.-H. A normally latched, large-stroke, inchworm microactuator. *J. Micromechanics Microengineering* **2007**, *17*, 8. [CrossRef]

21. Kloub, H. Design Concepts of Multistage Multistable Cooperative Electrostatic Actuation System with Scalable Stroke and Large Force Capability. In Proceedings of the ACTUATOR—International Conference and Exhibition on New Actuator Systems and Applications, Online, 17–19 February 2021; pp. 1–4.

22. Teal, D.; Gomez, H.C.; Schindler, C.B.; Pister, K.S.J.; Teal, D. Robust electrostatic inchworm motors for macroscopic manipulation and movement. In Proceedings of the 21st International Conference on Solid-State Sensors, Actuators and Microsystems (Transducers), online virtual conference, 20–25 June 2021.

23. Romasanta, L.J.; López-Manchado, M.A.; Verdejo, R. Increasing the performance of dielectric elastomer actuators: A review from the materials perspective. *Prog. Polym. Sci.* **2015**, *51*, 188–211. [CrossRef]

24. Schomburg, W.K. *Introduction to Microsystem Design*; Springer: Berlin/Heidelberg, Germany, 2011.

25. Gao, Y.; You, Z.; Zhao, J. Electrostatic comb-drive actuator for MEMS relays/switches with double-tilt comb fingers and tilted parallelogram beams. *J. Micromechanics Microengineering* **2015**, *25*, 45003. [CrossRef]

26. Legtenberg, R.; Groeneveld, A.W.; Elwenspoek, M. Comb-drive actuators for large displacements. *J. Micromech. Microeng.* **1996**, *6*, 320–329. [CrossRef]

27. Leadenham, S.; Erturk, A. M-shaped asymmetric nonlinear oscillator for broadband vibration energy harvesting: Harmonic balance analysis and experimental validation. *J. Sound Vib.* **2014**, *333*, 6209–6223. [CrossRef]

28. Schmitt, P.; Schmitt, L.; Tsivin, N.; Hoffmann, M. Highly Selective Guiding Springs for Large Displacements in Surface MEMS. *J. Microelectromechanical Syst.* **2021**, *30*, 597–611. [CrossRef]

29. Qiu, J.; Lang, J.; Slocum, A. A centrally-clamped parallel-beam bistable MEMS mechanism. Technical Digest. MEMS 2001. In Proceedings of the 14th IEEE International Conference on Micro Electro Mechanical Systems, Interlaken, Switzerland, 25 January 2001. [CrossRef]

30. Qiu, J.; Lang, J.; Slocum, A. A Curved-Beam Bistable Mechanism. *J. Microelectromechanical Syst.* **2004**, *13*, 137–146. [CrossRef]

31. Vysotskyi, B.; Parrain, F.; Aubry, D.; Gaucher, P.; Le Roux, X.; Lefeuvre, E. Engineering the Structural Nonlinearity Using Multimodal-Shaped Springs in MEMS. *J. Microelectromechanical Syst.* **2017**, *27*, 40–46. [CrossRef]

32. Vysotskyi, B.; Parrain, F.; Aubry, D.; Gaucher, P.; Lefeuvre, E. Innovative Energy Harvester Design Using Bistable Mechanism With Compensational Springs In Gravity Field. *J. Physics Conf. Ser.* **2016**, *773*, 012064. [CrossRef]

33. Li, B.; Li, G.; Lin, W.; Xu, P. Design and constant force control of a parallel polishing machine. In Proceedings of the 2014 4th IEEE International Conference on Information Science and Technology, Shenzhen, China, 26–28 April 2014; pp. 324–328.

34. Erlbacher, E. Method for Applying Constant Force with Nonlinear Feedback Control and Constant Force Device using Same. US Patent 5,448,146, 5 September 1995.

35. Boudaoud, M.; Haddab, Y.; Le Gor, Y. Modeling and optimal force control of a nonlinear electrostatic microgripper. *IEEE/ASME Trans. Mechatron.* **2012**, *18*, 1130–1139. [CrossRef]

36. Yang, S.; Xu, Q. Design and simulation of a passive-type constant-force MEMS microgripper. In Proceedings of the 2017 IEEE International Conference on Robotics and Biomimetics (ROBIO), Macau, China, 5–8 December 2017; pp. 1100–1105. [CrossRef]

37. Wang, P.; Xu, Q. Design and modeling of constant-force mechanisms: A survey. *Mech. Mach. Theory* **2018**, *119*, 1–21. [CrossRef]

38. Shahan, D.; Fulcher, B.; Seepersad, C.C. Robust design of negative stiffness elements fabricated by selective laser sintering. In Proceedings of the 2011 International Solid Freeform Fabrication Symposium, Austin, TX, USA, 8–10 August 2011.

39. Thewes, A.C.; Schmitt, P.; Löhler, P.; Hoffmann, M. Design and Characterization of an Electrostatic Constant-Force Actuator Based on a Non-Linear Spring System. *Actuators* **2021**, *10*, 192. [CrossRef]

40. Zhang, X.; Wang, G.; Xu, Q. Design, Analysis and Testing of a New Compliant Compound Constant-Force Mechanism. *Actuators* **2018**, *7*, 65. [CrossRef]

41. Guo; Tatar, E.; Fedder, G.K. Large-displacement parametric resonance using a shaped comb drive. In Proceedings of the IEEE 26th International Conference on Micro Electro Mechanical Systems (MEMS), Taipei, Taiwan, 20–24 January 2013.

42. Engelen, J.B.C.; Abelmann, L.; Elwenspoek, M.C. Optimized comb-drive finger shape for shock-resistant actuation. *J. Micromechanics Microengineering* **2010**, *20*, 105003. [CrossRef]

43. Ye, W.; Mukherjee, S.; MacDonald, N. Optimal shape design of an electrostatic comb drive in microelectromechanical systems. *J. Microelectromechanical Syst.* **1998**, *7*, 16–26. [CrossRef]

44. Schmitt, L.; Schmitt, P.; Hoffmann, M. Optimization of Electrostatic Bending-Plate Actuators: Increasing the Displacement and Adjusting the Actuator Stiffness. In Proceedings of the ACTUATOR 2022: International Conference and Exhibition on New Actuator Systems and Applications, Mannheim, Germany, 29–30 June 2021.

45. Toshiyoshi, H. Electrostatic Actuation. In *Comprehensive Microsystems*; Yogesh, B., Ed.; Gianchandani, Osamu Tabata, Hans Zappe: Paris, France, 2008; pp. 1–38.

46. Schmit, L.; Schmitt, P.; Barowski, J.; Hoffmann, J. Stepwise Electrostatic Actuator System for THz Reflect Arrays, GMM-Fachbericht 98: ACTUATOR 2021. In Proceedings of the International Conference and Exhibition on New Actuator Systems and Applications, Online, 17–19 February 2021.

47. Sarajlic, E.; Collard, D.; Toshiyoshi, H.; Fujita, H. 12-bit microelectromechanical digital-to-analog converter of displacement: Design, fabrication and char-acterization. In Proceedings of the IEEE 20th International Conference on Micro Electro Mechanical Systems (MEMS), Kyoto, Japan, 21–25 January 2007.

48. Toshiyoshi, H.; Kobayashi, D.; Mita, M.; Hashiguchi, G.; Fujita, H.; Endo, J.; Wada, Y. Microelectromechanical digital-to-analog converters of displacement for step motion actuators. *J. Microelectromechanical Syst.* **2000**, *9*, 218–225. [CrossRef]

49. Sarajlic, E.; Collard, D.; Toshiyoshi, H.; Fujita, H. Design and modeling of compliant micromechanism for mechanical digital-to-analog conversion of displacement. In Proceedings of the IEEJ Transactions on Electrical and Electronic Engineering, Online, 15–20 June 2007; pp. 357–364.

50. Schmitt, L.; Schmitt, P.; Hoffmann, M. Mechanischer 3-Bit Digital-Analog-Wandler (DAC) mit großem Stellweg. In *Proceedings of the MikroSystemTechnik Kongress, Germany, Ludwigsburg, 8–10 November 2021*; VDE-Verlag: Stuttgart-Ludwigsburg, Germany, 2021; ISBN 978-3-8007-5656-8.

51. Schmitt, L.; Schmitt, P.; Hoffmann, M. 3-Bit Digital-to-Analog Converter with Mechanical Amplifier for Binary Encoded Large Displacements. *Actuators* **2021**, *10*, 182. [CrossRef]

52. Schmitt, L.; Liu, X.; Schmitt, P.; Czylwik, A.; Hoffmann, M. Large Displacement Actuators With Multi-Point Stability for a MEMS-Driven THz Beam Steering Concept. *J. Microelectromechanical Syst.* **2023**, 1–13. [CrossRef]

53. Cho, S.K.; Moon, H.; Kim, C.-J. Creating, transporting, cutting, and merging liquid droplets by electrowetting-based actuation for digital microfluidic circuits. *J. Microelectromech. Syst.* **2003**, *12*, 70–80. [CrossRef]

54. Berthier, J. Electrowetting Theory. In *Micro and Nano Technologies, Micro-Drops and Digital Microfluidics*, 2nd ed.; William Andrew Publishing: Waltham, MA, USA, 2013; pp. 162–222. [CrossRef]

55. Samiei, E.; Tabrizian, M.; Hoorfar, M. A review of digital microfluidics as portable platforms for lab-on a-chip applications. *Lab Chip* **2016**, *16*, 2376–2396. [CrossRef]

56. Zhang, Y.; Liu, Y. Advances in integrated digital microfluidic platforms for point-of-care diagnosis: A review. *Sensors Diagn.* **2022**, *1*, 648–672. [CrossRef]

57. Berthier, J. EWOD Microsystems. In *Micro and Nano Technologies, Micro-Drops and Digital Microfluidics*, 2nd ed.; William Andrew Publishing: Waltham, MA, USA, 2013; pp. 225–324. [CrossRef]

58. Mugele, F.; Baret, J. C. Electrowetting: From basics to applications. *J. Physics: Condens. Matter* **2005**, *17*, R705–R774. [CrossRef]

59. Jones, T. An electromechanical interpretation of electrowetting. *J. Micromech. Microeng.* **2005**, *15*, 1184–1187. [CrossRef]

60. Chan, M.L.; Yoxall, B.; Park, H.; Kang, Z.; Izyumin, I.; Chou, J.; Megens, M.M.; Wu, M.C.; Boser, B.E.; Horsley, D.A. Design and characterization of MEMS micromotor supported on low friction liquid bearing. *Sensors Actuators A Phys.* **2012**, *177*, 1–9. [CrossRef]

61. Takei, A.; Binh-Khiem, N.; Iwase, E.; Matsumoto, K.; Shimoyama, I. Liquid motor driven by electrowetting. In Proceedings of the in IEEE 21st International Conference on Micro Electro Mechanical Systems, Tucson, AZ, USA, 13–17 January 2008.

62. Kang, H.-H.; Kim, J. EWOD (Electrowetting-on-Dielectric) Actuated Optical Micromirror. In Proceedings of the 19th IEEE International Conference on Micro Electro Mechanical Systems, Istanbul, Turkey, 22–26 January 2006.

63. Kopp, A.; Conrad, P.; Hoffmann, M.; Ament, C. System level modeling and closed loop control for a droplet-based micro-actuator. In Proceedings of the SICE International Symposium on Control Systems 2023 (Part of the 10th SICE Multi-Symposium on Control Systems), Kusatsu, Japan, 9–11 March 2023.

64. Conrad, P.; Berdnykov, A.; Hoffmann, M. Design of a Liquid Dielectrophoresis-driven Platform with Cooperative Actuation. In Proceedings of the ACTUATOR 2022, International Conference and Exhibition on New Actuator Systems and Applications, Mannheim, Germany, 28–30 June 2022.

Article

Multistage Micropump System towards Vacuum Pressure

Martin Richter * , **Daniel Anheuer** , **Axel Wille** , **Yuecel Congar and Martin Wackerle**

Fraunhofer Institute for Electronic Microsystems and Solid State Technologies EMFT, 80686 Munich, Germany; daniel.anheuer@emft.fraunhofer.de (D.A.); axel.wille@emft.fraunhofer.de (A.W.); yuecel.congar@emft.fraunhofer.de (Y.C.); martin.wackerle@emft.fraunhofer.de (M.W.)
* Correspondence: martin.richter@emft.fraunhofer.de; Tel.: +49-89-54759-455

Abstract: Fraunhofer EMFT's research and manufacturing portfolio includes piezoelectrically actuated silicon micro diaphragm pumps with passive flap valves. Research and development in the field of microfluidics have been dedicated for many years to the use of micropumps for generating positive and negative pressures, as well as delivering various media. However, for some applications, only small amounts of fluid need to be pumped, compressed, or evacuated, and until now, only macroscopic pumps with high power consumption have been able to achieve the necessary flow rate and pressure, especially for compressible media such as air. To address these requirements, one potential approach is to use a multistage of high-performing micropumps optimized to negative pressure. In this paper, we present several possible ways to cascade piezoelectric silicon micropumps with passive flap valves to achieve these stringent requirements. Initially, simulations are conducted to generate negative pressures with different cascading methods. The first multistage option assumes pressure equalization over the piezo-actuator by the upstream pump, while for the second case, the actuator diaphragm operates against atmospheric pressure. Subsequently, measurement results for the generation of negative gas pressures down to -82.1 kPa relative to atmospheric pressure (19.2 kPa absolute) with a multistage of three micropumps are presented. This research enables further miniaturization of many applications with high-performance requirements for micropumps, achievable with these multistage systems.

Keywords: silicon; micro diaphragm pump; vacuum; negative pressure; microfluidic; multistage

Citation: Richter, M.; Anheuer, D.; Wille, A.; Congar, Y.; Wackerle, M. Multistage Micropump System towards Vacuum Pressure. *Actuators* **2023**, *12*, 227. https://doi.org/10.3390/act12060227

Academic Editors: Manfred Kohl, Stefan Seelecke and Stephan Wulfinghoff

Received: 4 April 2023
Revised: 24 May 2023
Accepted: 26 May 2023
Published: 31 May 2023

1. Introduction

This article presents a technique for generating negative pressures via serially connecting piezoelectric silicon micromembrane pumps. The development of micropumps for various applications has aimed to generate high-pressure differences using different operating principles while minimizing the system's size and energy consumption. Through cascading several pumps, the entire system can be significantly reduced in size while maintaining performance, allowing for the implementation of complex MEMS systems in a compact package.

High positive pressures are, for instance, required for the bubble-tolerant, safe, and reliable delivery of fluids in medical applications such as insulin dosing or the delivery of pain medication [1]. Further potential areas of operation for micro-compressors include fuel cells [2] as well as micro-coolers, such as those for electronic devices [3]. The generation of low to high vacuum levels is a key technology for many industrial applications and processes, whereby often only a relatively small volume has to be evacuated [4]. Due to leakages or degassing over the lifetime of a device, vacuum sealing is not possible for all applications. Examples are portable miniaturized gas analysis systems for the investigation of gas mixtures in different industries [5] as well as electron optical systems for scanning electron microscopes with multiple beams [6] or micro degassing systems. The various systems have diverse requirements regarding the pressure level to be achieved along with the necessary flow rates [7]. The scientific community has demonstrated an interest in micro vacuum technologies for various applications, as seen in programs such as DARPA's Chip-Scale Vacuum Micro Pumps (CSVMP) [8].

Coarse vacuum is defined as pressures below 0.1 kPa absolute and is typically generated using mechanical pumps, starting at atmospheric pressure. For pressures below this, physical principles such as ion getter pumps are used to achieve high vacuum. Typically, a combination of mechanical pumps and physical pump principles are used to generate high vacuum. The development of mechanical MEMS micro-pumps for generating coarse vacuum is motivated by two factors:

1. The first factor motivating the development of mechanical MEMS micro-pumps for generating coarse vacuum is the high-power consumption of conventional mechanical pumps required to achieve coarse vacuum. Depending on the evacuated volume, such pumps require between ~50 W up to several kW of electrical power and weigh 2 kg or more [9]. In many cases, these vacuum pumps have to operate perpetually. In contrast, a MEMS system with cascaded piezo micromembrane pumps requires less than 1 W of power. However, it is worth noting that the evacuated volume is significantly smaller compared to mechanical vacuum pumps.
2. In addition to use cases for high vacuum, there are also applications that require portable systems with low absolute pressures. One example is the ion mobility spectrometer (IMS), which requires absolute pressures of about 10 kPa or below [10]. Another potential area of operation is the re-calibration of gas sensors on a portable device, where it is advantageous to apply an absolute pressure of approximately 20 kPa to the gas sensor [11]. Furthermore, several portable and battery-powered sensor systems that require vacuum are available, but no commercially available solutions exist in terms of compact size, weight, and energy consumption to enable handheld devices.

Actuation mechanisms for MEMS vacuum pumps are diverse and can be classified into two main categories: mechanical and non-mechanical. Mechanical mechanisms are divided into micro blowers and diaphragm pumps with valves. Micro blowers exploit a fluidic resonance effect at high frequencies and achieve an overpressure of 60 kPa [12]; nevertheless, there is no specification for the minimum negative pressure. Diaphragm pumps with valves include membrane pumps such as electrostatically actuated [13–15] or piezoelectric-driven [16,17] pumps, rotary pumps [18,19], and ejector-driven pumps [20,21]. Non-mechanical mechanisms are based on diffusion or ion-sorption. These various mechanisms are extensively researched. Non-mechanical pumps, such as ion-getter pumps [22,23], Knudsen pumps [24,25], and others, have also been introduced and are aimed towards micro vacuum pump applications.

The limitations of existing piezo-driven MEMS micropumps to achieve strong negative pressures rely on the weak compression ratio of those pumps (ratio between the stroke volume of the actuator to dead volume in the pump chamber). As the maximization of the stroke volume is given according to the actuator properties [26], the reduction of dead volume is limited by both the tolerances of the manufacturing processes as well as the nature of piezo physics, as 85% of the stroke is provided towards the pump chamber bottom. In this approach, a silicon MEMS for dead volume reduction via accurate etching and bonding technologies (silicon fusion bond) is applied. Furthermore, a pretension method (Figure 1) to avoid the drawbacks of piezo physics and without bending of the pump chip is used. With that, a compression ratio which is not yet known for MEMS micropumps is achieved, which is an important precondition to generate a large under pressure with air.

The operating conditions and achieved pressures of numerous micropumps with different actuation principles are very inhomogeneous [7]. Pressure requirements for certain applications are very high and, thus, not achievable with only one micropump; therefore, the combination of several pumps in a series is a promising and straightforward solution. Cascades with electrostatically driven micropumps are shown from Besharatian et al. with 2 to 24 stages with different configurations, where the minimal pressure of 24 stages is reported as ~97 kPa absolute [27–30]. Kim et al. report on 2, 4, and 18 stages with a maximum pressure difference of 17.5 kPa [14]. A piezo-driven two-stage micropump for air compression is published by Le at al. [31]. Non-mechanical Knudsen micropumps based on diffusion (thermal transpiration) for vacuum generation with 1 to 3 stages achieved 46.6 kPa absolute pressure in 1 stage [32]. Another Knudsen pump with 48 stages and a minimum pressure of 6.6 kPa absolute is presented [24].

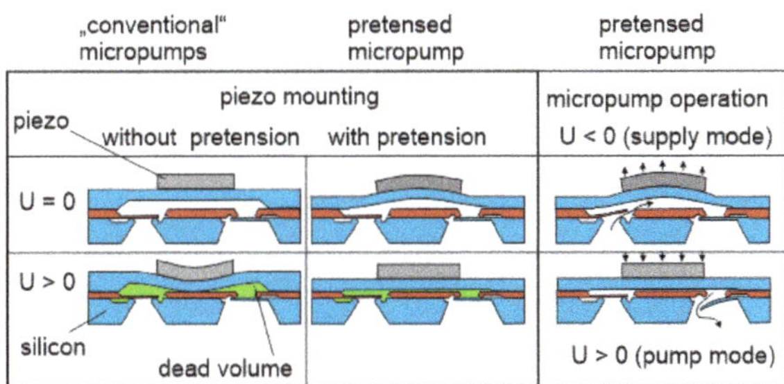

Figure 1. Schematic cross section of micropumps with reduced dead volume V_0 due to an electric pretension of the piezo during the gluing process. The dead volume V_0 (green) is defined as that volume, which remains in the pump chamber, if the actuator is in its lowest position.

The design of the micropumps as well as the pump cycles of each pump in the multistage is discussed in the beginning. The focus of the design section is to explain the actuator properties including stroke volume and blocking pressure with the pump cycle in a p-V diagram in order to approximate the suction pressure $p_{min,gas}$. As mentioned previously, a single pump cannot achieve coarse vacuum due to the compression ratio limitation of MEMS pumps. Therefore, in addition to optimizing a single pump, the cascading of micropumps is chosen as a method to enhance negative pressure.

Based on the design of a single micropump, two different possibilities of cascading are modelled and simulated, including coupling effects, to explore the possibility to achieve smaller absolute pressures. In this publication, two different methods of cascading with silicon micro diaphragm pumps are introduced. The design and simulation of the system is demonstrated using piezo-driven silicon micropumps with passive flap valves. Both cascading methods are applicable for overpressures as well as for negative pressures. In the pressure-balanced method, the generated pressure of the upstream pump also acts on the actuator of the adjacent stage. In the case of serial connection without pressure compensation, the atmospheric pressure applies to the actuator diaphragm of each cascading stage. In an ideal scenario, each pump of the cascade is optimized for the specific position according to its particular geometric properties. Finally, measurements with realized systems of two and three stages of 7 × 7 mm² piezo-actuated silicon micro-membrane pumps with similar design parameters are presented and compared to the simulation.

2. Materials and Methods

This section provides a detailed overview about the micropump design and simulation including different multistage setup options for micropump systems.

2.1. Micropump Design Optimized for Small Absolute Pressures

The micropumps in this study are made of three silicon layers and a PZT disk lead zirconate titanate (from PI Ceramics, type PI 151) acting as a piezoelectric actuator that is glued on top of the diaphragm (Figure 2). The silicon layers are structured using double-sided lithography and anisotropic KOH etching. A cross section with relevant design parameters is shown in Figure 3, and a detailed SEM image of a part of the micromachined valve seat is presented in Figure 4. The top layer forms the actuation diaphragm, which includes the pump chamber, while the two bottom layers form passive cantilever check valves. When a negative voltage is applied to the piezoelectric actuator, the diaphragm moves upwards and expands the pump chamber, resulting in gas being sucked from the inlet valve into the pump chamber, known as the supply mode. Conversely, a positive voltage leads to a downward movement of the actuator, compressing the gas in the pump chamber and causing fluid flow through the outlet valve into the outlet periphery, which is called the pump mode.

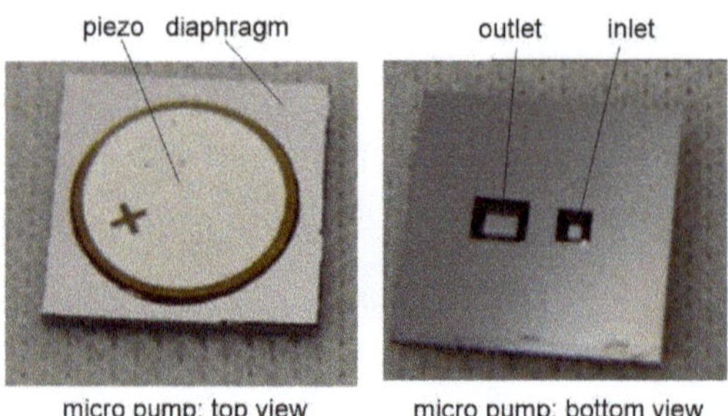

Figure 2. Top and bottom view of the $7 \times 7 \times 0.7$ mm^3 silicon micropump chip with mounted PZT and visible inlet and outlet ports.

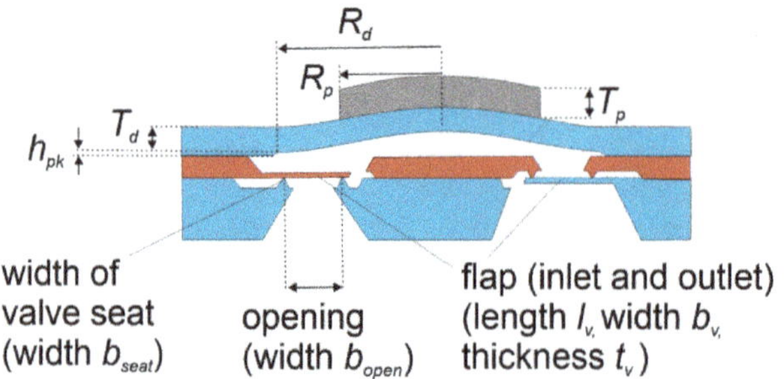

Figure 3. Cross section of the micropump with pretensed diaphragm explaining relevant design parameters. h_{pk} is the pump chamber height, T_d is the diaphragm thickness, T_p is the piezo actuator thickness, R_p is the piezo actuator radius, and R_d is the diaphragm radius.

Figure 4. SEM image of silicon valve sealing lip (width $b_{seat} = 3$ µm) to reduce surface contact area with the silicon flap in order to reduce sticking. In the right bottom corner, there is the opening. At the edge of the valve seat the compensation structures to protect convex corners during KOH etching are shown.

The micropump design is optimized towards a large compression ratio, which is defined as the ratio between the maximum stroke volume ΔV_{max} to the dead volume V_0 of the micropump. In order to reduce the dead volume, the PZT is mounted using a specific pretension method, depicted in Figure 1 [26]. As a result, it is possible to reduce the pump chamber height to a minimum value of 1 μm. Table 1 shows all design parameters of the micropump used for the simulation.

Table 1. Simulation parameters including material data, design geometries, fluid properties, and electrical parameters. All experiments used air as fluid.

Description	Symbol	Value	Unit
Silicon Material Parameter	Y_d	1.6×10^{11}	Pa
	ν	2.5×10^{-1}	1
	α_{Si}	3×10^{-6}	1/K
Piezoelectric Actuator Material Parameter	Y_p	6.7×10^{10}	Pa
	α_{Piezo}	6×10^{-6}	1/K
	d_{31}	2.1×10^{-10}	m/V
	C	2×10^{-9}	F
Micropump Geometry Parameter	h_{pk}	1×10^{-6}	m
	T_d	7×10^{-5}	m
	R_d	$3.15 \cdot 10^{-3}$	m
Piezo Geometry Parameter	T_p	1.5×10^{-4}	m
	R_p	2.84×10^{-3}	m
Environment Properties	p_0	1.013×10^2	kPa
	T	20	°C
Fluid Properties	ρ	1.3	kg/m^3
	η	1.823×10^{-5}	Pa s
Operation Properties	U_+	225	V
	U_-	-60	V

Both passive silicon check valves (inlet valves and outlet valves) have a cantilever (length $l_v = 800$ μm, width $b_v = 460$ μm, thickness $t_v = 15$ μm, opening $b_{open} = 390$ μm) above a square valve seat (width of the valve seat $b_{seat} = 3$ μm). Although a hard-hard seal is achieved between the flap and valve seat, only relatively small leakages (0.06 mL/min at 50 kPa) can be observed when negative pressure is applied. This is due to the polished surfaces of the flap and valve seat, which have a roughness below 1 nm. Next, no plastic deformation behaviour or fatigue can be observed due to the excellent properties of single-crystal silicon. The silicon flap valves are capable of withstanding pressures of up to at least 200 kPa in both directions. At very small positive pressure differences (with our design, below 3.7 kPa), the flow resistance is dominated by laminar gap flow; at higher pressure differences (above 3.7 kPa), the flow resistance is dominated by orifice flow. The behaviour and properties of the passive check valves made of silicon have been previously discussed in [33] and are not further discussed in this paper.

2.2. Pump Cycle of Micropumps Operated with Gases

The relation between the displaced volume V and the pressure p inside the pump chamber in dependence to the voltage U can be approximated as a linear behaviour, depicted in Equation (1). Neglecting the piezoceramic hysteresis and assuming Kirchhoff plate theory, the simplified formula is denoted [26]:

$$V(p, U_-) = C_p(p - p_0) - C_E^*(U_+ - U_-) + V_0 = C_p(p - p_0) + \Delta V + V_0 \tag{1}$$

$$V(p, U_+) - C_p(p - p_0) + V_0. \tag{2}$$

The coefficients C_p and C_E^* are derived analytically in [26,34] and assuming round geometry for piezo as well as diaphragm. It should be mentioned that C_p as well as

C_E^* are extended analytically expressions shown in the annex of reference [28]. They depend on the thicknesses (T_d, T_p) and radii (R_d, R_p) of piezo and diaphragm as well as Young's moduli (Y_d, Y_p) and Poisson ratio ν of the silicon diaphragm. Additionally, C_E^* is direct proportional to the transverse piezo coefficient d_{31}. This coefficient describes the relative lateral shrinking of the piezo ceramics, if an electrical field is applied in vertical direction. The transverse piezo coefficient d_{31} as well as C_E^* is always negative. The voltage stroke $U_+ - U_-$ multiplied with $-C_E^*$ represent the maximal actuator stroke volume ΔV without back pressure. The thermal expansion coefficients of silicon α_{Si} and piezo α_{Piezo} are necessary to estimate the vertical shift of the actuator diaphragm with a changing temperature. This temperature change influences the dead volume V_0 and occurs either during the manufacturing process (e.g., if the glue is hardened at a higher temperature) or during operation (if the pump is not operated at ambient temperature).

$$\Delta V = -C_E^*(U_+ - U_-) \tag{3}$$

The primary obstacle for using piezo-driven MEMS pumps as vacuum pumps is optimizing the compression ratio through increasing the stroke volume and decreasing the dead volume. To illustrate this challenge, we propose a p-V diagram in which the x-axis represents the absolute pressure in the pump chamber and the y-axis represents the volume of the pump chamber, which is enclosed by the actuation diaphragm and check valves. This diagram clearly illustrates the significant impact of both the stroke volume and dead volume on the target parameter $p_{min,gas}$.

With that, the pump cycle is described in a p-V diagram, depicted in Figure 5. Linear behaviour for the negative voltage U_- and the positive voltage U_+ is assumed. The calculated dead volume V_0 of the micropump consists of the remaining pump chamber volume, when the diaphragm is at the lower returning point as well as a certain dead volume from the flap valves. The model assumes an operation of the micropump at atmosphere pressure p_0 with a rapid voltage switch between U_- and U_+ with a certain operation frequency f. A is the starting point of the cycle with a positive electrical voltage U_+ and assuming a stable actuator position after the end of the supply mode. While the actuation voltage decreases, the gas inside the pump chamber is compressed rapidly, for instance, during a time range of 1 ms. As depicted in the p-V diagram (Figure 5), the pump follows according to its equation of state compressing the gas volume until the actuation characteristics of state B is reached. During that short time no significant flow through the valve occurs, resulting in a maximum over pressure $p_{max,gas}$ achieved via the compression. The outlet valve opens and gas flows during a longer time scale towards the outlet, representing the pump mode. Figure 5 depicts this situation during the pressure and volume decrease from state B to state C. At state C, the pump outlet valve is pressure balanced with the periphery and the entire stroke volume is pumped to the outlet, assuming a neglectable leakage flow through the inlet valve. It is considered a stable position after the end of the pump mode.

Defined by the operation frequency at $t = T/2 = 1/(2f)$, the negative voltage U_- is applied rapidly, resulting in an upward movement of the diaphragm towards state D.

The pump chamber volume is increasing, so the gas is expanding to a minimum pressure. The corresponding pressure $p_{min,gas}$ results in a gas flow through the opened inlet valve inside the pump chamber. This phase is named supply mode, until the inlet valve is pressure balanced and state A is reached. The pump cycle is now completed.

The volume difference between the two actuator positions A and C is defined as stroke volume ΔV. Without back pressure, the pump cycle transports this volume from the inlet to the outlet. An operational frequency f causes the pump cycle to be repeated, which is assumed to be sufficiently small to perform the entire stroke with a stroke volume ΔV. The result is an average flow rate of $Q = \Delta V \cdot f$ achieved by the pump. The remaining volume in the pump chamber and the valve unit is defined as dead volume V_0. It is calculated with the diaphragm actuator in its lowest position at U_+ (Figure 5). The compression ratio ε can be defined as the ratio between the maximal stroke volume ΔV and the dead volume V_0.

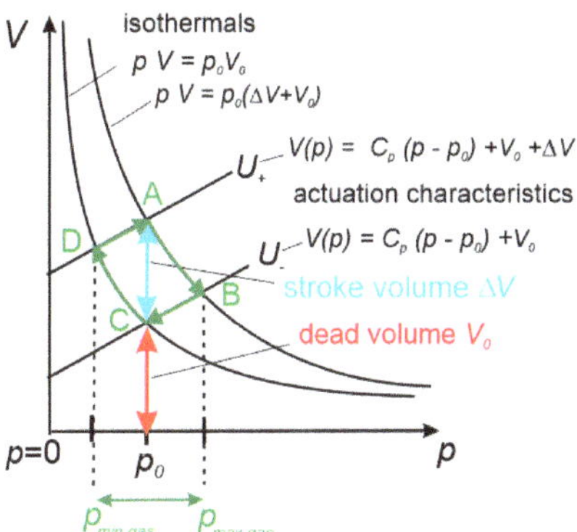

Figure 5. p-V diagram of the pump cycle of a micropump operating at atmosphere pressure p_0 pumping gas.

Combining the actuator characteristics with the isothermals according to Figure 5, the pressures $p_{max,gas}$ and $p_{min,gas}$ can be calculated analytically, depicted in Equations (4) and (5).

$$\text{Pump mode}: \ p_{max,gas} = \frac{C_p p_0 - V_0 + \sqrt{\left(C_p p_0 + V_0\right)^2 + 4C_p p_0 \Delta V}}{2C_p} \tag{4}$$

$$\text{Supply mode}: \ p_{min,gas} = \frac{C_p p_0 - V_0 - \Delta V_{max} \pm \sqrt{\left(C_p p_0 - V_0 - \Delta V\right)^2 - 4C_p p_0 V_0}}{2C_p} \tag{5}$$

In order to achieve the smallest possible absolute pressure, it is important to minimize the dead volume V_0 and to maximize the stroke volume ΔV. It should be mentioned that the hyperbolic shape of the isothermals has a larger gradient at smaller pressures. With that, a given volume stroke is achieving a smaller pressure stroke at the supply mode compared to the pump mode:

$$p_0 - p_{min,gas} < p_{max,gas} - p_0 \tag{6}$$

In this investigation, isothermal equations of state are assumed. In reality, the behaviour can be adiabatic or polytropic, in which case the fundamental conclusion is unaffected. With adiabatic behaviour, the radian is actually smaller, which causes higher pressure peaks.

Hysteresis occurs when the piezo is driven in big signal mode. When the external electrical field applied to the PZT is changed, the atoms react through displacing the load on a very short timescale (less than one microsecond), which is referred to as effect 1. However, the Weiss domains also change their size on a much slower timescale (above one millisecond and longer), referred to as effect 2. This second effect is associated with energy losses. Piezo creeping is a known effect that occurs after a step voltage is applied to the PZT. Initially, the piezo changes about 97% of the stroke immediately (due to effect 1), while the remaining 3% occurs on a logarithmic timescale (due to effect 2) and can take several seconds to achieve the full stroke [35].

To account for these effects, the micropump was operated continuously at a frequency of 100 Hz. This frequency is sufficiently low to avoid inducing piezo heating due to hysteresis losses. Heating of piezo ceramics are not observed until operating frequencies reach several kHz, and such high frequencies result in stroke loss, allowing the piezo creeping effect to be ignored. Additionally, because the voltage always starts and ends at the same level, the same level, the actuator is in a defined position.

2.3. Simulation of Performance Parameters of a Single Micropump

Through combining the input parameters from Table 1 and the theoretical explanation provided in the previous Section 2.2, we have calculated the output parameters for a single micropump, as shown in Table 2.

Table 2. Simulation of performance parameters of a single micropump optimized for low pressure applications.

Description	Symbol	Value	Unit
Performance parameter	ΔV	149	nL
	Δp_{block}	322	kPa
	$p_{max,gas}$	189.5	kPa
	$p_{min,gas}$	41.3	kPa
Design parameter	V_0	84	nL
	ε	1.79	1
Simulation output	h_-	11.58	µm
	h_+	0.96	µm

This micropump is optimized for low-pressure gas applications due to its high compression ratio ε (stroke volume: $\Delta V = 149$ nL, dead volume: $V_0 = 84$ nL). To minimize the dead volume, it is crucial to cycle the actuation diaphragm in close proximity to the pump chamber bottom, resulting in a small value of h_+. The symbols h_- and h_+ describe the distances from the pump chamber bottom to the actuator in the diaphragm centre at negative and positive voltage, respectively. Absolute actuator positions are calculated via considering additional effects such as the different thermal expansion coefficients of silicon and PZT (α_{Si} and α_{Piezo}, respectively), the influence of the mounting voltage, and the operating temperature of the micropump (room temperature) in the simulation.

The most relevant outcome of the simulation is a $p_{min,gas}$ of 41.3 kPa (absolute pressure), which would outperform the state of the art for a single stage micropump. Ideal valves with no leakages as well as ideal bending of the actuator diaphragm are assumed; the real $p_{min,gas}$ is not expected to achieve that value.

2.4. Cascading Micropumps to Achieve Small Absolute Pressures

In this section, different methods for cascading micropumps in a series to enhance the performance of a single micropump and achieve lower absolute pressures are investigated. In order to achieve vacuum pressures, the outlet of stage $n + 1$ is connected to the inlet of stage n (Figure 6). Three different methods to cascade pumps are discussed:

- Cascading without pressure balance, where the micropumps are "just connected" in a series;
- Cascading with pressure balanced outlet, where the reference pressure above the piezo of stage $n + 1$ is connected to the achieved pressure of stage n; and
- Cascading with pressure balanced inlet, where the reference pressure above the piezo of stage n is connected to the achieved pressure of stage n.

It needs to be demonstrated which concept is most promising to achieve the smallest absolute pressure.

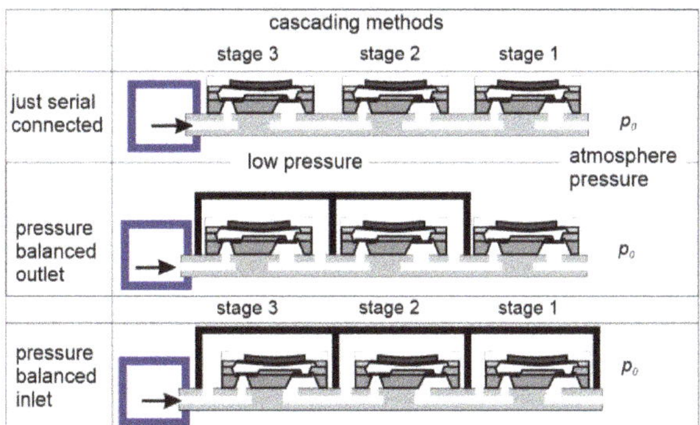

Figure 6. Different methods of cascading micropumps with 3 stages each to achieve vacuum pressure in the blue framed chamber compared to atmosphere pressure p_0. In the just serial connected case (case 1), the actuators of every stage are working against atmospheric pressure. Pressure balanced outlet (case 2a) shows a housing on top of the actuator connected to the micropump outlet to a achieve a pressure balance between outlet and actuator. In the pressure balanced inlet (case 2b), the housing is connected to the micropump inlet of each stage.

2.4.1. Under Pressure with Micropumps: Just Serially Connected (Case 1)

Fluidic capacitance reduces at pressures below atmosphere pressure. Considering that every added stage after the initial atmosphere stage with the same stroke volume result in a smaller gain in under pressure. Especially if the absolute pressure is getting closer to vacuum, the gradient of the isothermals increases dramatically. For this situation a very high compression ratio (high stroke volume ΔV and small dead volume V_0) is crucial to progress towards vacuum pressures.

The reduction of the actuator starting position of the stage 2 diaphragm due to the under pressure achieved at stage 1 does not result in an analogous reduction in dead volume, as the actuation diaphragm gets in contact with the pump chamber bottom (Figure 7). Furthermore, the stroke volume is reduced due to this touch down. In summary, a serial connection alone without pressure balancing is not useful to achieve small absolute pressures.

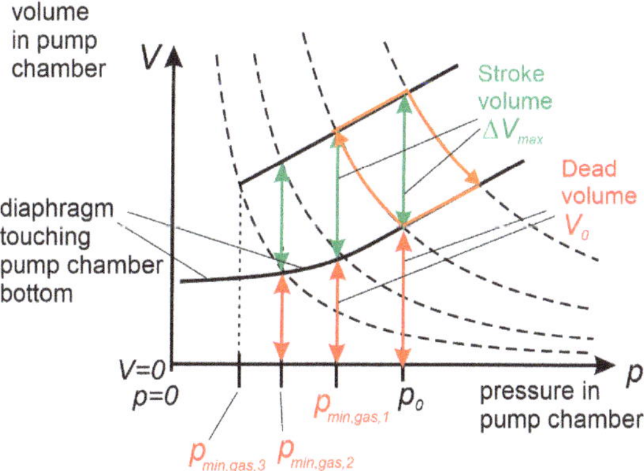

Figure 7. Pump cycles of 3 consecutive stages for serial connection of micropumps, just serial connected without pressure compensation.

2.4.2. Under Pressure with Micropumps: Pressure Balanced (Case 2)

Cascading the micropumps via pressure balance ensures a constant dead volume as well as a constant stroke volume at higher stages to achieve negative pressure. The drawbacks regarding a reduction of the actuator stroke volume through touch down can be avoided (Figure 8). The increased fluidic capacitance due to the high gradients of the isothermal state equations are limiting the progress towards vacuum. However, it is evident that micropumps have to be pressure balanced if a small absolute pressure has to be achieved.

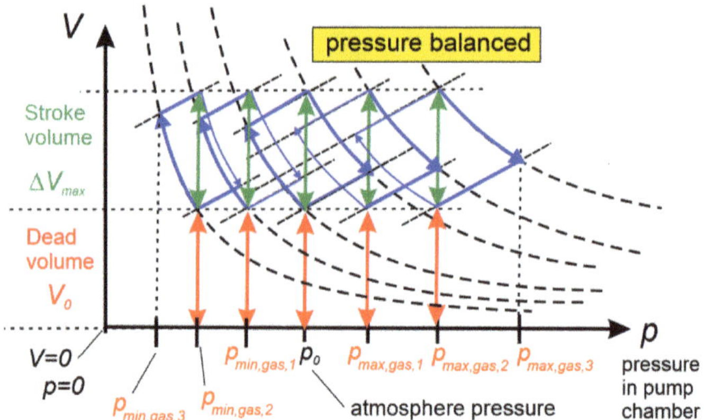

Figure 8. Pressure-balanced configuration of a multistage of three micropumps to achieve negative pressure.

It is assumed that all micropumps of the multistage have the same design properties and the actuation signal is the same. In the following section, simulations are carried out to calculate the negative pressure of these configurations.

2.5. Simulation of a Pressure-Balanced Multistage Micropump for Negative Pressure

Based on the input parameters of the micropump (Table 1), the coefficients C_p and C_E^* are calculated using the analytical expression in the annex of ref. [28]. The stroke volume ΔV is calculated using Equation (3), and the dead volume V_0 is calculated using the geometry of the pump chamber, if the actuation diaphragm is in the lower position. Finally, the theoretical negative pressure $p_{min,gas}$ of the micropump is calculated using Equation (6). In this approach, the simulation of stage 2 takes the pressure of stage 1 $p_{min,gas,1}$ as a new reference pressure for stage 2 (instead of atmospheric pressure p_0). The pressure $p_{min,gas,2}$ achieved at stage 2 is used as a reference pressure for stage 3. To calculate the actuator position for the pump cycle, the parameters from Table 1 are used. According to Figure 9, a minimum pressure $p_{min,gas,1}$ of 41.3 kPa absolute for stage 1 is calculated. Stage 2 results in a $p_{min,gas,2}$ amount of 15.6 kPa, and for stage 3, a theoretical value $p_{min,gas,3} = 5.7$ kPa is derived. As the gradient of the isothermals increases at small absolute pressures, an expansion of a volume ΔV during the supply mode results in a smaller pressure decrease per stage.

It is important to mention that no valve leakages are being considered in this simulation. Furthermore, ideal symmetric bending of the piezo-diaphragm is assumed. In addition to that, the electrically pretensed mounted actuator removes nearly all volume from the pump chamber during pump mode. In reality, due to inhomogeneities in the piezo ceramics, such as gas bubbles after sintering and inhomogeneous powder mixing before sintering, the bending curve of the actuators shows deviations from the symmetric bending curve. This results in an increased dead volume and reduced compression ratio. As a result, the experimental values of $p_{min,gas,1}$, $p_{min,gas,2}$, and $p_{min,gas,3}$ are expected to not achieve the same performance compared to these simulation results.

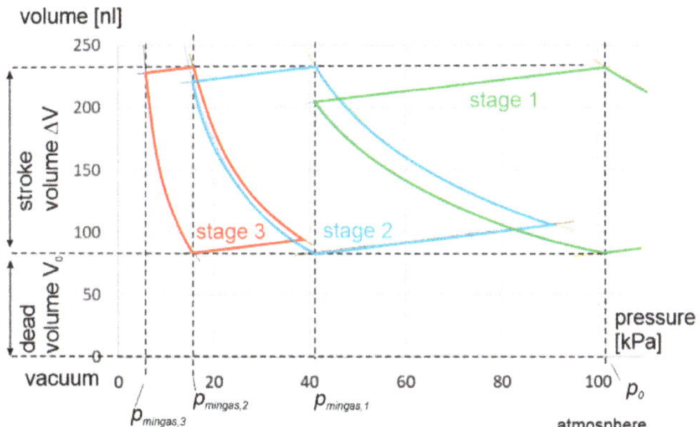

Figure 9. Simulation result of a multistage with three pressure balanced silicon micropumps. All micropumps are identical in design, and have the same stroke volume and dead volume.

3. Measurements

In the following section, measurements with piezoelectric actuated silicon micromembrane pumps to achieve vacuum pressures are conducted. The results of initial tests using single pumps and subsequent measurements using various multistage combinations are presented.

3.1. Stackable Housing Concept for Pressure-Balanced Micropump

Three micropumps were selected and assembled into micropump modules for carrying out the measurements. The modules, which can be stacked, include the micropump, sealing, and electrical connections, as shown in the picture on the right-hand side of Figure 10. The pump outlet is connected to a channel that leads directly to the unstructured bottom side of the module, while the inlet and the free space above the pump device are also connected. A silicone seal is integrated into the top side of the module to ensure an air-tight fluidic connection to the next module. The left-hand side picture of Figure 10 shows an example stack of three cascaded pump modules used for the measurements. The micropumps achieve the lowest pressure at the application or sensor 1 connector, and air is pumped from the application connector through the modules to the ambience or discharge connector.

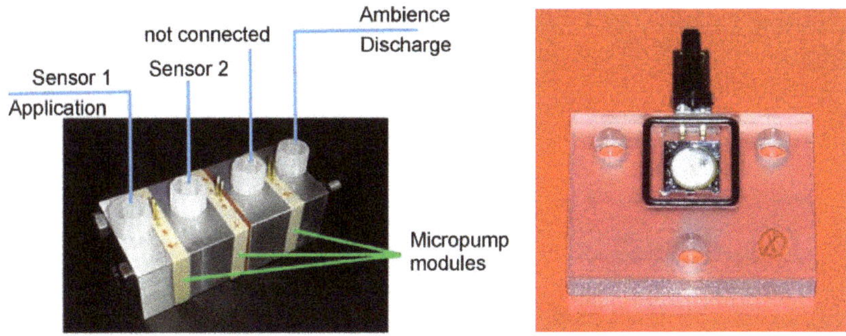

Figure 10. Micropump module in stackable housing with silicone sealing (**right**) and stack of three cascaded micropump modules including connectors to pressure sensors (**left**).

During the measurements of the stacked modules, the micropumps were actuated by a piezo amplifier connected to a waveform generator. Differential pressure sensors connected to the two left connectors were read out with an analogue-to-digital converter module from National Instruments and a measurement software. The data were recorded with a sampling rate of 1 Hz and the pumps were assembled and tested as seen in Figure 11.

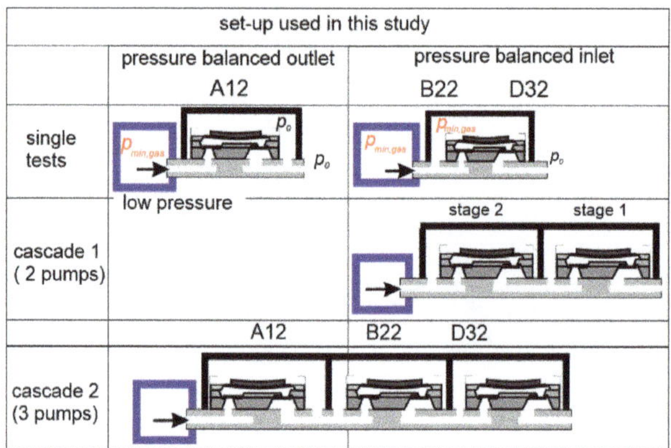

Figure 11. Different setups used in this study for single-pump tests of micropumps A12, B22, and D32 and multistage tests with two and three micropumps.

3.2. Minimum Gas Pressure for Single Pumps

As explained in Section 3.2, the separated silicon pump chips are assembled to pump modules (Figure 10). The modules with integrated pumps B22 and D32 are pressure balanced to the inlet pressure, so the actuator pressure is balanced to the minimum air pressure achieved by the respective micropump. Meanwhile the module with pump A12 is pressure balanced to the outlet atmosphere pressure. Figure 11 depicts those configurations schematically.

As a first standard characterization, the static actuator stroke of the separated pumps is measured. The piezoelectric actuator is pressure balanced to atmosphere pressure. Figure 12 illustrates the stroke from -80 V to 300 V. With a particular actuation voltage between 200 V and 230 V, the bottom of the pump chamber bottom gets touched by all actuators with a comparable stroke.

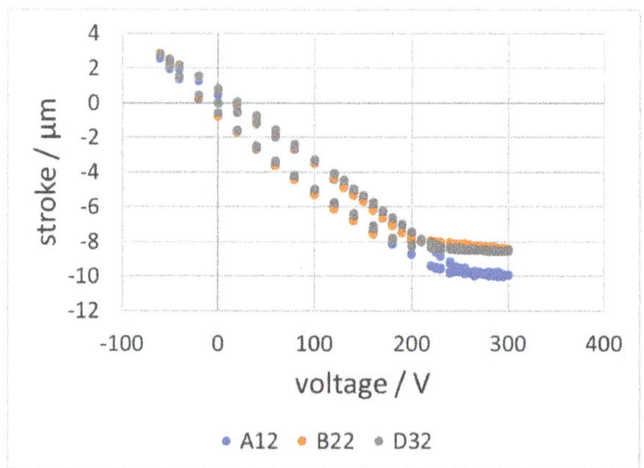

Figure 12. Static actuator stroke of the selected three micropumps (A12, B22, D32). Actuator dimensions are depicted in Table 1. (PZT from PI Ceramics, type PI 151).

The absolute displacement of the actuator is mainly influenced by the actuation voltage and the pressure difference between the top and bottom side of the diaphragm. Therefore, the upper actuator position at minimum voltage during supply mode is stable for inlet pressure-balanced pumps. For outlet pressure-balanced pumps, the lower actuator position at positive voltage during pump mode is stable.

Figure 13 shows the time-dependent pressure which the single pump modules achieve while evacuating air from a test volume connected to a pressure sensor. The sensor measures the differential pressure regarding the atmosphere. After a certain time of 250 s, the pump B22 achieved a pressure $p_{min,gas,B22}$ of -50 kPa (according to 51.3 kPa absolute), the pump D32 achieved $p_{min,gas,D32} = -47$ kPa (according to 54.3 kPa absolute), and pump A12 achieved $p_{min,gas,A12} = -35$ kPa (according to 66.3 kPa absolute).

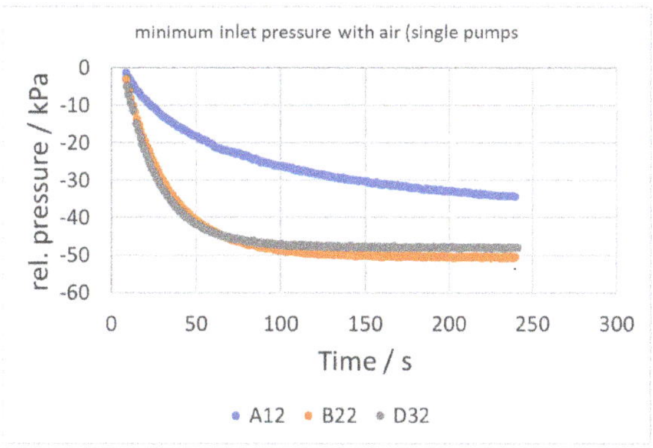

Figure 13. Measurement results of $p_{min,gas}$ from the selected three micropumps (A12, B22, D32).

As shown in Figure 13, none of the micropumps were able to achieve the theoretically simulated value of 41.3 kPa absolute pressure (equivalent to -60.0 kPa relative to atmospheric pressure). Nevertheless, the performance required to achieve a negative pressure of approximately -50 kPa is much better than that of other piezo-driven micropumps with passive check valves, and even better than that of the best micro blowers. It can be assumed that the different performance from pumps B22 and D32 compared to pump D32 can be explained partly by leakage losses of the valve. The gradient of the characteristics in the beginning is proportional to the stroke volume ΔV. As the micropump A12 shows a smaller gradient, it can be assumed from this measurement that the stroke volume of pump A12 is also much smaller compared to the pumps B22 and D32. Those three micropumps are cascaded and characterized in the following section.

3.3. Minimum Gas Pressure for a Multistage of Two Micropumps (Pressure Balanced)

In the following measurement, depicted in Figure 14, the micropumps D32 and B22 are deployed in the multistage. The pressure sensor measurement data show the $p_{min,gas}$ of pump D32, used for stage 1, as well as pump B22, employed in stage 2. After approximately 4 min, the pressure level of -73 kPa is reached, which corresponds to 28.3 kPa absolute pressure. The measurement begins at ambient pressure of around 100 kPa and drops exponentially below the application pressure.

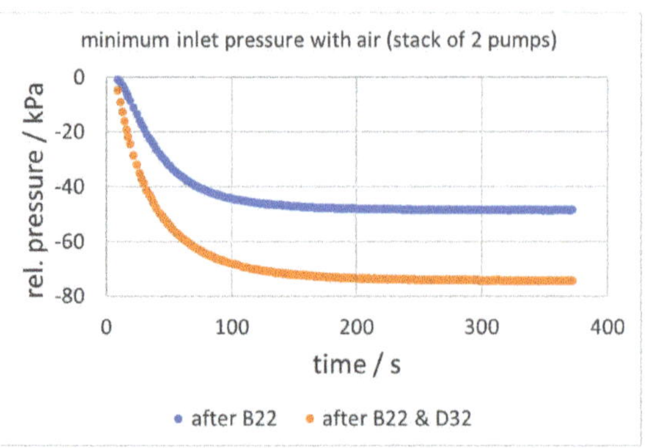

Figure 14. Time- and pressure-dependent cascaded micropumps. Two cascaded micropumps (B22, D32) achieve an absolute pressure of 28.3 kPa.

3.4. Minimum Gas Pressure for a Multistage of Three Micropumps (Pressure Balanced)

Figure 15 depicts the measurement after including micropump A12 to the cascade, forming a three-micropump pressure-balanced multistage. Pump D32 is used for stage 1, B22 for stage 2, and A12 is utilized for stage 3. Although the single-pump performance of A12 is significantly lower compared to B22 and D32 (Figure 13), an improvement of the minimum vacuum gas pressure $p_{min,gas}$ to a level of 19.2 kPa absolute is achieved.

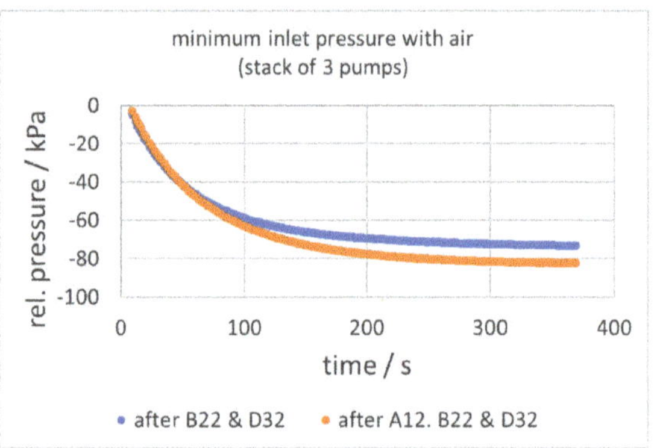

Figure 15. Minimum air inlet pressure measurement result in dependence of time with a multistage of three micropumps. D32 in stage 1, B22 in stage 2, and A12 for stage 3 achieved an absolute pressure of 19.2 kPa in approximately 200 s.

4. Discussion and Outlook

The single micropumps achieved low absolute gas pressures $p_{min,gas}$ of 51.3 kPa, 54.3 kPa, and 66.3 kPa. Although this result is superior compared to the state of the art, in theory, a value of 41.3 kPa absolute should be feasible. The multistage configuration with two micropumps achieved an absolute gas pressure $p_{min,gas}$ of 28.3 kPa, and the multistage configuration with three pumps reached an absolute gas pressure $p_{min,gas}$ of 19.2 kPa. Although, to our knowledge, this is the first time for a mechanical micropump device to generate such a low absolute gas pressure. Theoretically, an absolute gas pressure $p_{min,gas,2}$ of 15.6 kPa and, for stage 3, an absolute gas pressure $p_{min,gas,3}$ of 5.7 kPa, are calculated.

4.1. Discussion

The difference between simulation and measurement can be partly explained in a qualitative way using the following reasons:

- Model simplifications: The pressure-balanced configuration model assumes no influence of the pump actuation characteristics through generating negative pressure. This assumption is not true, as the negative pressure generated by the pump lifts the diaphragm upwards away from the pump chamber bottom. As a result, the dead volume of the micropump increases and the compression ratio decreases. This effect reduces the ability to achieve high negative pressures.
- Tolerances in material properties: In the PZT ceramics calculation, a homogeneous material and a homogenous d_{31} coefficient is assumed. This assumption is inaccurate due to trapped bubbles during the sintering process as well as inhomogeneous powder mixing that may occur in the PZT ceramics production process. Considering that, the bending characteristics of an actuation diaphragm are not symmetric. This influence may reduce the actuator stroke volume and increase the dead volume. Both parameters have a significant influence on the compression ratio, which is the key parameter to achieve low gas pressures.
- Valve leakages: The passive silicon flap valve forms a hard-hard seal between the valve cantilever and valve seat. Although the silicon flap valve is accurately placed above the valve seat and the sealing surface is polished silicon without plastic deformations during bonding, there exist several possibilities for gaps at the valve seat leading to leakages:

    1. After removing a sacrificial layer (nitride/oxide) of 150 nm in the valve manufacturing process, there is an initial gap between cantilever and valve seat.
    2. As the valve seat has a squared shape due to KOH etching (Figures 3 and 4), the corners of the cantilever can bend upwards in the opposite direction from the valve seat, resulting in additional remaining gaps.

    Both issues are a source of additional leakages, reducing the vacuum pump performance.

- Squeeze film damping: The micropump has been operated at a frequency of 100 Hz. In order to reduce the micropump dead volume, the pretension of the diaphragm was adjusted so the diaphragm nearly touches the pump chamber bottom (a distance below 1 µm was envisaged). This is important to reduce the dead volume V_0. However, at these small pump chamber heights, squeeze film damping occurs, and the time available for supply mode and pump mode might not be sufficient to push the gas out of the region between diaphragm and pump chamber bottom, which reduces the effective stroke volume.

4.2. Outlook

This study discussed how to design a micropump assembly to achieve low gas pressures. For further improvements in the direction of coarse vacuum, the following steps are envisaged:

- Reduce leakages of the microvalves: Different design modifications to reduce both discussed gaps in hard-hard sealing are realized.
- Optimize valve bending at small absolute pressures: Mechanical cantilever valves open and close according to the pressure difference generated by the actuator movement. In a multistage with multiple micropumps upstream, just a few kPa of pressure difference is available to open the valve. In an optimized version of this multistage configuration, every stage has its own valve geometry adapted for each pressure regime.
- Research in physical properties at small absolute pressures: With very small remaining gap heights in the micrometer range, the leakage rate is not only defined by the convection with Navier–Stokes equations, but also by self-diffusion. The theory describing the gas leakages of microvalves has to be extended in order to optimize the design of the valves adapted for these self-diffusion properties.

- Model adaption: The model will be extended with an implementation of the actuator diaphragm lifting effect for the pressure-balanced micropump due to the generated negative pressure.
- Actuator optimization: Another potential optimization strategy involves adapting the micropump design as well as the actuation voltage for the individual stages. Especially for stages closer to the vacuum, micropumps with large compression ratios and a reduced blocking pressure are advantageous. The design of the piezo-diaphragm actuator has a trade-off between stroke volume (proportional to the compression ratio) and blocking pressure. Both applications rely on micropumps optimized for these requirements. For these optimizations, the p-V diagram provides a guideline.

Author Contributions: Conceptualization, M.R. and D.A.; methodology, M.R. and M.W.; validation, M.W. and A.W.; formal analysis, M.R.; investigation, D.A. and Y.C.; resources, D.A.; data curation, M.W.; writing—original draft preparation, M.R. and D.A.; writing—review and editing, D.A.; visualization, M.R. and D.A.; supervision, M.R.; project administration, D.A.; funding acquisition, A.W. and M.R. All authors have read and agreed to the published version of the manuscript.

Funding: This research was funded by Deutsche Forschungsgemeinschaft (German Research Foundation) with grant number KU 3410/1-1.

Institutional Review Board Statement: Not applicable.

Informed Consent Statement: Not applicable.

Data Availability Statement: If data is required, please get in touch with the corresponding author for access.

Acknowledgments: We kindly thank the Fraunhofer EMFT silicon micropump group for the supply of micropump samples.

Conflicts of Interest: The authors declare no conflict of interest.

References

1. Bußmann, A.B.; Grünerbel, L.M.; Durasiewicz, C.P.; Thalhofer, T.A.; Wille, A.; Richter, M. Microdosing for drug delivery application—A review. *Sens. Actuators A Phys.* **2021**, *330*, 112820. [CrossRef]
2. Herz, M.; Kibler, S.; Söllner, M.; Scheufele, B.; Richter, M.; Lueth, T.C.; Bock, K. Entwicklung einer energieeffizienten piezoelektrischen Hochfluss-Mikropumpe für Methanol-Brennstoffzellen. *Mikrosystemtechnik-Kongress* **2011**, *4*, 74–77.
3. Heppner, J.D.; Walther, D.C.; Pisano, A.P. The design of ARCTIC: A rotary compressor thermally insulated μcooler. *Sens. Actuators A Phys.* **2007**, *134*, 47–56. [CrossRef]
4. Jousten, K. *Handbuch Vakuumtechnik*; Springer Fachmedien Wiesbaden: Wiesbaden, Germany, 2018.
5. Tassetti, C.-M.; Mahieu, R.; Danel, J.-S.; Peyssonneaux, O.; Progent, F.; Polizzi, J.-P.; Machuron-Mandard, X.; Duraffourg, L. A MEMS electron impact ion source integrated in a microtime-of-flight mass spectrometer. *Sens. Actuators B Chem.* **2013**, *189*, 173–178. [CrossRef]
6. van Someren, B.; van Bruggen, M.J.; Zhang, Y.; Hagen, C.W.; Kruit, P. Multibeam Electron Source using MEMS Electron Optical Components. *J. Phys. Conf. Ser.* **2006**, *34*, 1092–1097. [CrossRef]
7. Grzebyk, T. MEMS Vacuum Pumps. *J. Microelectromech. Syst.* **2017**, *26*, 705–717. [CrossRef]
8. Defense Advanced Research Projects Agency. Mighty Micropumps: Small but Powerful Vacuum Pumps Demonstrated: DARPA Creates Microscale Pumps to Evacuate Tiny Vacuum Chambers. Available online: https://www.darpa.mil/news-events/2013-06-04 (accessed on 9 May 2023).
9. Reichelt Chemietechnik GmbH + Co. Mini-Vakuum-Membranpumpe für Gasförmige Medien und Reinstmedien. Available online: https://www.rct-online.de/de/pumpen/gaspumpen/mini-vakuum-membranpumpe-fuer-gasfoermige-medien-und-reinstmedien (accessed on 11 May 2023).
10. Cumeras, R.; Figueras, E.; Davis, C.E.; Baumbach, J.I.; Gràcia, I. Review on Ion Mobility Spectrometry. Part 1: Current instrumentation. *Analyst* **2015**, *140*, 1376–1390. [CrossRef] [PubMed]
11. Bäther, W.; Raupers, B.; Lehmann, S. Gas-Measuring Device: Patent Application Publication. U.S. Patent US 2016/0327532 A1, 9 January 2015.
12. Murata Manufacturing Co., Ltd. Microblower MZB3004T04: Microblower (Air Pump). Available online: https://www.murata.com/en-eu/products/mechatronics/fluid/overview/lineup/microblower_mzb3004t04 (accessed on 17 May 2023).
13. Astle, A.; Paige, A.; Bernal, L.P.; Munfakh, J.; Kim, H.; Najafi, K. Analysis and Design of Multistage Electrostatically-Actuated Micro Vacuum Pumps. In *Microelectromechanical Systems*; ASMEDC: New York, NY, USA, 2002; pp. 477–486.

14. Kim, H.; Astle, A.A.; Najafi, K.; Bernal, L.P.; Washabaugh, P.D. An Integrated Electrostatic Peristaltic 18-Stage Gas Micropump With Active Microvalves. *J. Microelectromech. Syst.* **2015**, *24*, 192–206. [CrossRef]
15. Zengerle, R.; Ulrich, J.; Kluge, S.; Richter, M.; Richter, A. A bidirectional silicon micropump. *Sens. Actuators A Phys.* **1995**, *50*, 81–86. [CrossRef]
16. Leistner, H.; Wackerle, M.; Congar, Y.; Anheuer, D.; Roehl, S.; Richter, M. Robust Silicon Micropump of Chip Size $5 \times 5 \times 0.6$ mm^3 with 4 mL/min Air and 0.5 mL/min Water Flow Rate for Medical and Consumer Applications. In *Actuator 2021, Proceedings of the International Conference and Exhibition on New Actuator Systems and Applications: GMM Conference, Online Event, 17–19 February 2021*; Schlaak, H., Ed.; VDE Verlag GmbH: Berlin, Germany; Offenbach, Germany, 2021; pp. 113–116.
17. Richter, M.; Wackerle, M.; Kibler, S.; Biehl, M.; Koch, T.; Müller, C.; Zeiter, O.; Nuffer, J.; Halter, R. Miniaturized drug delivery system TUDOS with accurate metering of microliter volumes. In Proceedings of the SENSOR, International Conference on Sensors and Measurement Technology, 16, AMA Conferences, Nürnberg, Germany, 14–16 May 2013; pp. 420–425.
18. Du, M.; Ma, Z.; Ye, X.; Zhou, Z. On-chip fast mixing by a rotary peristaltic micropump with a single structural layer. *Sci. China Technol. Sci.* **2013**, *56*, 1047–1054. [CrossRef]
19. Pankhurst, P.; Abdollahi, Z.M. Evaluation of a novel portable micro-pump and infusion system for drug delivery. *IEEE Eng. Med. Biol. Soc. Annu. Conf.* **2016**, *2016*, 465–468.
20. Tanaka, S.; Tsukamoto, H.; Miyazaki, K. Development of Diffuser/Nozzle Based Valveless Micropump. *J. Fluid Sci. Technol.* **2008**, *3*, 999–1007. [CrossRef]
21. Wong, C.C.; Aeschliman, D.P.; Henfling, J.F.; Sniegowski, J.J.; Rodgers, M.S. Development of an Ejector-Driven Micro-Vacuum Pump. In *Micro-Electro-Mechanical Systems (MEMS)*; American Society of Mechanical Engineers: New York, NY, USA, 2001; pp. 485–493.
22. Grzebyk, T.; Górecka-Drzazga, A.; Dziuban, J.A. Glow-discharge ion-sorption micropump for vacuum MEMS. *Sens. Actuators A Phys.* **2014**, *208*, 113–119. [CrossRef]
23. Green, S.R.; Malhotra, R.; Gianchandani, Y.B. Sub-Torr Chip-Scale Sputter-Ion Pump Based on a Penning Cell Array Architecture. *J. Microelectromech. Syst.* **2013**, *22*, 309–317. [CrossRef]
24. Gupta, N.K.; An, S.; Gianchandani, Y.B. A monolithic 48-stage Si-micromachined Knudsen pump for high compression ratios. In Proceedings of the 2012 IEEE 25th International Conference on Micro Electro Mechanical Systems (MEMS), Paris, France, 29 January–2 February 2012; IEEE: Piscataway, NJ, USA, 2012; pp. 152–155.
25. Vargo, S.E.; Muntz, E.P. Initial results from the first MEMS fabricated thermal transpiration-driven vacuum pump. In Proceedings of the AIP Conference Proceedings, Sydney, Australia, 9–14 July 2000; pp. 502–509.
26. Herz, M.; Horsch, D.; Wachutka, G.; Lueth, T.C.; Richter, M. Design of ideal circular bending actuators for high performance micropumps. *Sens. Actuators A Phys.* **2010**, *163*, 231–239. [CrossRef]
27. Lee, S.; Yee, S.Y.; Besharatian, A.; Kim, H.; Bernal, L.P.; Najafi, K. Adaptive gas pumping by controlled timing of active microvalves in peristaltic micropumps. In Proceedings of the TRANSDUCERS 2009—2009 International Solid-State Sensors, Actuators and Microsystems Conference, Denver, CO, USA, 21–25 June 2009; IEEE: Piscataway, NJ, USA, 2009; pp. 2294–2297.
28. Sandoughsaz, A.; Besharatian, A.; Bernal, L.P.; Najafi, K. Modular stacked variable-compression ratio multi-stage gas micropump. In Proceedings of the 2015 Transducers—2015 18th International Conference on Solid-State Sensors, Actuators and Microsystems (Transducers), Anchorage, AK, USA, 21–25 June 2015; IEEE: Piscataway, NJ, USA, 2015; pp. 704–707.
29. Besharatian, A.; Kumar, K.; Peterson, R.L.; Bernal, L.P.; Najafi, K. Valve-only pumping in mechanical gas micropumps. In Proceedings of the 2013 Transducers & Eurosensors XXVII: The 17th International Conference on Solid-State Sensors, Actuators and Microsystems (Transducers & Eurosensors XXVII), Barcelona, Spain, 16–20 June 2013; IEEE: Piscataway, NJ, USA, 2013; pp. 2640–2643.
30. Besharatian, A.; Kumar, K.; Peterson, R.L.; Bernal, L.P.; Najafi, K. A Scalable, modular, multi-stage, peristaltic, electrostatic gas micro-pump. In Proceedings of the 2012 IEEE 25th International Conference on Micro Electro Mechanical Systems (MEMS 2012), Paris, France, 29 January–2 February 2012; pp. 1001–1004.
31. Le, S.; Hegab, H. Investigation of a multistage micro gas compressor cascaded in series for increase pressure rise. *Sens. Actuators A Phys.* **2017**, *256*, 66–76. [CrossRef]
32. McNamara, S.; Gianchandani, Y.B. A micromachined Knudsen pump for on-chip vacuum. In Proceedings of the TRANSDUCERS '03, 12th International Conference on Solid-State Sensors, Actuators and Microsystems, Digest of Technical Papers (Cat. No.03TH8664), Boston, MA, USA, 8–12 June 2003; IEEE: Piscataway, NJ, USA, 2003; pp. 1919–1922.
33. Richter, M. Modellierung und Experimentelle Charakterisierung von Mikrofluidsystemen und deren Komponenten. Ph.D. Thesis, Universität der Bundeswehr München, München, Germany, 1998.
34. Herz, M. Optimierung der Förderrate einer Piezoelektrischen Hochleistungs-Mikropumpe. Ph.D. Thesis, Technische Universität München, München, Germany, 2011.
35. PI Ceramic GmbH. Piezoelectric Ceramic Products: Fundamentals, Characteristics and Applications. *Catalogue* **2016**, *CAT125E*, R3.

 *actuators*

Article

Maskless Writing of Surface-Attached Micro-Magnets by Two-Photon Crosslinking

Nicolas Geid [1,2], Jan Ulrich Leutner [1], Oswald Prucker [1] and Jürgen Rühe [1,2,*]

[1] Department of Microsystems Engineering (IMTEK), University of Freiburg, Georges-Köhler-Allee 103, D-79110 Freiburg, Germany
[2] Cluster of Excellence livMatS @ FIT–Freiburg Center for Interactive Materials and Bioinspired Technologies, Georges-Köhler-Allee 105, D-79110 Freiburg, Germany
* Correspondence: ruehe@imtek.de

Abstract: Surface-bound 3D micro-magnets are fabricated from photoreactive copolymers filled with magnetic nanoparticles by maskless 3D writing. The structures are generated by 2-photon crosslinking (2PC), which allows direct writing into solid films of composites consisting of magnetic particles and a photoreactive elastomer precursor. With this strategy, it is possible to directly write complex, surface-bound magnetic actuator structures, which generates new opportunities in the fields of microfluidics and bioanalytical systems. Compared to the common 2-photon polymerization, in which the writing process takes place in a liquid resin, the direct writing based on the 2PC method takes place in a solid polymer film (i.e., in the glassy state).

Keywords: micro-actuators; two-photon lithography; two-photon crosslinking; photoreactive polymers; micro-magnets

Citation: Geid, N.; Leutner, J.U.; Prucker, O.; Rühe, J. Maskless Writing of Surface-Attached Micro-Magnets by Two-Photon Crosslinking. *Actuators* **2023**, *12*, 124. https://doi.org/10.3390/act12030124

Academic Editor: Richard Yongqing Fu

Received: 7 February 2023
Revised: 6 March 2023
Accepted: 10 March 2023
Published: 15 March 2023

1. Introduction

Micro-actuators are very interesting systems both from an academic as well as from an application-oriented point of view as they allow the generation of large fields of actuators which can work in concert. They can be used to generate coordinated and cooperative movements and, thus, allow the development of novel devices and systems [1,2]. Such systems are particularly attractive for microfluidic pumping and mixing and even more so for complex active stimulation of biological systems, including those that involve cell stimulation [3]. A strong inspiration for the design of micro-actuators has frequently been ciliated organisms in nature, which use the movement of tiny hairs to move liquid. Many different methods have been developed to mimic such ciliated organisms using micro-actuators and actuator fields, frequently called artificial cilia [4–7]. These synthetically generated actuators react to different external stimuli, e.g., to electrostatic [8], light [9,10], piezo [11] and magnetic actuation [6,12–15]. Magnetic actuation is particularly suitable because the process can be well controlled and the weak interaction of magnetic fields with biological materials usually causing only minor perturbations [16,17]. Magnetic micro-actuators generally consist of magnetic particles which are incorporated in elastomers. In most cases, mask- or mold-based processes are employed to form the magnetic microstructures [14,18,19]. Published work on artificial cilia shows that high pumping efficiencies and controlled particle transport can be achieved by actuation with an external magnetic field [20,21]. In addition to mold-based processes, a mask-based two-color lithography process based on C,H insertion crosslinking has recently been described for the generation of magnetic microflaps, which have been incorporated into microfluidic chips. These actuators achieved average flow velocities of hundreds of µm/s, which shows that effective pumping rates can also be achieved with such a materials system [14,22,23].

Micro-molding [24] and mask-based techniques are standard technologies and allow the generation of large actuator fields. However, they are limited when it comes to the generation of more complex geometries that vary locally, which is required in more demanding applications, such as when cooperative action is needed. Molding processes, for example, cannot produce undercut or other complex structures, such as spherical objects, where demolding is difficult or even impossible. Further, in such processes, it is difficult to implement local variations in the chemical composition or mechanical properties of the actuators, e.g., produce structures with variations in stiffness. Additionally, for all design changes of the structure, a new mold needs to be generated which makes complex studies with frequent structural redesign tedious. In contrast to this, a direct writing process allows the flexible selection of 3D designs for actuators, giving full control in the manufacturing of complex and eventually cooperative actuator systems.

A direct writing process that has received much attention in recent years is 2-photon lithography [25–27], in which high-resolution microstructures are usually written by 2-photon polymerization using a femtosecond laser [28–32]. In such processes, the 2-photon activation is used to locally initiate a polymerization reaction leading to the formation of polymers or a polymer network in the illuminated volume element (voxel) [33]. A variety of freely swimming micro-actuators generated using 2-photon polymerization has been described. Such systems are of great interest, especially in the field of medical micro-robots [34–36]. Recent publications have also considered the writing of surface-attached magnetic micro-springs using 2-photon polymerization, which could be moved with the help of a magnetic field. In this process, 2.1 wt.% Fe_3O_4 nanoparticles were introduced in a monomer resin [37,38].

Recently, a new technique in 2-photon lithography called 2-photon crosslinking has been introduced, which represents an attractive alternative to the commonly used 2-photon polymerization process [39]. In this method, copolymers are used which are equipped with a photochemically reactive group and can be simultaneously crosslinked and surface-bonded by 2-photon excitation in the glassy state using C,H insertion chemistry [39]. Compared to 2-photon polymerization, several layers of different polymers can be applied on top of each other, resulting in a multifunctional material [40]. This technique uses no monomeric compounds, which is very attractive from a safety point of view for working in an optics laboratory. Additionally, as the polymer can be thoroughly purified before use, the final structures are monomer-free, which is very important for any biological or biomedical application. Quality control with respect to the contents of residual monomers in additive manufacturing processes based on polymerization reactions is difficult to ensure as only single objects are generated. This is potentially a serious problem as the monomer content depends very strongly on the details of the conditions under which the writing process is performed [39,41].

Following up on this concept, we present here a novel method for the direct writing of micro-magnets using 2-photon crosslinking (2PC), based on a photoactive elastomeric copolymer with magnetically embedded nanoparticles. A schematic illustration of this process is shown in Figure 1b,c. As described, the process is based on 2-photon absorption in a small volume around the focal point (voxel), which eventually leads to formation of a polymer network via a C,H insertion crosslinking (CHic) reaction. The very same reaction leads to a surface-attached layer if the surface of the substrate is first decorated with an alkyl silane layer [40–43]. The non-crosslinked polymer is easily removed using a suitable solvent, and the written microstructures remain on the surface as they are covalently linked to the substrate. We will also show how the resulting layers can be actuated using a rotating magnet which is operated below the substrate and discuss the resulting deflections of the micro-actuators based on simple Euler–Bernoulli beam theory [44,45].

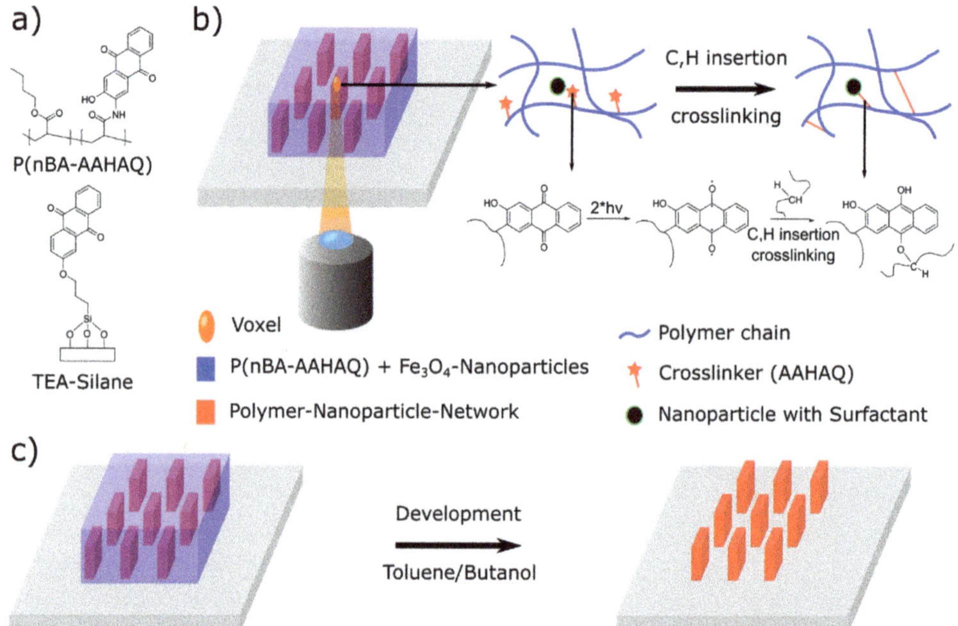

Figure 1. (**a**) Molecular structures of the copolymer P(nBA-AAHAQ), and the triethoxy anthraquinone silane (TEA-silane) used for a covalent attachment to the glass slide. (**b**) Schematic representation of the 2-photon crosslinking process. The femtosecond laser triggers a C,H insertion crosslinking reaction within a small volume (voxel) in the solid, nanoparticle containing, polymer layer. The formation of a polymer network and the incorporation of the nanoparticles result in the writing of three-dimensional structures from bottom to top. (**c**) By developing in a suitable solvent (e.g., toluene/butanol), the non-crosslinked polymer chains are washed out and the micro-actuators are obtained.

2. Materials and Methods

Materials: 2-Amino-3-hydroxyanthraquinone was purchased from TCI, Germany. All other chemicals were purchased from Sigma Aldrich, Germany. Dimethyl acrylamide and n-butylacrylate (nBA) were purified by filtration through basic aluminum oxide and distillation. 2,2′-Azobis(2-methylpropionitrile) (V70; Fujifilm Wako Chemicals Europe, Neuss, Germany) was recrystallized from ethanol. All other chemicals and solvents were used as received.

Photoreactive monomer: 2-Amino-3-hydroxyanthraquinone (2.39 g, 10.0 mmol, 1.0 eq) was dissolved in an inert gas atmosphere in 1,4-dioxane (100 mL). A solution of acrylic acid chloride (0.68 mL, 0.72 g, 8.0 mmol, 0.8 eq) in dioxane (20 mL) was added while stirring at 0 °C. After complete addition, the reaction mixture was heated for 6 h to reflux. Afterwards the reaction solution was filtered and the filter cake was taken up in ethanol (3 × 150 mL) and heated to reflux. The solid product was filtered off and dried under vacuum, resulting in a yellow product with a yield of 84% (1.97 g, 6.7 mmol).

^{1}H-NMR (250 MHz, DMSO-d6, δ): 5.80 (dd, J = 2 Hz, 10 Hz, 1H), 6.35 (dd, J = 2 Hz, 17 Hz, 1H), 6.87 (dd, J = 10 Hz, 17 Hz, 1H), 7.63 (s, 1H, Ar-H), 7.82–7.94 (m, 2H, Ar-H), 8.10–8.22 (m, 2H, Ar-H), 9.05 (s, 1H, Ar-H), 9.81 (s, 1H, OH), 11.71 (s, 1H, NH).

^{13}C-NMR (250 MHz, DMSO-d6, δ): 112.51 (C3), 119.90 (C6), 126.71 (C5), 127.34 (C11), 127.49 (C14), 128.84 (C17), 130.77 (C13), 132.51 (C12), 133.09 (C1), 133.96 (C16), 134.15 (C8), 134.88 (C9), 135.17 (C2), 153.38 (C4), 164.92 (C15), 182.36 (C7), 182.75 C10).

Photocurable copolymer: The copolymer was synthesized via free radical polymerization with 92.5 mol% nBA and 7.5 mol% 2-amino-3-hydroxyanthraquinone (AAHAQ) at 30 °C with the V70 initiator. Reaction time: 24 h, yield 70%. The AAHAQ content was determined to be 7.4% according to NMR spectroscopy based on the integration of the signals due to the aromatic protons (7.6–7.9 ppm). According to the GPC data, the molecular masses of the obtained copolymers were M_n = 93.000 g/mol and M_W = 241.000 g/mol, so that the polymers carried, on average, approximately 70 AAHAQ repeat units per polymer chain.

Magnetic resin: The photo-cross-linkable magnetic composite was obtained by mixing 25 µL of commercial Fe_3O_4 superparamagnetic ferrofluid (EMG 905 containing oleic acid as stabilizer, Ferrotec, NH, Bedford, MA, USA) in a solution of 100 mg photocurable copolymer P(nBA-7.5%AAHAQ) in 0.975 mL toluene to obtain a homogenous dispersion.

Fabrication process of the micro-actuators: For the fabrication of the magnetic micro-actuators, glass slides (22 × 22 mm, 170 µm thick) were first silanized with a 50 mM triethoxy anthraquinone silane ((TEA)-silane) in toluene. The synthesis of the TEA-silane was carried out according to [39]. The solution (100 µL) was applied to a glass slide and spin-coated at 1000 rpm for 30 s. For the surface attachment, it was heated to 120 °C for 30 min and the non-attached silane was washed out with toluene. To obtain a polymer nanoparticle layer of about 80–100 µm, 50–100 µL of the dispersion described above was drop-cast onto the treated glass slide. The toluene was slowly evaporated over 2 h at room temperature to avoid bubble formation. As soon as the layer had solidified, the remaining toluene could be removed by heating to 100 °C for 1 h. A 100 µm thick glassy polymer composite layer with approx. 10 wt.% nanoparticles was obtained. After completion of the 2-PC process, the substrates were placed in n-butanol and developed for at least 2 h, thus washing out the unreacted polymer.

Design of the micro-actuators: The beam-like micro-actuators were designed to have a height of 60–75 µm and a length of 20 µm. The thickness varied from 3 µm to 8 µm depending on the structure. The distance between two adjacent bars was 80 µm.

Direct writing of the magnetic micro-actuators: All structures were designed using Solidworks or Think3D computer programs to generate a stereolithography file (stl.). The 2-photon lithographs were performed using the Nanoone setup (UpNano GmbH, Vienna, Austria). It was equipped with a Ti-sapphire laser having a wavelength of 780 nm and a laser power of 500 mW. This printer uses a galvanometer scanning method. The generated stl. files were loaded into the Think3D version 1.7.3 software (Nanoone, UpNano GmbH, Vienna, Austria). For writing, a 20× magnification lens with 0.7 NA was used. Printing was performed in bottom-up mode and the lens was immersed into water. Slicing was set to 0.5 µm and hatching to 0.15 µm; a laser power of 15 mw and a scan speed of 100 mm/s were used.

Magnetic actuation and optical microscopy: A rotating permanent magnet was used to actuate the structures. This magnet had a field strength of 1.4 T (N42). The magnet was rotated by a motor at 30 rpm installed 1.5 cm below the stage of an optical microscope. With the help of a magnetometer, a maximum field strength of 130 mT was measured at the sample position. In this way, the structures could be made to move by the gradient of the magnetic field strength which ranged from 0 to 130 mT.

SEM: To prevent collapse, the structures kept in ethanol were dried using a critical point dryer (Leica EM CPD300, Wetzlar, Germany). The parameters chosen for the CPD were a cooling temperature of 10 °C, CO_2 in medium velocity, 16 exchange cycles, a critical temperature of 38 °C, and gas out medium velocity. Subsequently, scanning electron micrographs (5 kV, 0.1 nA) could be obtained using an FEI Scios 2 HiVac (Thermo Fisher, Waltham, MA, USA). The samples were gold-sputtered with a Cressington Sputter Coater 108auto. The images were taken at a 45° angle.

3. Results and Discussion

The material systems for the generation of the micro magnetic structures by 2-photon crosslinking processes consisted of superparamagnetic particles and a copolymer, which was composed of a matrix component and a photoreactive unit for crosslinking. As photoreactive group 2-amino-3-hydroxyanthraquinone (AAHAQ) developed by Schwärzle et al. was chosen, which exhibits pronounced 2-photon absorption due to its designed electronic structure (i.e., conjugated π-system, planarity, donor and acceptor groups) [39]. Briefly, the photoreactive group AAHAQ was obtained in a one-step synthesis from 2-amino-3-hydroxyanthraquinone and acrylic acid chloride. The desired P(nBA-AAHAQ) copolymer (see Figure 1a) was prepared by free radical polymerization of N-buylacrylate (nBA) and AAHAQ at 30 °C using a low-temperature initiator (2,2′-azobis(4-methoxy-2,4-dimethylvaleronitrile (V70).

For the incorporation of the nanoparticles, particular attention must be paid to the compatibility of the particles and the copolymer. N-butylacrylate is a suitable soft component and an attractive component for biological applications, especially cell experiments [46]. Due to the hydrophobic nature of the nBA polymer, hydrophobically coated (i.e., oleic acid stabilized) iron nanoparticles were chosen, to allow sufficient miscibility. The copolymer and the nanoparticles were dispersed in toluene to form a homogeneous dispersion, which was stable between 0–20 wt.% content of nanoparticles. In order to obtain comparable results, in the following, we focus on dispersions with 10 wt.% nanoparticles.

For attaching the forming structures to the surfaces, a triethoxy anthraquinone silane (TEA-silane, see Figure 1a) was prepared and a self-assembled monolayer was formed on the glass surface. For the writing process, the dispersion was drop-cast onto the substrate to form a thin film of about 100 µm. Upon 2-photon illumination, the chromophore in the copolymer was excited into a triplet state and the polymer crosslinked through C,H insertion reactions (CHic). At the same time, the crosslinker units in those voxels in direct contact with the silane monolayer were also excited and attached to the forming network, so that the entire forming structure became covalently bound to the glass substrate. Moreover, the crosslinker molecules could probably also insert into the C,H bonds of the oleic acid molecules on the surface of the Fe_3O_4-nanoparticles, thus firmly binding the nanoparticles into the network. After writing the latent 3D image into the film, it was developed using n-butanol, a theta-solvent for the polymer. If a good solvent, such as toluene, was chosen instead, the structures formed became too strongly swollen, so that, in some cases, the swelling pressure became so great that they were torn off from the surface and the yield of perfectly formed structures was lower than that in the case where a solvent for the polymer was used as the developer, which was still capable of dissolving the polymer, but was of lower solvent quality. The micro-actuators were then transferred to ethanol in which the actuation took place. To dry the structures, i.e., for SEM imaging, in the case of high-aspect-ratio structures, critical point drying with CO_2 was employed. Since no meniscus formed during the drying process, a collapse of the structures was completely prevented, and large fields of perfectly shaped actuators were obtained.

To characterize the movement of micro-actuators written with the 2PC process, simple beam structures were investigated. A 20× water immersion objective with 0.7 NA was used for writing the micro-actuators. The writing parameters chosen for this material were a laser power of 15 mW, a writing speed of 100 mm/s, a hatching distance of 0.15 µm and a slicing distance of 0.5 µm. These parameters were used for all experiments described in this paper. Figure 2a shows a schematic representation of the deflection of the micro-actuators. A permanent magnet (N42) was rotated with a speed of 30 rpm under the micro-actuators, which responded to the gradient of the field lines. The magnetic flux density B in the y- and z-directions at the actuator position as a function of the rotation angle is shown in Figure 2b. The micro-actuators experienced a maximum magnetic flux density of 130 mT, which led to the desired deflection of the beams. Figure 2d shows the resulting deflection of the beams under investigation. The actuators not deflected are shown in the upper part of the figure, and the maximum deflected actuators are shown in the lower part of the figure. The beams shown here were printed with a width of 20 µm, a thickness of 3 µm and a height of 70 µm. In ethanol, the micro-actuators swelled by a factor of 1.2 but remained firmly attached to the substrate as they were covalently bound there. The actuator fields were practically free of defects, illustrating the high reproducibility of this manufacturing technique. Figure 2c shows scanning electron microscope images that demonstrate the high resolution of the beam structures. As the specimens dried, they de-swelled, causing a slight decrease in volume. The drying process was also the reason for the slight rounding of the structures at the edges.

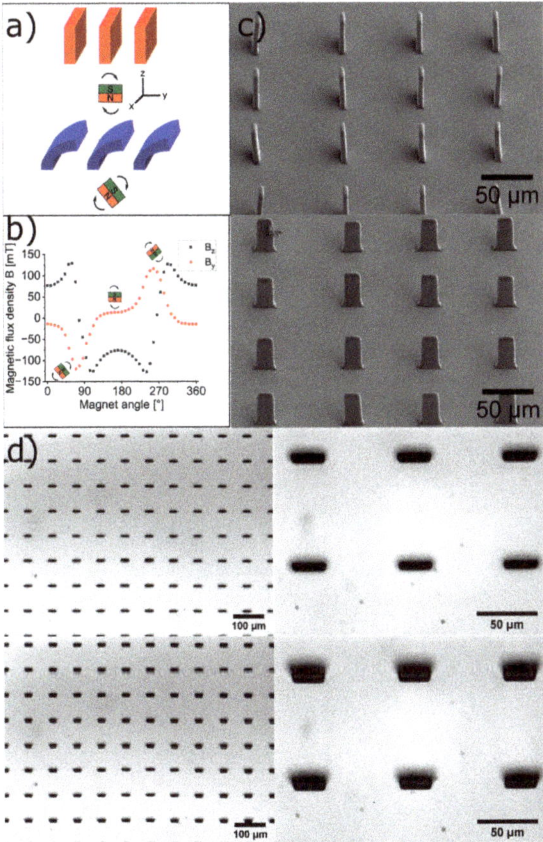

Figure 2. (**a**) Schematic representation of the deflection of beam structures by a rotating permanent magnet. (**b**) Graphical representation of the magnetic flux density B as a function of the rotation angle of the permanent magnet (N42). The permanent magnet was placed at a distance of 1.5 cm between the center of the magnet and the sample position. (**c**) Scanning electron microscopy images of the beam structures after critical point drying. (**d**) The corresponding light microscopic images of the written beam structures in the undeflected (**top**) and maximally deflected states (**bottom**). In the field shown, 49 structures (one printing field) were written in 6 min.

2PC allows for a rather simple variation of the actuator dimensions or aspect ratios. We have investigated the latter and compared our results to simple theoretical predictions using the Euler–Bernoulli beam theory [45]. The written beams have one end fixed to the surface and one free end. By neglecting the torsional motions, a simple cantilever model can be used. Since the nanoparticles are homogeneously distributed in the beam, a uniformly distributed volume load can be assumed. For simplification, this volume load is projected onto a line load q, which is shown schematically in Figure 3a. Thus, the model used now corresponds to a cantilever with a uniformly distributed load. The aim is to determine the maximum deflection at the free end of the beam as a function of the beam structure, i.e., the beam height. An expression for the maximum deflection of the beam can be obtained by integrating twice the ordinary differential equation of the deflection curve. Here, it is assumed that the system is in the linear elastic range (Hooke's law) and the equation is therefore only valid for small deflections. For the maximum deflection δ of the beam the following formula is obtained:

$$\delta = \frac{q \cdot h^4}{8} \cdot E \cdot I$$

Thus, the maximum deflection of a beam depends on the fourth power of the height of the beam. The dependence of the deflection on the Young's modulus E reflects the material properties. The area moment of inertia I involves the structural properties of the beams and can be described according to:

$$I = \frac{b \cdot a^3}{12}$$

In this expression, the moment of inertia of the area is linearly related to the width b and depends on the cube of the thickness a of the beam. However, changing the width or thickness would also increase the load because the nanoparticles are homogeneously distributed throughout the volume. Since the equation uses the uniformly distributed line load, increasing the thickness or width will not increase the load in the equation. Therefore, in order to obtain the proper dependency of these parameters on deflection, the increase in load due to a change in width or thickness must be projected onto the linear load, as shown in Figure 3b,c. As a result, the deflection δ is independent of the width and inversely proportional to the square of the thickness.

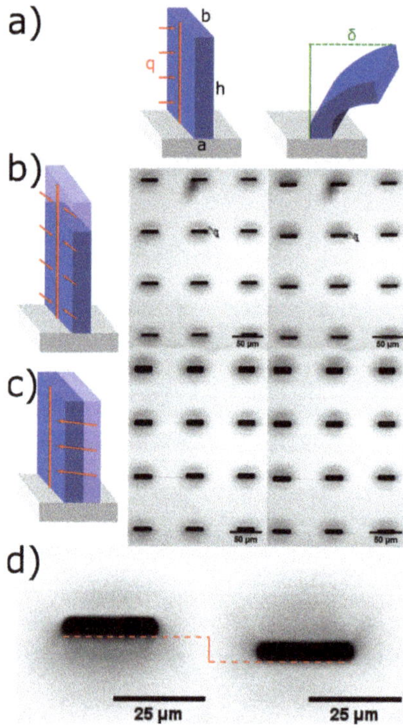

Figure 3. (**a**) Schematic representation of a cantilever beam with a uniformly distributed load and the resulting deflection. The position of the maximum deflection is shown by the green line. Schematic representation and light microscope images of beams which were (**b**) varied in height, increasing the height from 60 to 75 µm in 5 µm steps, and (**c**) varied in thickness, increasing the thickness from 5 to 8 µm in 1 µm steps. The not-deflected structures are shown on the left and the deflected structures on the right. (**d**) Zoomed-in image showing the deflection of a beam with a height of 65 µm.

To verify the results obtained from beam theory, the deflections were determined using light microscopy images (Figure 3b,c). For better visualisation of the actuation, Figure 3c shows a zoomed-in image of the deflection of a beam with a height of 65 μm; in the supporting information videos of the actuation process are shown in Figure 3b,c. (Movies S1 and S2). The obtained deflections were plotted against the respective height h or thickness a. These were then fitted as functions of h and a. Figure 4 shows the fit curves obtained. A good agreement between the measured deflections and the calculated dependencies (solid lines) can be seen. Even for larger deflections in the range of 10 μm (>10% of the height of the structure), the measured deflection agreed well with theory, demonstrating that the simple beam theory is suitable to describe the deflection behavior.

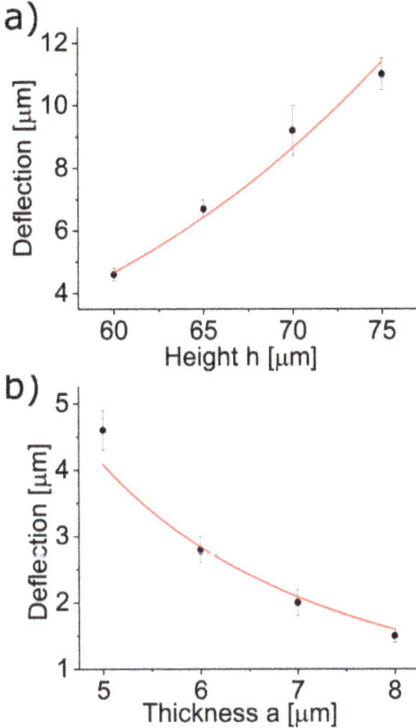

Figure 4. Plot of the deflection against (**a**) the height and (**b**) the width of the beam. The height was fitted proportional to h^4 and the thickness anti-proportional to a^2 according to beam theory.

4. Conclusions

Two-photon crosslinking of a composite consisting of a precursor of an elastic polymer and superparamagnetic particles allows direct writing of flexible 3D micro-magnets of almost any three-dimensional shape with high resolution. A unique feature of this method is that the magnetic micro-actuators are not printed from a liquid monomer containing solution, but from a polymer in the glassy state. This allows the writing of complex systems and large arrays of micro magnets in a very simple way in one step with high yield. Additionally, simple beam theory can be used to predict the deflection behavior of such architectures as a function of actuator dimensions and mechanical properties.

We believe that these findings pave the way for writing of cooperative actuator fields, which allow, among other things, controlled pushing of particles by combinations of movable and non-movable actuators, as well as the generation of metachronal waves and multicomponent micro-actuators.

Supplementary Materials: The following supporting information can be downloaded at: https://www.mdpi.com/article/10.3390/act12030124/s1, Movie S1: Support_Figure3_1.gif, Movie S2: Support_Figure3_2.gif.

Author Contributions: Conceptualization, J.R. and O.P.; methodology, N.G. and J.U.L.; investigation, N.G. and J.U.L.; resources, J.R.; writing—original draft preparation, N.G.; writing—review and editing, N.G., O.P. and J.R.; visualization, N.G. and J.U.L.; supervision, J.R.; funding acquisition, J.R. All authors have read and agreed to the published version of the manuscript.

Funding: This research was funded by Deutsche Forschungsgemeinschaft (DFG, German Research Foundation) under Priority Programme SPP 2206–KOMMMA and partially by the Deutsche Forschungsgemeinschaft (DFG, German Research Foundation) under Germany's Excellence Strategy–EXC-2193/1–390951807.

Data Availability Statement: The data presented in this study are available upon request from the corresponding author. The data are not publicly available because they are part of ongoing research.

Acknowledgments: Natalia Schatz and Martin Schönstein are thanked for valuable technical assistance.

Conflicts of Interest: The authors declare no conflict of interest. The funders had no role in the design of the study; in the collection, analyses, or interpretation of data; in the writing of the manuscript; or in the decision to publish the results.

References

1. den Toonder, J.M.J.; Onck, P.R. Microfluidic manipulation with artificial/bioinspired cilia. *Trends Biotechnol.* **2013**, *31*, 85–91. [CrossRef]
2. Zhang, S.; Wang, Y.; Onck, P.; den Toonder, J.M.J. A concise review of microfluidic particle manipulation methods. *Microfluid. Nanofluidics* **2020**, *24*, 24. [CrossRef]
3. Zhou, H.; Mayorga-Martinez, C.C.; Pané, S.; Zhang, L.; Pumera, M. Magnetically Driven Micro and Nanorobots. *Chem. Rev.* **2021**, *121*, 4999–5041. [CrossRef] [PubMed]
4. Jager, E.W.; Smela, E.; Inganäs, O. Microfabricating Conjugated Polymer Actuators. *Science* **2000**, *290*, 1540–1545. [CrossRef] [PubMed]
5. Sahadevan, V.; Panigrahi, B.; Chen, C.-Y. Microfluidic Applications of Artificial Cilia: Recent Progress, Demonstration, and Future Perspectives. *Micromachines* **2022**, *13*, 735. [CrossRef]
6. Piraux, L. Magnetic Nanowires. *Appl. Sci.* **2020**, *10*, 1832. [CrossRef]
7. Venkata Kamalakar, M.; Raychaudhuri, A.K. Resistance anomaly near phase transition in confined ferromagnetic nanowires. *Phys. Rev. B* **2010**, *82*, 195425. [CrossRef]
8. Toonder, J.M.J.; Bos, F.; Broer, D.; Filippini, L.; Gillies, M.; de Goede, J.; Mol, T.; Reijme, M.; Talen, W.; Wilderbeek, H.; et al. Artificial cilia for active micro-fluidic mixing. *Lab Chip* **2008**, *8*, 533–541. [CrossRef]
9. van Oosten, C.L.; Bastiaansen, C.W.M.; Broer, D.J. Printed artificial cilia from liquid-crystal network actuators modularly driven by light. *Nat. Mater.* **2009**, *8*, 677. [CrossRef]
10. Li, M.; Kim, T.; Guidetti, G.; Wang, Y.; Omenetto, F.G. Optomechanically Actuated Microcilia for Locally Reconfigurable Surfaces. *Adv. Mater.* **2020**, *32*, 2004147. [CrossRef]
11. Oh, K.; Chung, J.-H.; Devasia, S.; Riley, J.J. Bio-mimetic silicone cilia for microfluidic manipulation. *Lab Chip* **2009**, *9*, 1561. [CrossRef]
12. Timonen, J.V.I.; Johans, C.; Kontturi, K.; Walther, A.; Ikkala, O.; Ras, R.H.A. A Facile Template-Free Approach to Magnetodriven, Multifunctional Artificial Cilia. *ACS Appl. Mater. Interfaces* **2010**, *2*, 2226–2230. [CrossRef] [PubMed]
13. Islam, T.U.; Bellouard, Y.; den Toonder, J.M.J. Highly motile nanoscale magnetic artificial cilia. *Proc. Natl. Acad. Sci. USA* **2021**, *118*, e2104930118. [CrossRef] [PubMed]
14. Belardi, J.; Schorr, N.; Prucker, O.; Rühe, J. Artificial Cilia: Generation of Magnetic Actuators in Microfluidic Systems. *Adv. Funct. Mater.* **2011**, *21*, 3314–3320. [CrossRef]
15. Hanasoge, S.; Ballard, M.; Hesketh, P.J.; Alexeev, A. Asymmetric motion of magnetically actuated artificial cilia. *Lab Chip* **2017**, *17*, 3138–3145. [CrossRef] [PubMed]
16. Zhang, S.; Zhang, R.; Wang, Y.; Onck, P.R.; den Toonder, J.M.J. Controlled Multidirectional Particle Transportation by Magnetic Artificial Cilia. *ACS Nano* **2020**, *14*, 10313. [CrossRef] [PubMed]
17. Judith, R.M.; Fisher, J.K.; Spero, R.C.; Fiser, B.L.; Turner, A.; Oberhardt, B.; Taylor, R.M.; Falvo, R.; Superfine, R. Micro-elastometry on whole blood clots using actuated surface-attached posts (ASAPs). *Lab Chip* **2015**, *15*, 1385. [CrossRef] [PubMed]
18. Fahrni, F.; Prins, M.W.J.; van Ijzendoorn, L.J. Micro-fluidic actuation using magnetic artificial cilia. *Lab Chip* **2009**, *9*, 3413–3421. [CrossRef]
19. Evans, B.A.; Shields, A.R.; Carroll, R.L.; Washburn, S.; Falvo, M.R.; Superfine, R. Magnetically actuated nanorod arrays as biomimetic cilia. *Nano Lett.* **2007**, *7*, 1428–1434. [CrossRef]
20. Khaderi, S.N.; den Toonder, J.M.J.; Onck, P.R. Magnetically Actuated Artificial Cilia: The Effect of Fluid Inertia. *Langmuir* **2012**, *28*, 7921–7937. [CrossRef]
21. Zhou, Z.-g.; Liu, Z.-w. Biomimetic Cilia Based on MEMS Technology. *J. Bionic. Eng.* **2008**, *5*, 358. [CrossRef]

22. Hussong, J.; Schorr, N.; Belardi, J.; Prucker, O.; Rühe, J.; Westerweel, J. Experimental investigation of the flow induced by artificial cilia. *Lab Chip* **2011**, *11*, 2017–2022. [CrossRef] [PubMed]

23. Khaderi, S.N.; Craus, C.B.; Hussong, J.; Schorr, N.; Belardi, J.; Westerweel, J.; Prucker, O.; Rühe, J.; den Toonder, J.M.J.; Onck, P.R. Magnetically-actuated artificial cilia for microfluidic propulsion. *Lab Chip* **2011**, *11*, 2002–2010. [CrossRef] [PubMed]

24. Heckele, M.; Schomburg, W.K. Review on micro molding of thermoplastic polymers. *J. Micromech. Microeng.* **2004**, *14*, R1–R14. [CrossRef]

25. Pawlicki, M.; Collins, H.A.; Denning, R.G.; Anderson, H.L. Two-Photon Absorption and the Design of Two-Photon Dyes. *Angew. Chem.* **2009**, *121*, 3292. [CrossRef]

26. LaFratta, C.N.; Fourkas, J.T.; Baldacchini, T.; Farrer, R.A. Multiphoton fabrication. *Angew. Chem. Int. Ed.* **2007**, *46*, 6238. [CrossRef]

27. Park, S.-H.; Yang, D.-Y.; Lee, K.-S. Two-photon stereolithography for realizing ultraprecise three-dimensional nano/microdevices. *Laser Photon- Rev.* **2009**, *3*, 1–11. [CrossRef]

28. Coenjarts, C.A.; Ober, C.K. Two-Photon Three-Dimensional Microfabrication of Poly(Dimethylsiloxane) Elastomers. *Chem. Mater.* **2004**, *16*, 5556–5558. [CrossRef]

29. Montemayor, L.C.; Meza, L.R.; Greer, J.R. Design and Fabrication of Hollow Rigid Nanolattices via Two-Photon Lithography. *Adv. Eng. Mater.* **2014**, *16*, 184–189. [CrossRef]

30. Dayan, C.B.; Chun, S.; Krishna-Subbaiah, N.; Drotlef, D.-M.; Akolpoglu, M.B.; Sitti, M. 3D Printing of Elastomeric Bioinspired Complex Adhesive Microstructures. *Adv. Mater.* **2021**, *33*, e2103826. [CrossRef]

31. Wu, S.; Serbin, J.; Gu, M. Two-photon polymerisation for three-dimensional micro-fabrication. *J. Photochem. Photobiol. A* **2006**, *181*, 1–11. [CrossRef]

32. Faraji Rad, Z.; Prewett, P.D.; Davies, G.J. High-resolution two-photon polymerization: The most versatile technique for the fabrication of microneedle arrays. *Microsyst. Nanoeng.* **2021**, *7*, 71. [CrossRef] [PubMed]

33. Lee, K.-S.; Kim, R.H.; Yang, D.-Y.; Park, S.H. Advances in 3D nano/microfabrication using two-photon initiated polymerization. *Prog. Polym. Sci.* **2008**, *33*, 631. [CrossRef]

34. Wang, X.; Qin, X.-H.; Hu, C.; Terzopoulou, A.; Chen, X.-Z.; Huang, T.-Y.; Maniura-Weber, K.; Pané, S.; Nelson, B.J. 3D Printed Enzymatically Biodegradable Soft Helical Microswimmers. *Adv. Funct. Mater.* **2018**, *28*, 1804107. [CrossRef]

35. Giltinan, J.; Sridhar, V.; Bozuyuk, U.; Sheehan, D.; Sitti, M. 3D Microprinting of Iron Platinum Nanoparticle-Based Magnetic Mobile Microrobots. *Adv. Intell. Syst.* **2021**, *3*, 2000204. [CrossRef]

36. Peters, C.; Ergeneman, O.; García, P.D.W.; Müller, M.; Pané, S.; Nelson, B.J.; Hierold, C. Superparamagnetic Twist-Type Actuators with Shape-Independent Magnetic Properties and Surface Functionalization for Advanced Biomedical Applications. *Adv. Funct. Mater.* **2014**, *24*, 5269. [CrossRef]

37. Xia, H.; Wang, J.; Tian, Y.; Chen, Q.-D.; Du, X.-B.; Zhang, Y.-L.; He, Y.; Sun, H.-B. Ferrofluids for Fabrication of Remotely Controllable Micro-Nanomachines by Two-Photon Polymerization. *Adv. Mater.* **2010**, *22*, 3204. [CrossRef]

38. Wang, J.; Xia, H.; Xu, B.-B.; Niu, L.-G.; Wu, D.; Chen, Q.-D.; Sun, H.-B. Remote manipulation of micronanomachines containing magnetic nanoparticles. *Opt. Lett.* **2009**, *34*, 581–583. [CrossRef]

39. Schwärzle, D.; Hou, X.; Prucker, O.; Rühe, J. Polymer Microstructures through Two-Photon Crosslinking. *Adv. Mater.* **2017**, *29*, 1703469. [CrossRef]

40. Schuh, K.; Prucker, O.; Rühe, J. Tailor-Made Polymer Multilayers. *Adv. Funct. Mater.* **2013**, *23*, 6019–6023. [CrossRef]

41. Prucker, O.; Brandstetter, T.; Rühe, J. Surface-attached hydrogel coatings via C,H insertion crosslinking for biomedical and bioanalytical applications (Review). *Biointerphases* **2017**, *13*, 10801. [CrossRef] [PubMed]

42. Kanokwijitsilp, T.; Körner, M.; Prucker, O.; Anton, A.; Lübke, J.; Rühe, J. Kinetics of Photocrosslinking and Surface Attachment of Thick Polymer Films. *Macromolecules* **2021**, *54*, 6238. [CrossRef]

43. Kotrade, P.F.; Rühe, J. Malonic Acid Diazoesters for C−H Insertion Crosslinking (CHic) Reactions: A Versatile Method for the Generation of Tailor-Made Surfaces. *Angew. Chem. Int. Ed.* **2017**, *56*, 14405–14410. [CrossRef]

44. Li, X.-F. A unified approach for analyzing static and dynamic behaviors of functionally graded Timoshenko and Euler Bernoulli beams. *J. Sound Vib.* **2008**, *318*, 1210–1229. [CrossRef]

45. Gere, J.M.; Goddno, B.J. *Mechanics of Materials, Brief Edition*; Cengage Learning: Stamford, CT, USA, 2012; pp. 480–492.

46. Grespan, E.; Martewicz, S.; Serena, E.; Le Houerou, V.; Rühe, J.; Elvassore, N. Analysis of calcium transients and uniaxial contraction force in single human embryonic stem cell-derived cardiomyocytes on microstructured elastic substrate with spatially controlled surface chemistries. *Langmuir* **2016**, *32*, 12190–12201. [CrossRef] [PubMed]

actuators

Article

Model Order Reduction of Microactuators: Theory and Application

Arwed Schütz [1,2,*] and Tamara Bechtold [1]

1   Department of Engineering, Research Group Modelling and Simulation of Mechatronic Systems, Jade University of Applied Sciences, Friedrich-Paffrath-Str. 101, 26389 Wilhelmshaven, Germany
2   Chair of Control Engineering, University of Augsburg, Am Technologiezentrum 8, 86159 Augsburg, Germany
*   Correspondence: arwed.schuetz@jade-hs.de

Abstract: This paper provides an overview of techniques of compact modeling via model order reduction (MOR), emphasizing their application to cooperative microactuators. MOR creates highly efficient yet accurate surrogate models, facilitating design studies, optimization, closed-loop control and analyses of interacting components. This is particularly important for microactuators due to the variety of physical effects employed, their short time constants and the many nonlinear effects. Different approaches for linear, parametric and nonlinear dynamical systems are summarized. Three numerical case studies for selected methods complement the paper. The described case studies emerged from the *Kick and Catch* research project and within a framework of the German Research Foundation's Priority Program, *Cooperative Multistable Multistage Microactuator Systems (KOMMMA)*.

Keywords: model order reduction; finite element method; microactuators; multiphysics; MEMS

Citation: Schütz, A.; Bechtold, T. Model Order Reduction of Microactuators: Theory and Application. *Actuators* **2023**, *12*, 235. https://doi.org/10.3390/act12060235

Academic Editor: Micky Rakotondrabe

Received: 29 April 2023
Revised: 31 May 2023
Accepted: 1 June 2023
Published: 7 June 2023

1. Introduction

Microactuators are the hidden facilitators of everyday life and enable devices ranging from smartphones, printers and automotives to industrial facilities. Similar diversity is found in the physical effects deployed for actuation, ranging from shape memory alloys to electrostatics. In general, these devices convert energy into mechanical motion. However, regardless of the specific design, the manufacturing of microactuators requires designated processes and takes several months. In addition, the built hardware usually cannot be repaired or modified; it can only be replaced. For these reasons, a device should be engineered to the highest extent possible prior to production. Another level of complexity is added in the case of cooperative microactuators due to the higher number of actuators and potential cross-coupling. Therefore, reliable models are crucial for this task as they allow us to study excitations, to design control schemes and to optimize the design. Numerically investigating these models is a challenging process due to their computational complexity, nonlinearities and the small time constants inherent to microactuators. This issue is addressed by methods of compact modeling, which aim for computationally efficient yet accurate surrogate models. This methodology is applicable to various physics, nonlinearities and coupling as found in cooperative microactuators. Focusing on microactuators and their potential cooperation, this paper provides an overview of a prominent branch of compact modeling: methods of model order reduction (MOR). MOR generates significantly smaller surrogate models on the basis of large-scale dynamical systems as arising from, e.g., the finite element method (FEM). It has been widely applied in the simulation of microelectromechanical systems (MEMS) and to enhance traditional simulation tools [1]. To reduce a dynamical system, it is projected onto a low-dimensional subspace that captures most of its dynamics. In a physical sense, the state vector, e.g., a displacement field, is approximated by a linear combination of inherent patterns, also known as modes, shapes or the reduced basis.

1.1. State of the Art: Projection-Based Linear Model Order Reduction for Microactuators

Extensive research has investigated this linear MOR process [2] and several methods have been proposed [3]. These methods mainly differ in how they identify the patterns, i.e., how the reduced basis and the projection are computed. More specifically, a reduced basis may be a local or global approximation, may guarantee certain system-theoretic properties or may be limited to original systems of small dimension due to the method's computational complexity. All these reductions can be nested sequentially to combine different methods. The interested reader is referred to [4] for an intuitive overview and to [5–7] for a comprehensive handbook covering methods and a variety of applications. The following paragraphs consider the most established classes of these methods, namely *modal truncation, substructuring, balanced truncation, Krylov subspaces* and *proper orthogonal decomposition (POD)*. Special emphasis is placed on applications to microactuators to provide starting points for interested readers.

One of the oldest methods is *modal truncation* [8]. Well established in structural dynamics, it approximates displacements via the superposition of vibration modes. Unlike many other methods, these modes are not purely numerical constructs but have actual physical meaning: when excited with an eigenfrequency, the device vibrates in the corresponding eigenmode. This reduced basis forms a global approximation and leads to a diagonal reduced order model (ROM), which allows for even faster computations. The consideration of modal derivatives qualifies the concept as nonlinear MOR [9,10]. Modal truncation has been widely used for MEMS, e.g., for electromechanic RF microswitches with geometrical nonlinearities [11] and fluid structural interactions [12]. Further examples include micromirror arrays [13] and MEMS gyroscopes [14–17].

Another early branch of MOR from structural mechanics is *substructuring* [18,19], which includes *Guyan reduction* or *static condensation* [18], *Craig-Bampton reduction* [20] and *component mode synthesis* [21–23]. The general idea is to decompose the domain into substructures, which are reduced individually. These reduced substructures might be collected in a library and coupled to represent a full system. Hence, this methodology is well suited for cooperative microactuators as it emphasizes coupling. For each substructure, only the degrees of freedom (DOFs) not contributing to the coupling interfaces are reduced. Therefore, the corresponding ROM's DOFs comprise two sets: the reduced coordinates and all interface-related DOFs of the original substructure. As a result, large ROMs are required for good accuracy [24]. This concept has been applied to MEMS to investigate gas sensors [25], the failure modes of RF microswitches [26], electrothermomechanical microgrippers [27] or gyroscopes [28].

A noteworthy system-theoretic method is *balanced truncation* [29–32], which guarantees an optimal global reduction and, as a rather distinctive property, features an a priori error bound. Based on the control-theoretic concepts of controllabilty and observability, the system is transformed into a balanced realization. The transformed states are sorted by their Hankel singular values, which can be interpreted as the states' energies. Truncating insignificant states achieves the reduction. While this method ensures desirable properties, it carries high computational costs and is therefore limited to small models [33]. Consequently, little research has investigated purely balanced truncation for MEMS, e.g., for a gyroscope [34]. Instead, it has often been applied as a second reduction step in combination with, e.g., Krylov-subspace-based methods [35,36].

Methods based on *Krylov subspaces* [37–39] are among the most efficient and often the only choice for large-scale models [40]. They are also known as *rational interpolation* or *moment matching* and utilize the concept of transfer functionsThey ensure that the Taylor-expanded transfer functions of both the original and the reduced model match for the first r moments. Hence, a corresponding reduced basis forms a local approximation around an expansion point s_0 in the frequency domain. The default expansion point of $s_0 = 0$ focuses on the steady-state behavior, while dynamic responses require higher expansion points. Specialized variants have been proposed, e.g., for second-order systems [41] or for the optimal choice of expansion points [42]. This class of methods has been deployed in numerous

MEMS-related research articles, especially due to its synergy with high-dimensional FEM models. Examples include simple microstructures [43], electro-thermal MEMS [44–48], thermomechanical microgrippers [27], piezoelectric devices [49–51], electromechanical actuators [52], electromagnetic systems [53,54], gyroscopes [55] and accelerometers [56].

Another approach to the construction of a reduced basis is data-driven methods such as *POD*. Based on simulated data, POD finds a reduced basis via statistical methods that cover the variance in these data. While this reduced basis is limited to its training, it is easy to implement. In addition, it is commonly used for nonlinear systems as concepts from system theory might be inapplicable. An early application of POD to microactuators investigated a microswitch with squeeze-film damping [43]. Further examples include MEMS beams [57,58], resonators [59] and micromirrors [60].

Please note that it is also possible to create parametric ROMs [61,62] to conduct efficient design studies. Parametric influences might originate from, e.g., material properties or geometry. A parameter's effect is usually either captured by an affine expression [63,64] or approximated by interpolation [65]. Additionally, the reduced basis needs to be extended to capture parametric changes. This methodology has facilitated the design process of microswitches [66], RF resonators [64], gyroscopes [64,67,68], anemometers [67], microthrusters [63,68] or thermoelectric generators [48].

1.2. State of the Art: Projection-Based Nonlinear Model Order Reduction for Microactuators

However, all these methods of projection-based MOR are limited to linear systems. In the case of nonlinearities, they cannot reduce the nonlinear terms or are not even applicable at all. This poses a major bottleneck because MEMS are often subject to nonlinear effects, ranging from large deformations, electrostatic forces, hysteresis for piezoelectric devices or shape memory alloys, to mechanical contact. A remedy is provided by additionally approximating the nonlinear forces in an efficient way. This second approximation step is also known as *hyper-reduction*. The following paragraphs briefly introduce relevant hyper-reduction methods, such as an *approach for systems with few nonlinearities*, the *trajectory piecewise-linear (TPWL)* approximation, *polynomial tensors*, *discrete empirical interpolation method (DEIM)* and *energy conserving mesh sampling and weighting (ECSW)*. The first two methods are explained in detail in Section 2.3 and applied to numerical test cases in Section 3. Please note that while these methods achieve great results, they require individual treatment for each model and are often limited to load cases considered in their training. Another common challenge is to obtain the data needed from commercial simulation software, often limiting the choices.

An *approach for systems with few nonlinearities* is to transform them into artificial inputs [24,69]. This method is straightforward to implement, preserves the physical meaning and does not rely on training data. Further, larger numbers of nonlinearities can be lumped into fewer terms to achieve compatibility. However, this approach is limited to nonlinearities that depend on a single or a few DOFs at most, e.g., to one-dimensional electrostatic forces or mechanical contact. MEMS-related applications range from RF switches [69] and scanning-probe data storage [24] to electromechanical beam actuators [52].

The TPWL approximation is a robust method for general nonlinear systems [70]. A combination of linearized systems sampled along a training trajectory approximate a nonlinear system. The combination weights depend on the reduced state and change throughout the simulation. Therefore, a TPWL-approximated system is still nonlinear, but the corresponding terms are few and are efficient to evaluate. While the approach is robust and only requires easily obtainable data, its accuracy strongly depends on the weighting scheme and the sampling strategy. Furthermore, it relies on data generated by extensive simulation of the original model. The main work proposing this approach featured an electromechanical MEMS as a case study [70]. Later work reduced thermal actuators [71], thermal switches [47,72,73] or solenoid actuators [74] based on TPWL.

Polynomial tensors are another intuitive approach for hyper-reduction. The concept is to approximate the reduced nonlinear forces by a Taylor expansion [75,76]. However, the Taylor series' coefficients are tensors of increasing order and, thus, the method quickly becomes inapplicable due to the amount of entries. Therefore, polynomial tensors effectively only suit nonlinearities that can be approximated by low-order Taylor expansions. One such nonlinearity is St. Venant–Kirchhoff materials, which describe linear–elastic systems at large deformations [4]. This method may be deployed to incorporate geometrical nonlinearities into substructures, as introduced in Section 2.2. A direct application to MEMS is found for an electromechanical actuator with squeeze-film damping [75].

A commonly used hyper-reduction method is the *DEIM* [77], which evaluates only a few of the original nonlinearities. These evaluated nonlinearities serve as weights for precomputed force patterns to approximate the whole nonlinear force vector. The approximated force vector is subsequently projected back onto the reduced space. Hence, the nonlinear vector is approximated by a reduced force basis with state-dependent weights. This scheme does not ensure stability and leads to asymmetry [4]. Constructing the reduced force basis and choosing the subset of nonlinearities to evaluate are data-driven processes and rely on training data, limiting the prediction quality. Applications to MEMS or to models featuring the same physics comprise electrothermal microgrippers [78], MEMS switches [79] and transistors [79].

A recently introduced hyper-reduction method is the *ECSW* procedure [80–82]. This approach considers the virtual work of the reduced forces over all finite elements. A subset of elements is determined so that the combination of their weighted energies approximates the original total work. The weights compensate for the energies of the numerous excluded elements [81]. This approach has similarities to the transfer of loads from a fine FEM mesh to a coarser one [81]. In contrast to the DEIM and many other hyper-reduction methods, stability and symmetry are preserved. Furthermore, the reduced force vector is approximated directly, instead of approximating the full vector and subsequently projecting it to the reduced space [81]. Again, training data are needed and this limits the prediction quality, even though there is some robustness. To our knowledge, this method is yet to be extended to the microactuator community.

1.3. Alternatives to Projection-Based Model Order Reduction

While projection-based MOR achieves highly efficient and accurate surrogate models and also preserves the original model's structure, alternative methodologies exist. These approaches are also suitable to obtain compact models of microactuators and vary drastically in their complexity and performance. Commonly deployed methodologies for compact modeling are *look-up tables, meta-modeling, generalized Kirchhoffian network (GKN)* and *machine-learning*-based or *data-driven* approaches. Please note that these techniques can often be combined with MOR, e.g., to update a nonlinear stiffness matrix via a look-up table [17] or to approximate nonlinear forces via artificial neural networks (ANNs) [83,84].

Look-up tables are the most basic solution and consist of precomputed outputs for sampled input combinations. While they are robust and easy to implement, they strongly depend on the sampling strategy and, potentially, an interpolation scheme. Furthermore, they are unsuitable for dynamical systems and drastically lose accuracy as the number of parameters increases.

Meta-modeling or *response surfaces* extend the previous approach of look-up tables with regression analysis [85]. While they also require several sampled solutions of the original models, they achieve higher accuracy and some extrapolation quality. However, their performance depends on how well the basic function matches the relation to be modeled. Usually, they are not deployed to approximate dynamical systems but relations between design variables and outputs for design optimization.

A prominent branch of compact modeling is GKNs, which transfer the concept of electrical Kirchhoffian networks to other physical domains [86]. Therefore, a microactuator might be represented by a network of lumped elements. This methodology covers multiphysical problems as well as nonlinear effects while preserving the basic structure of the original model. Another advantage is the physical meaning of all components and the computational efficiency, but accuracy might be sacrificed due to the lumped nature. Furthermore, it requires expert knowledge to divide a structure into lumped elements and to fill them with an appropriate mathematical model. Applications to microsystems are common [87] and include capacitive MEMS transducers [88] and acoustic ultrasonic MEMS transducers [89], as well as magnetic, electric and acoustic transducers [87].

A novel and promising branch is *machine-learning*-based and *data-driven* approaches. This includes methods such as ANNs [90] or data-driven MOR via operator inference [91–93]. These methods have in common that they rely on vast amounts of training data. In the context of simulation, these data are easy to obtain without noise or outliers and are available from commercial software. Solving the original model numerous times to obtain data and subsequent training leads to high computational costs. In addition, some methods require expert knowledge to adjust the hyperparameters or choose architectures. Furthermore, the structure of the problem and the physical interpretability might not be preserved. Nevertheless, this class of methods is suitable for a wide range of problems and synergizes well with simulated data.

1.4. Outline of the Article

This paper reviews the methodology of MOR and emphasizes its application to microactuators. The aim is to further establish MOR in the microactuator community. Therefore, a brief methodological overview tailored to the microactuator community is provided. References for different actuators are intended as starting points for interested readers. Moreover, three extensive case studies demonstrate the potential of MOR.

The remainder of the paper is structured as follows. Section 2 provides the theory, describing the process from the numerical modeling of microactuators to reduced order models. Subsequently, Section 3 applies the theory from Section 2 to three microsystem-oriented case studies, covering several physics and nonlinear effects. Finally, Section 4 summarizes this work.

2. Compact Modeling by Means of Mathematical Model Order Reduction

This section describes how to derive a highly efficient surrogate model as illustrated in Figure 1. The starting point is a mathematical model given by the governing partial differential equation (PDE). Spatial discretization via, e.g., the FEM leads to an accurate but large-scale system of n ordinary differential equations (ODEs), as described in Section 2.1. Subsequently, MOR constructs a highly efficient surrogate model, i.e., an ODE system of the same form, but with much smaller dimensions $r << n$, as described in Section 2.2.

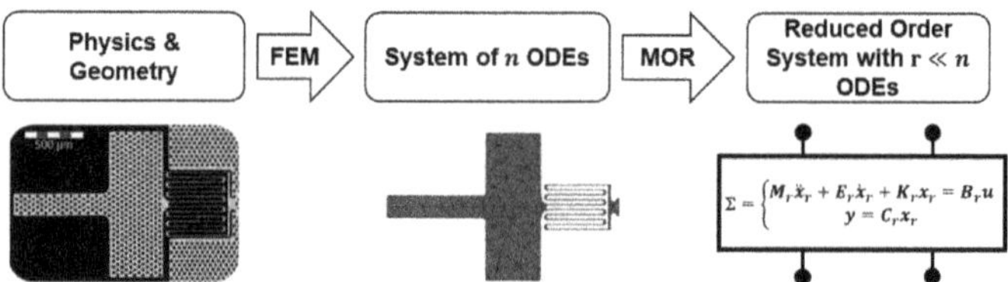

Figure 1. Workflow of MOR-based compact modeling. Starting from a microsystem and its governing physics, the FEM assembles a high-dimensional dynamical system. This system is then reduced by methods of MOR, resulting in a surrogate model of drastically smaller dimensions.

2.1. Mathematical Modeling of Microactuators

Physical laws dictate the behavior of all microactuators. At the continuum level, these laws can be described by mathematical models (PDEs), which usually comprise two components: conservation laws and constitutive relations. The former arise from the universal conservation of quantities, such as energy or mass; the latter introduce experimentally confirmed material relations. This concept applies to numerous physical domains, including structural, thermal, acoustic and electric, as well as to their multiphysical coupling. This section deploys linear elastic dynamics as an example, since it is relevant to all microactuators. The governing PDE is Newton's second law, which is denoted by the following second-order PDE for time-independent density:

$$\nabla \cdot \sigma + f = \rho \ddot{x}. \tag{1}$$

Here, σ is the stress tensor, f the body force per volume, ρ the density and x the displacement vector. The relevant constitutive relation is Hooke's law given by

$$\sigma = C\varepsilon, \tag{2}$$

where C is the stiffness tensor and ε is the strain tensor. The following strain–displacement equation from infinitesimal strain theory completes the mathematical description:

$$\varepsilon = \frac{1}{2}\left[\nabla x + (\nabla x)^T\right]. \tag{3}$$

While the above equations provide the complete mathematical model, they can only be solved analytically for the most basic scenarios. An established approach to finding a remedy is numerical methods such as the FEM. The FEM subdivides the computational domain into smaller subdomains called finite elements and approximates the solution with element-wise polynomial shape functions. Mathematically, this spatial discretization converts the initial PDE into a system of linear ODEs Σ in the form of

$$\Sigma = \begin{cases} M\ddot{x} + E\dot{x} + Kx = Bu \\ y = Cx + Du \end{cases}, \tag{4}$$

where M, E and $K \in \mathbb{R}^{n\times n}$ are the system matrices and $x \in \mathbb{R}^n$ is the state vector. In contrast to Equations (1) and (3), x comprises numerous nodal displacements at different positions and not continuous displacement functions. The input vector is denoted by $u \in \mathbb{R}^p$, the user-defined output vector by $y \in \mathbb{R}^q$. The inputs are distributed by $B \in \mathbb{R}^{n\times p}$ and the outputs are computed from the state vector by $C \in \mathbb{R}^{q\times n}$. The feedthrough matrix $D \in \mathbb{R}^{q\times p}$ includes the direct effect that inputs may have on outputs. As the system is linear and all matrices remain constant over time, it is referred to as linear time-invariant. Systems of this form are common in control theory and several concepts for further analysis apply. One such concept is the transfer function $H(s)$, which is an equivalent system representation. It relates Laplace transformations of input and output functions

$$Y(s) = H(s)\,U(s), \tag{5}$$

where s is the complex frequency variable. For the system in Equation (4), the transfer function is given by

$$H(s) = C\left(s^2 M + s E + K\right)^{-1} B + D. \tag{6}$$

2.2. Projection-Based Linear Model Order Reduction

Although numerical methods such as the FEM are capable of solving sophisticated multiphysical problems, they suffer under high computational costs. These high computational costs arise from the fact that FEM-generated dynamical systems reach large dimensions up to $10^6 \ldots 10^8$. Therefore, these models' dimensions impede efficient design studies and prevent application in control circuits, especially considering the small time scales of microactuators. A well-established approach to tackling this challenge is projection-based MOR [94], which creates surrogate models of the same structure but significantly smaller dimension. These surrogates enable fast prediction, more extensive analysis options, parametric investigations and feedback control.

The basic idea behind MOR is to decompose the solutions into patterns. Restricting the solution space to the most important of these patterns results in a low-dimensional surrogate model. These patterns are also known as modes, shapes or reduced basis vectors. In a mathematical sense, the state vector x becomes a linear combination of predefined patterns as illustrated in Figure 2. After orthonormalizing the r most important patterns for numerical reasons, they are assembled as columns of a projection matrix $V \in \mathbb{R}^{n \times r}$. The reduced state vector $x_r \in \mathbb{R}^r$ comprises the weights of all these patterns. Omitting most patterns as they barely contribute introduces an approximation error x_ε, and it holds that

$$x = V x_r + x_\varepsilon. \tag{7}$$

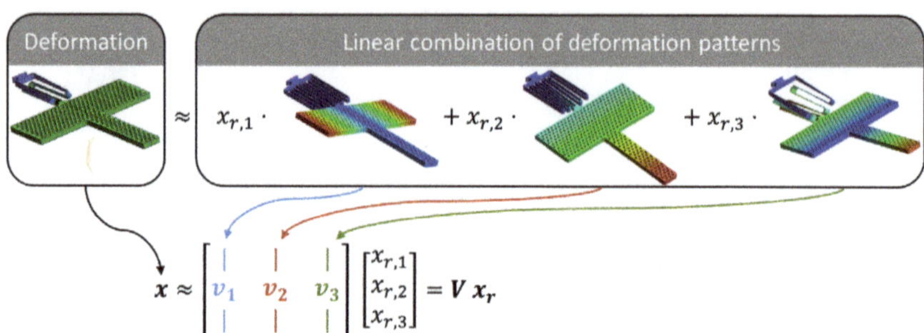

Figure 2. General idea of MOR: approximating the state as a combination of few relevant patterns. In this example, the state corresponds to the deformation of an electrostatic beam actuator and it is approximated by three eigenmodes. Here, the color indicates deformation magnitude. The Eigenmodes are assembled as columnvectors V. The corresponding weights are collected within the reduced state vector x_r. In general, the specific vectors in V depend on the method chosen for MOR.

However, substituting this approximation into the system in Equation (4) results in an overdetermined system, which also includes an approximation error. To obtain a unique solution and to eliminate the error from the equation, the system is projected onto V along $null(W^T \in \mathbb{R}^{r \times n})$. The pattern-based approximation in Equation (7) and an appropriate projection reduce the system in Equation (4) to

$$\Sigma_r = \begin{cases} \overbrace{W^T M V}^{M_r} \ddot{x}_r + \overbrace{W^T E V}^{E_r} \dot{x}_r + \overbrace{W^T K V}^{K_r} x_r = \overbrace{W^T B}^{B_r} u \\ y = \underbrace{C V}_{C_r} x_r + D u \end{cases}, \tag{8}$$

where $M_r, E_r, K_r \in \mathbb{R}^{r \times r}$, $B_r \in \mathbb{R}^{r \times p}$ and $C_r \in \mathbb{R}^{q \times r}$ are the reduced system matrices. These reduced matrices only need to be computed once and can be subsequently deployed in applications. Please note that the inputs u and outputs y remain unchanged. This ROM contains multiple orders of magnitude less ODEs than the original FEM system in Equation (4) as $r \ll n$. As a result, all subsequent computations are significantly faster. Specific methods to construct the reduced basis are presented and discussed in Section 1, including modal truncation, substructuring, balanced truncation, Krylov subspaces and POD.

2.3. Projection-Based Nonlinear Model Order Reduction

In general, real-world physics are nonlinear. In some cases, the nonlinearities barely contribute within the operating conditions of interest, and a linear model provides sufficient accuracy. However, the field of MEMS features several potential nonlinear effects, such as large deformations, electrostatic forces, hysteresis for piezoelectric devices or shape memory alloys or mechanical contact. Therefore, nonlinear approaches are inevitable for microactuators. In mathematical terms, a nonlinear problem depends on its solution. Hence, the solution process follows an iterative scheme until convergence criteria are eventually satisfied. Due to these iterations, nonlinearities significantly inflate the omputational demands. Considering a system Σ_{nl} as in Equation (4) but with nonlinear restoring forces $f(x) \in \mathbb{R}^n$ leads to

$$\Sigma_{nl} = \begin{cases} M\ddot{x} + E\dot{x} + f(x) = Bu \\ y = Cx + Du \end{cases}. \tag{9}$$

Such a nonlinear system as in Equation (9) entails two great challenges for projection-based MOR. Firstly, many methods to identify reduced bases introduced in Section 2.2 leave their range of validity. Most of the related research deploys the POD because it does not rely on system-theoretic concepts, which might only be defined for linear systems. The second challenge is that the nonlinear term cannot be reduced: approximating the state by patterns and projecting the system as in Equation (8) leads to

$$\Sigma_{nl,r} = \begin{cases} M_r\ddot{x}_r + E_r\dot{x}_r + W^T f(Vx_r) = B_r u \\ y = C_r x_r + Du \end{cases}. \tag{10}$$

Evaluating the nonlinear term $W^T f(Vx_r)$ requires us to project the reduced state vector x_r into the original high-dimensional space, evaluating the full set of nonlinearities, and then project them back onto the reduced space. Obviously, this process is less efficient than evaluating the original model, especially with respect to the iterative solution scheme. A well-established approach is an efficient approximation of the nonlinear term, which is also known as *hyper-reduction* [95]. In preparation for the numerical case studies in Sections 3.2 and 3.3, an approach for systems with few nonlinearities [69] and the TPWL [96] approximation are described.

If a system has only a few nonlinearities, a robust and straightforward approach is to isolate them and to handle them as additional inputs [24,69]. In mathematical terms, the nonlinear term is decomposed into

$$f(x) = Kx - B_F u_F(\hat{C}x), \tag{11}$$

where B_F is an additional input matrix that distributes nonlinear forces scaled by the nonlinear inputs $u_F(\hat{C}x)$. To evaluate the nonlinearities in u_F, only a small subset $\hat{C}x$ of the full state vector is required. Hence, this approach performs best when the nonlinearities depend on only a few DOFs, which is the case for, e.g., electrostatic forces or simple mechanical contact. Inserting this decomposition into the system in Equation (10) gives

$$M_r\ddot{x}_r + E_r\dot{x}_r + K_r x_r = \underbrace{W^T \begin{bmatrix} B & B_F \end{bmatrix}}_{B_r^*} \underbrace{\begin{bmatrix} u \\ u_F(\hat{C}_r x_r) \end{bmatrix}}_{u^*(\hat{C}_r x_r)}, \tag{12}$$

where B_r^* collects the input matrices and $u^*(\hat{C}_r x_r)$ summarizes the inputs. Even though this system is still nonlinear, the relocation into inputs enables the use of linear MOR methods [69]. Another noteworthy advantage is that the approximation is not limited to training data. In the case of a higher number of nonlinearities, they can be condensed by grouping schemes to require significantly fewer nonlinear evaluations. Although this step introduces another approximation, it minimizes the computational demand. Specific grouping schemes are, e.g., to approximate distributed electrostatic forces by a single lumped force [69] or by a force distribution scaled by a single nonlinear term [52].

An early and reliable approach for general nonlinear systems is the TPWL approximation [70]. The idea is to approximate a nonlinear system by a combination of linearized systems that are sampled along a training trajectory. The weights of each system depend on the reduced state and change in the course of the simulation. Due to these state-dependent weighting scheme, the TPWL-approximated system still contains nonlinearities, but they are few and are efficient to evaluate. The nonlinear force vector is expressed as a weighted sum of N linearizations and reads

$$f(x) \approx \sum_{i=1}^{N} \Big(w_i(x) \, (-f_i + K_i \, x) \Big), \tag{13}$$

where the subscript i indicates the ith sampling point, $w_i(x)$ the state-dependent weight and f_i and K_i the linearization. Please note that the sampled systems are linear and, therefore, can be reduced by methods of linear MOR. However, a global reduced basis for all systems is required to achieve compatibility. Combining the approximation in Equation (13) with the system in Equation (10) results in a TPWL-reduced system given by

$$M_r \, \ddot{x}_r + E_r \, \dot{x}_r + \underbrace{\sum_{i=1}^{N} \Big(w_i(x_r) \, \overbrace{W^T K_i \, V}^{K_{i,r}} \Big) \, x_r}_{K_r(x_r)} = \underbrace{\left[\sum_{i=1}^{N} \Big(w_i(x_r) \, \overbrace{W^T f_i}^{f_{i,r}} \Big) \quad B_r \right]}_{B_r^*(x_r)} \underbrace{\begin{bmatrix} 1 \\ u \end{bmatrix}}_{u^*}. \tag{14}$$

The remaining nonlinear terms are the N weights $w_i(x_r)$, which only depend on the reduced state vector x_r according to Algorithm 1. First, the distance between the current state and the sampled states is computed. Based on the minimum distance m, preliminary weights $\hat{w}_i$ are calculated. To obtain the final weights, these are normalized by their sum.

Algorithm 1 Weighting scheme for TPWL.

---

for $i = 1, \ldots, N$ **do**
$\quad d_i \leftarrow \| x_r - x_{r,i} \|$
$m \leftarrow \min_{i=1,..,N} d_i$
for $i = 1, \ldots, N$ **do**
$\quad \hat{w}_i \leftarrow e^{-\beta \frac{d_i}{m}}$
$S \leftarrow \sum_i^N \hat{w}_i$
for $i = 1, \ldots, N$ **do**
$\quad w_i \leftarrow \dfrac{\hat{w}_i}{S}$

---

3. Exemplary Applications of MOR to Microactuators

This section introduces three numerical case studies of different physical domains with both linear and nonlinear setups. The first case study in Section 3.1 describes the linear MOR of a piezoelectric chip actuator based on earlier work [97]. The multiphysical coupling gives rise to a unique challenge as it potentially introduces instability to the corresponding ROMs. Applying the designated methods preserves or reintroduces stability. Section 3.2 presents the second case study, which reduces an electromechanical microactuator as demonstrated in [52]. This device corresponds to a cantilever beam actuated into electrostatic pull-in. Electrostatic forces and mechanical contact render the model nonlinear. A novel third case study in Section 3.3 reduces a geometrically nonlinear beam model via TPWL. The modeled actuator is the same as in Section 3.2, but the FEM model is coarser and exclusively composed of three-dimensional elements. All case studies deploy *Ansys® Academic Research Mechanical, Release 2022 R2* for FEM modeling to compute reference solutions and to obtain system matrices. The process of MOR either uses *Model Reduction inside Ansys* [98] by CADFEM® and/or a Python-based implementation [99,100].

3.1. Piezoelectric Chip Actuator

This numerical case study investigates the PA3JEA piezoelectric chip actuator and is based on earlier work [97] that contains a more detailed description. Figure 3 depicts the actuator and its symmetry-exploiting FEM model. The ceramic coating houses 33 piezoelectric layers. The modeled geometry excludes the interdigitated silver electrodes since their effect on the system is insignificant. However, their electrical behavior is included and the electrical potentials of each electrode layer are coupled. Further, the cathodes are grounded. Therefore, the model is composed exclusively of THP51 ceramic [97]. Mechanical boundary conditions prevent rigid body motion and vertical deformation of the actuator's base. The load case considered is a force acting vertically at the top surface's center. The vertical displacement at this position and the voltage at the anode constitute the desired outputs. This model comprises 3395 nodes and corresponds to a system of 9892 differential algebraic equations (DAEs).

Figure 3. The PA3JEA piezoelectric chip actuator and its symmetry-exploiting FEM model [97].

MOR of piezoelectric devices introduces two additional challenges: the potential loss of stability and rounding errors sensitive to the chosen unit system. The first challenge is to preserve the original system's stability. Instability translates into a potentially unbounded output for a bounded input and, thus, renders reduced systems useless. Methods to preserve or to reintroduce stability are *Schur after MOR* [101], *MOR after Schur* [101], *MOR after implicit Schur* [102] and *multiphysics structure-preserving MOR* [38,49,101]. They differ in their approach and vary significantly in their computational efficiency. Recommended methods are either Schur after MOR or multiphysics structure-preserving MOR. The second challenge is the chosen unit system as it affects the matrices' condition numbers, especially for multiphysical studies. For this reason, preliminary studies are recommended to determine the best-conditioned setting in order to minimize rounding errors.

The original system of DAEs representing the FEM model in Figure 3 is reduced in several settings to compare the four stability-preserving algorithms. All reductions deploy Krylov subspaces to compute the reduced basis. Further, all reduced order models are evaluated in a harmonic analysis in a frequency range of 0 kHz to 500 kHz for a unit force. The settings to be varied include the expansion point and the reduced model's dimension. The former comprises the values 0 Hz, 250 kHz and 500 kHz, while the latter is either set to 60 or 120. This setup results in six different combinations, which are evaluated for all four algorithms.

A comparison of the anode's voltage for a reduced dimension of 60 and an expansion point of 0 Hz is shown in Figure 4. As the curves cannot be distinguished, the figure is extended by corresponding relative errors. In general, the relative error barely surpasses 10^{-7}, but increases towards higher frequencies due to the low expansion point. The least accurate but most efficient method is Schur after MOR. The best accuracy is achieved by multiphysics structure-preserving MOR.

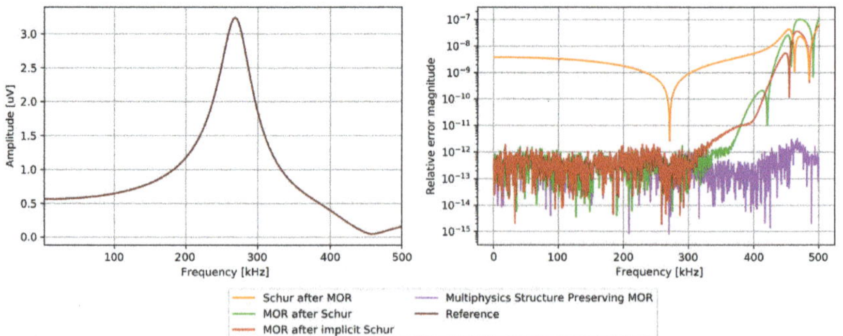

Figure 4. Response of the anode's voltage to a harmonic unit force computed by the FEM for reference and four reduced order models stabilized by different algorithms (**left**). The corresponding relative errors allow more detailed conclusions (**right**). This evaluation is one of six setups to compare the four stability-preserving algorithms [97].

Figure 4 compares the methods in one of the six settings and only for one of the two outputs. The remaining information is summarized in Figure 5. This plot condenses the frequency-dependent relative errors as in Figure 4 into a single number by taking the average magnitude. Although this procedure strongly depends on the chosen frequency range, it is constant for each combination and, therefore, constitutes a valid procedure. The findings coincide with the ones for Figure 4. In addition, it can be observed that the mechanical output is approximated significantly better than the electrical one.

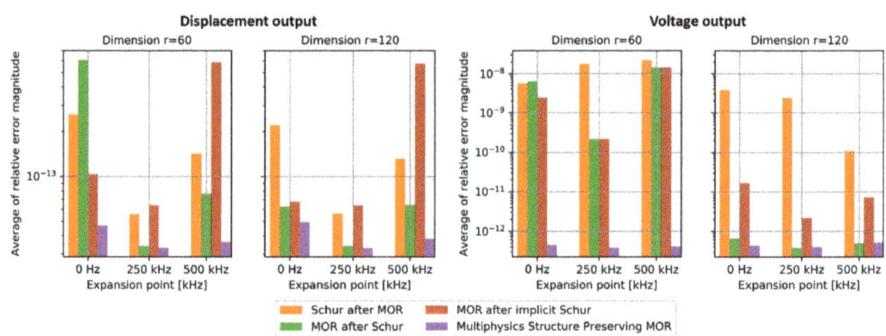

Figure 5. Comparing the average relative error magnitude for both outputs and all six reduction settings. Multiphysics structure-preserving MOR achieves the best accuracy. MOR after Schur benefits most from increasing the reduced dimension. In general, a central expansion point leads to the best approximation quality [97].

The times required for each method with respect to the original model's dimension are given in Figure 6. For a more detailed analysis, both the total time and the MOR-exclusive part are given. The absolute time demand as well as the scaling with model dimension vary significantly between the methods. Schur after MOR is the most efficient method, taking the least amount of time and scaling well with larger dimensions. In contrast, MOR after Schur is limited to small-sized models, as it leads to dense system matrices and high computational costs. Multiphysics structure-preserving MOR and MOR after implicit Schur perform similarly, but require more time than Schur after MOR in their stability-preserving computations. All computations were performed on an Intel® Core™ CPU $4 \times 3.0\,$GHz and 64 GB RAM.

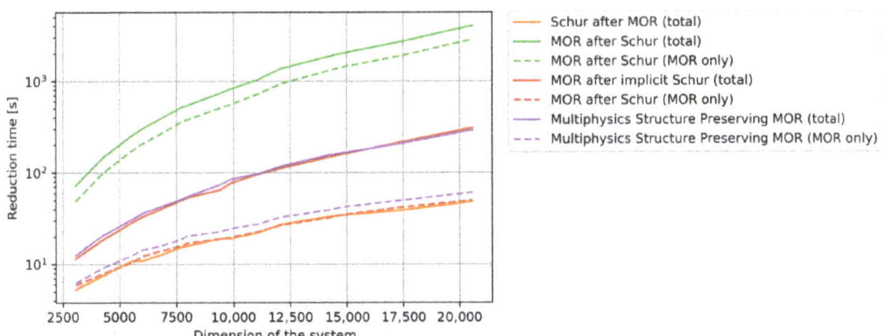

Figure 6. The four method's computational times vs. the original model's dimension n. Dashed lines correspond to elapsed time due to MOR, whereas full lines provide the total demand [97].

In conclusion, MOR significantly reduces the dimension of the original model and, thus, introduces significant computational benefits. Further, all four methods preserve the original model's stability and even the worst achieved accuracy suffices for most applications. However, Schur after MOR and multiphysics structure-preserving MOR are the methods to recommend as they are efficient and reliable. In the case of large-scale FEM models, the recommendation narrows down to Schur after MOR. Further information can be found in [97].

3.2. Electromechanical Beam Actuator

The subject of this numerical case study is a single actuator of the cooperative microsystem shown in Figure 7. An extensive description of this study is available in [52]. The design has been developed within the *Kick and Catch* project [52,103] and deploys multiple cooperating actuators to rotate a freely moving body. The overall goal is a multistable, quasistatic micromirror.

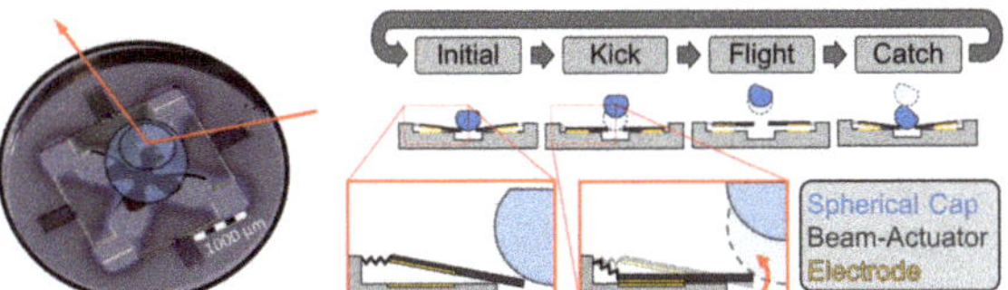

Figure 7. The *Kick and Catch* actuator system [103] (**left**) and its operating principle (**right**). On the left, the indicated spherical cap rests on the four electrostatic beam actuators and is deflecting an incident light ray. As illustrated on the right, these beam actuators are actuated into pull-in to launch the spherical cap. After a free flight phase, the sphere is caught and rests stably. Consecutive flight phases allow for a high deflection angle.

One of the four electrostatic microactuators with mechanical contact constitutes the numerical case study. This actuator is highly representative because its physical principles are commonly found in a large class of microactuators. Figure 8 presents the actuator's design, which is composed of three sections: a beam tip, an electrode and a compliant meander spring. The actuator is mounted 10 µm above its counter electrode, which attracts the beam when voltage is applied.

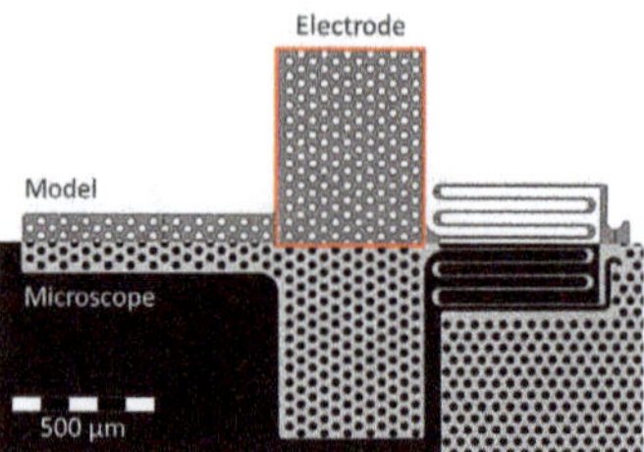

Figure 8. The electrostatic beam actuator and its symmetry-exploiting FEM model side by side. From left to right, the beam comprises three sections: a tip for leverage, an electrode area marked in red and a compliant meander spring that enables the pull-in motion. The electrode is subject to electrostatic forces, which are nonlinear because they depend on the beam's deformation [52].

In terms of FEM, the actuator is a linear mechanical system as in Equation (4) and of dimension $n = 25,134$. Lumped transducer elements below the electrode introduce electrostatic forces and mechanical contact. Due to the lumped nature, a nonlinear force acts on every node of the electrode area. These nonlinear forces depend on the node's out-of-plane displacement x_k and read

$$f^k = f^k_{el} + f^k_{cont}$$
$$\text{with: } f^k_{el} = \frac{\varepsilon A_k}{2} \frac{1}{(x_k + g_0)^2} V^2$$
$$f^k_{cont} = \begin{cases} k_n |x_k + g_0| & \text{if } (x_k + g_0) < 0 \\ 0 & \text{else.} \end{cases}$$

(15)

Hence, the force on node k is composed of an electrostatic part f^k_{el} and a contact force f^k_{cont}. The electrostatic force is based on permittivity ε, the node's effective area A_k, the initial gap g_0 and the applied voltage V. The contact force has a penalty-based structure and, thus, occurs upon penetration. Its magnitude depends on the amount of penetration and the contact stiffness k_n.

This setup is well suited for the nonlinear MOR technique for systems with few non-linearities. As the nonlinearities only depend on a single DOF, they can be evaluated with acceptable computational costs. Furthermore, the out-of-plane displacement of adjacent nodes barely deviates. Therefore, the nonlinear forces can be grouped to reduce the number of nonlinear computations. Each group summarizes its nodes' effective areas and their contact stiffness, respectively. In addition, a representative displacement per group is determined. This grouping procedure drastically reduces the number of nonlinear computations and can be achieved by geometric groups [69] or by clustering [52].

For the model in Figure 8, the linear part is reduced by a Krylov-subspace-based approach to a dimension of $r = 100$. Further, the nonlinear forces on the electrode are summarized into six groups by two approaches: geometric grouping and agglomerative clustering. The resulting two reduced order models are evaluated for a transient load case, which applies a step voltage of 30 V. As a result, the beam is actuated into pull-in within less than 400 µs. To assess the two methods' accuracy, the vertical tip displacement is tracked and compared to an FEM reference solution. This comparison is presented in Figure 9, in which the contact event is highlighted. Both reduced order models excellently match their reference and even provide convergence after the contact event. Although the behavior after contact differs, it cannot be compared as the FEM solution fails to converge. The speed-up factor due to MOR and the two grouping schemes is more than 250, reducing from 784 s for the FEM analysis to less than 3 s on an Intel® Core™ CPU 4 x 3.0 GHz and 64 GB RAM. These computational times only indicate the efficiency, because the evaluation did not use the same solver but a less optimized one.

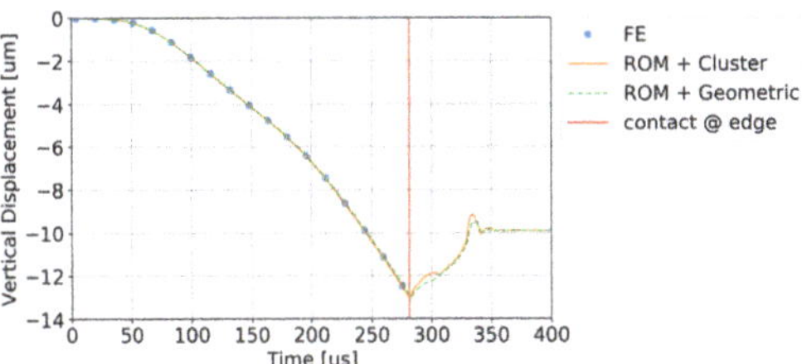

Figure 9. Vertical displacement of the actuator's tip after applying a step voltage of 30 V, computed by two reduced order models and the FEM for reference [52].

3.3. Geometrically Nonlinear Beam Actuator

Depending on the load case, analysis of the beam actuator in Section 3.2 requires us to consider large deformations. These geometrical nonlinearities are common for microactuators and, in contrast to Section 3.2, render the whole system nonlinear. This novel numerical case study deploys a static and purely mechanical load case of the isolated beam actuator shown in Figure 8. A new FEM mesh of the homogenized actuator geometry results in a system of $n = 843$ ODEs. A downward force on the beam's tip is gradually increased up to 500 μN in steps of 10 μN.

Two methods are combined to reduce this setup: POD to construct the reduced basis and TPWL to handle the nonlinearities. Both rely on sampled quantities of the original model: the former requires samples of the state vector, the latter linearized system matrices. TPWL dynamically composes the current set of system matrices from a pool of linearized matrices. State-dependent weights quantify the difference between the current state and previously sampled ones. Figure 10 visualizes this procedure, illustrating selected states as deformations of the beam. Please note that TPWL can also be applied independently of MOR. Hence, this study compares the TPWL-approximated original model and its reduced version to the FEM reference solution. As a result, sources of deviation can be identified more clearly.

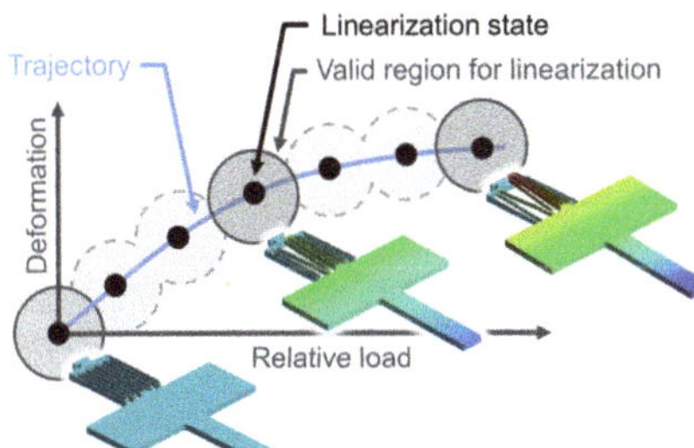

Figure 10. TPWL approximates a nonlinear system as a weighted sum of linearized ones. These linearizations are obtained at different states along the trajectory. States correspond to physical deformations, of which three are illustrated. The color corresponds to deformation magnitude. The linearized models are valid in the vicinity of the respective linearization states, as indicated by circles.

The specific TPWL approximation uses a weighting parameter of $\beta = 250$ and 11 uniformly distributed samples. The process of MOR via POD decomposes 51 snapshots obtained from the FEM reference solution. The 25 most dominant left-singular vectors form the reduced basis as the remaining singular values barely contain additional information, as illustrated in Figure 11. Therefore, they are truncated, resulting in a reduced dimension of $r = 25$.

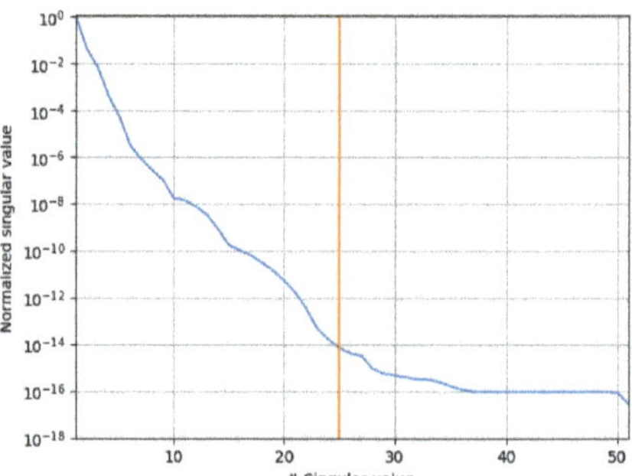

Figure 11. Descending singular values of the FEM study's 51 snapshots, normalized by the largest singular value. The reduced basis of left-singular vectors is truncated at $r = 25$ indicated by the vertical line.

The TPWL-approximated model and its reduced version are then evaluated for the same 51 load steps as the original model. The output quantity assessing their performance is the beam's tip displacement analogous to Section 3.2. Figure 12 provides the results and the relative errors of this analysis. Both models match the reference solution as the relative error barely surpasses 10^{-2}. As with most interpolation-based methods, the error peaks between samples. The reduction step introduces negligible additional errors when compared to the TPWL approximation. Both methods are computationally more efficient and reduce the original solution time of 1.157 s per load step to 36.68 ms and 443.7 µs, respectively. Therefore, the reduced order model is faster by a factor of more than 2500. Please note that the FEM analysis uses a more optimized solver. The study was conducted on an Intel® Core™ CPU (4 x 3.0 GHz) and 64 GB RAM.

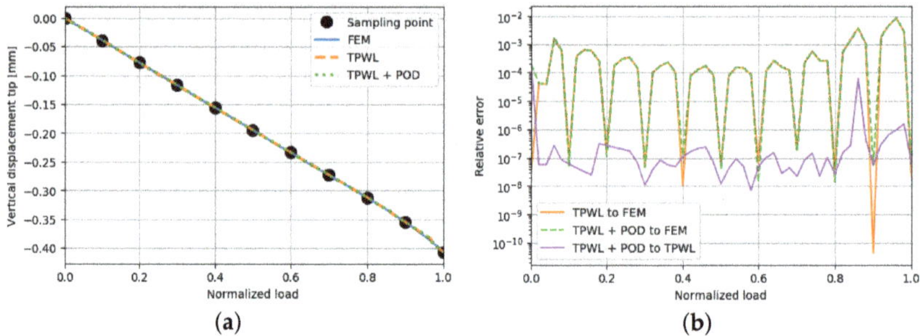

(a)   (b)

Figure 12. (**a**) Comparison of the vertical tip displacement computed by the TPWL-approximated model, its reduced version and the FEM for reference. The plot also indicates sampling positions for TPWL. (**b**) In addition, the relative errors are shown, including the deviation of the TPWL-approximated model and its reduced version to the FEM reference solution. Further, the error between the two TPWL approximations is provided. In general, the minimum error coincides with the sampling positions and reaches maximum values in between. The coarsely sampled TPWL approximation is the main source of deviation, in contrast to the excellent match that MOR achieves.

Both the TPWL approximation and its reduced version achieve noteworthy computational benefits while losing less than 1 % accuracy for this case study. This deviation predominantly originates from the TPWL approximation, while MOR to a dimension of only 25 induces a negligible error. Please note that a more sophisticated sampling strategy improves the accuracy and may even require fewer samples. Furthermore, the weighting scheme's settings bear potential for optimization and promise better accuracy at the cost of several parameter studies. However, the aim of this numerical study is to demonstrate the basic principle and its potential, without application-dependent finetuning.

4. Conclusions and Outlook

Cooperative actuators often lead to more complex models than their independent counterparts. As a result, the model dimensions grow and limit the potential for design studies or closed-loop control. This work emphasizes the solution that MOR provides to this challenge. This methodology generates highly efficient surrogate models, as demonstrated in three numerical case studies, reaching speed-up factors of more than 250 and excellent accuracy. The gain in efficiency facilitates extensive parameter studies and design optimization. Furthermore, numerous physical effects, parametric influences and even nonlinearities can be considered with appropriate MOR methods. Reduced order models can be coupled to investigate interactions within systems. In addition, they can be conveniently combined with different model types, such as look-up tables, lumped element models or neural networks. The reduction process also preserves the model's structure but encrypts the original model, protecting intellectual property when sharing models. Another advantage is more sophisticated closed-loop control, which becomes more feasible. However, MOR requires extra work and designated tools. Although several commercial solutions exist, consistent tool chains are rare. One reason is that most MOR methods are intrusive and need to access the underlying mathematical model. Nevertheless, the extensive literature on microsystem-related MOR suggests its potential for this field of research, accelerating if not enabling several applications.

Alternatives to MOR include look-up tables, meta-models, GKNs and data-driven approaches such as operator inference or ANNs. While look-up tables and meta-models are easy to implement, they are based on solutions of the original model. Therefore, the number of full samples grows exponentially with the number of parameters, often leading to coarse sampling. In addition, they are better suited to model relations between input parameters and outputs, rather than dynamical systems. However, these two approaches are robust and require significantly less expert knowledge. GKNs represent an original system as an equivalent network of lumped elements and are a viable alternative to projection-based MOR. They preserve the original model's structure and physical meaning, while being capable of multiphysics and nonlinear effects. Moreover, they are established in the field of microactuators. This technique requires expert knowledge and is less accurate and flexible than MOR, e.g., regarding additional outputs. Data-driven approaches such as operator inference or ANNs are suitable for numerous tasks, but their prediction quality is limited to scenarios included in their training. They do not preserve physical meaning or the structure of the original problem, but work well with simulated data without noise or outliers.

Current trends of MOR include their combination with data-driven approaches such as ANNs. This synergy enables highly efficient training on reduced data or grey-box models incorporating physical knowledge. The former approach uses MOR to drastically reduce the dimensions of training data, which leads to faster training, easier networks and the more efficient tuning of hyperparameters. The latter concept extends linear ROMs with ANNs to include parametric influences or nonlinear effects, e.g., by updating matrices for parametric changes or nonlinearities. The potential is yet to be exploited and diffused into the microactuator community.

Author Contributions: Conceptualization, T.B. and A.S.; investigation, A.S.; writing—original draft preparation, A.S.; writing—review and editing, T.B. and A.S.; visualization, A.S.; supervision, T.B.; funding acquisition, T.B. All authors have read and agreed to the published version of the manuscript.

Funding: This research was funded by Deutsche Forschungsgemeinschaft (German Research Foundation), grant number 424616052.

Data Availability Statement: The data can be provided upon reasonable request.

Acknowledgments: Siyang Hu and Chengdong Yuan are thanked for the fruitful discussions. Further, we thank Thorlabs GmbH for supplying additional information and for their kind permission to use pictures from their website in our work.

Conflicts of Interest: The authors declare no conflict of interest. The funding institutions had no role in the design of the study; in the collection, analyses, or interpretation of data; in the writing of the manuscript, or in the decision to publish the results.

Abbreviations

The following abbreviations are used in this manuscript:

ANN	artificial neural network
DAE	differential algebraic equation
DEIM	discrete empirical interpolation method
DOF	degree of freedom
ECSW	energy conserving mesh sampling and weighting
FEM	finite element method
GKN	generalized Kirchoffian network
MEMS	microelectromechanical system
MOR	model order reduction
ODE	ordinary differential equation
PDE	partial differential equation
POD	proper orthogonal decomposition
ROM	reduced order model
TPWL	trajectory piecewise-linear

References

1. Bechtold, T.; Schrag, G.; Feng, L. (Eds.) *System-Level Modeling of MEMS*; Wiley-VCH-Verl.: Weinheim, Germany, 2013; Advanced Micro and Nanosystems Volume 10.
2. Benner, P.; Ohlberger, M.; Cohen, A.; Willcox, K. *Model Reduction and Approximation*; Society for Industrial and Applied Mathematics: Philadelphia, PA, USA, 2017.
3. Baur, U.; Benner, P.; Feng, L. Model Order Reduction for Linear and Nonlinear Systems: A System-Theoretic Perspective. *Arch. Comput. Methods Eng.* **2014**, *21*, 331–358. [CrossRef]
4. Rutzmoser, J. Model Order Reduction for Nonlinear Structural Dynamics: Simulation-Free Approaches. Ph.D. Thesis, Technischen Universität München, Garching, Germany, 2018.
5. Benner, P.; Grivet-Talocia, S.; Quarteroni, A.; Rozza, G.; Schilders, W.; Silveira, L.M. (Eds.) *Model Order Reduction: Volume 1: System- and Data-Driven Methods and Algorithms*, 1st ed.; De Gruyter: Berlin, Germany, 2021; Volume 1.
6. Benner, P.; Grivet-Talocia, S.; Quarteroni, A.; Rozza, G.; Schilders, W.; Silveira, L.M. (Eds.) *Model Order Reduction: Volume 2: Snapshot-Based Methods and Algorithms*; De Gruyter: Berlin, Germany, 2021; Volume 2.
7. Benner, P.; Grivet-Talocia, S.; Quarteroni, A.; Rozza, G.; Schilders, W.; Silveira, L.M. (Eds.) *Model Order Reduction: Volume 3: Applications*, 1st ed.; De Gruyter: Berlin, Germany, 2021; Volume 3.
8. Lord Rayleigh, J. *The Theory of Sound*; Macmillan: New York, NY, USA, 1894.
9. Weeger, O.; Wever, U.; Simeon, B. On the use of modal derivatives for nonlinear model order reduction. *Int. J. Numer. Methods Eng.* **2016**, *108*, 1579–1602. [CrossRef]
10. Vizzaccaro, A.; Salles, L.; Touzé, C. Comparison of nonlinear mappings for reduced-order modelling of vibrating structures: Normal form theory and quadratic manifold method with modal derivatives. *Nonlinear Dyn.* **2021**, *103*, 3335–3370. [CrossRef]
11. Mehner, J.; Gabbay, L.D.; Senturia, S.D. Computer-aided generation of nonlinear reduced-order dynamic macromodels. II. Stress-stiffened case. *J. Microelectromech. Syst.* **2000**, *9*, 270–278. [CrossRef]
12. Mehner, J.; Doetzel, W.; Schauwecker, B.; Ostergaard, D. Reduced order modeling of fluid structural interactions in MEMS based on model projection techniques. In Proceedings of the Transducers '03 , Boston, MA, USA, 8–12 June 2003; pp. 1840–1843.
13. Bennini, F.; Mehner, J.; Dötzel, W. Computational Methods for Reduced Order Modeling of Coupled Domain Simulations. In *Transducers '01 Eurosensors XV*; Obermeier, E., Ed.; Springer: Berlin/Heidelberg, Germany, 2001; pp. 260–263.
14. Putnik, M.; Sniegucki, M.; Cardanobile, S.; Kehrberg, S.; Kuehnel, M.; Nagel, C.; Degenfeld-Schonburg, P.; Mehner, J. Incorporating geometrical nonlinearities in reduced order models for MEMS gyroscopes. In Proceedings of the IEEE Inertial Sensors 2017, Kauai, HI, USA, 28–30 March 2017; pp. 43–46.
15. Putnik, M.; Sniegucki, M.; Cardanobile, S.; Kuhnel, M.; Kehrberg, S.; Mehner, J. Predicting the Resonance Frequencies in Geometric Nonlinear Actuated MEMS. *J. Microelectromech. Syst.* **2018**, *27*, 954–962. [CrossRef]
16. Putnik, M.; Cardanobile, S.; Sniegucki, M.; Kehrberg, S.; Kuehnel, M.; Degenfeld-Schonburg, P.; Nagel, C.; Mehner, J. Simulation methods for generating reduced order models of MEMS sensors with geometric nonlinear drive motion. In Proceedings of the the 5th IEEE International Symposium on Inertial Sensors & Systems, Lake Como, Italy, 26–29 March 2018; pp. 1–4.
17. Putnik, M. Simulation Methods for the Mechanical Nonlinearity in MEMS Gyroscopes. Ph.D. Thesis, Fakultät für Elektrotechnik und Informationstechnik, Chemnitz, Germany, 2019.

18. Guyan, R.J. Reduction of Stiffness and Mass Matrices. *AIAA J.* **1964**, *3*, 380. [CrossRef]
19. de Klerk, D.; Rixen, D.J.; Voormeeren, S.N. General Framework for Dynamic Substructuring: History, Review and Classification of Techniques. *AIAA J.* **2008**, *46*, 1169–1181. [CrossRef]
20. Craig, R.R.; Bampton, M.C.C. Coupling of substructures for dynamic analyses. *AIAA J.* **1968**, *6*, 1313–1319. [CrossRef]
21. MacNeal, R.H. A hybrid method of component mode synthesis. *Comput. Struct.* **1971**, *1*, 581–601. [CrossRef]
22. Rubin, S. Improved Component-Mode Representation for Structural Dynamic Analysis. *AIAA J.* **1975**, *13*, 995–1006. [CrossRef]
23. Craig, R.R.; Ni, Z. Component mode synthesis for model order reduction of nonclassicallydamped systems. *J. Guid. Control. Dyn.* **1989**, *12*, 577–584. [CrossRef]
24. Lienemann, J. Complexity Reduction Techniques for Advanced MEMS Actuators Simulation. Ph.D. Thesis, Albert-Ludwigs-Universität Freiburg, Freiburg, Germany, 2006.
25. Sadek, K.; Moussa, W. Application of adaptive multilevel substructuring technique to model CMOS micromachined thermistor gas sensor, part (I): A feasibility study. In Proceedings of the International Conference on MEMS, NANO and Smart Systems, Banff, AB, Canada, 20–23 July 2003; pp. 279–284.
26. Sadek, K.; Lueke, J.; Moussa, W. A Coupled Field Multiphysics Modeling Approach to Investigate RF MEMS Switch Failure Modes under Various Operational Conditions. *Sensors* **2009**, *9*, 7988–8006. [CrossRef] [PubMed]
27. Binion, D.; Chen, X. Coupled electrothermal–mechanical analysis for MEMS via model order reduction. *Finite Elem. Anal. Des.* **2010**, *46*, 1068–1076. [CrossRef]
28. Giannini, D.; Bonaccorsi, G.; Braghin, F. Size optimization of MEMS gyroscopes using substructuring. *Eur. J. Mech. A/Solids* **2020**, *84*, 104045. [CrossRef]
29. Mullis, C.; Roberts, R. Synthesis of minimum roundoff noise fixed point digital filters. *IEEE Trans. Circuits Syst.* **1976**, *23*, 551–562. [CrossRef]
30. Moore, B. Principal component analysis in linear systems: Controllability, observability, and model reduction. *IEEE Trans. Autom. Control* **1981**, *26*, 17–32. [CrossRef]
31. Chahlaoui, Y.; Lemonnier, D.; Vandendorpe, A.; van Dooren, P. Second-order balanced truncation. *Linear Algebra Its Appl.* **2006**, *415*, 373–384. [CrossRef]
32. Reis, T.; Stykel, T. Balanced truncation model reduction of second-order systems. *Math. Comput. Model. Dyn. Syst.* **2008**, *14*, 391–406. [CrossRef]
33. Wolf, T.; Castagnotto, A.; Eid, R. *Moderne Methoden der Regelungstechnik 3-Teil B-Einführung in die Modellreduktion*; Technische Universität München: München, Germany, 2016.
34. M'Closkey, R.T.; Gibson, S.; Hui, J. System Identification of a MEMS Gyroscope. *J. Dyn. Syst. Meas. Control* **2001**, *123*, 201–210. [CrossRef]
35. Kamon, M.; Wang, F.; White, J. Generating nearly optimally compact models from Krylov-subspace based reduced-order models. *IEEE Trans. Circuits Syst. II Analog Digit. Signal Process.* **2000**, *47*, 239–248. [CrossRef]
36. Rudnyi, E.B.; Korvink, J.G. Review: Automatic Model Reduction for Transient Simulation of MEMS-based Devices. *Sens. Update* **2002**, *11*, 3–33. [CrossRef]
37. Grimme, E.J. Krylov Projection Methods for Model Reduction. Ph.D. Thesis, University of Illinois at Urbana-Champaign, Champaign, IL, USA, 1997.
38. Freund, R.W. Krylov-subspace methods for reduced-order modeling in circuit simulation. *J. Comput. Appl. Math.* **2000**, *123*, 395–421. [CrossRef]
39. Freund, R.W. Model reduction methods based on Krylov subspaces. *Acta Numer.* **2003**, *12*, 267–319. [CrossRef]
40. Rudnyi, E.B.; Korvink, J.G. Model Order Reduction for Large Scale Engineering Models Developed in ANSYS. In *Applied Parallel Computing. State of the Art in Scientific Computing*; Hutchison, D., Kanade, T., Kittler, J., Kleinberg, J.M., Mattern, F., Mitchell, J.C., Naor, M., Nierstrasz, O., Pandu Rangan, C., Steffen, B., et al., Eds.; Lecture Notes in Computer Science; Springer: Berlin/Heidelberg, Germany, 2006; Volume 3732, pp. 349–356.
41. Bai, Z.; Su, Y. SOAR: A Second-order Arnoldi Method for the Solution of the Quadratic Eigenvalue Problem. *SIAM J. Matrix Anal. Appl.* **2005**, *26*, 640–659. [CrossRef]
42. Gugercin, S.; Antoulas, A.C.; Beattie, C. H2 Model Reduction for Large-Scale Linear Dynamical Systems. *SIAM J. Matrix Anal. Appl.* **2008**, *30*, 609–638. [CrossRef]
43. Hung, E.S.; Yang, Y.J.; Senturia, S.D. Low-order models for fast dynamical simulation of MEMS microstructures. In Proceedings of the International Solid State Sensors and Actuators Conference (Transducers '97), Chicago, IL, USA, 19 June 1997; pp. 1101–1104.
44. Bechtold, T.; Rudnyi, E.B.; Korvink, J.G. Automatic order reduction of thermo-electric model for micro-ignition unit. In Proceedings of the International Conference on Simulation of Semiconductor Processes and Devices, Granada, Spain, 4–6 September 2002; pp. 131–134.
45. Bechtold, T. Model Order Reduction of Electro-Thermal MEMS. Ph.D. Thesis, Albert-Ludwigs-Universität Freiburg, Freiburg, Germany, 2005.
46. Bechtold, T.; Rudnyi, E.B.; Korvink, J.G. *Fast Simulation of Electro-Thermal MEMS*; Springer: Berlin/Heidelberg, Germany, 2007.
47. Liu, Y.; Yuan, W.; Chang, H.; Ma, B. Compact thermoelectric coupled models of micromachined thermal sensors using trajectory piecewise-linear model order reduction. *Microsyst. Technol.* **2014**, *20*, 73–82. [CrossRef]

48. Yuan, C.; Kreß, S.; Sadashivaiah, G.; Rudnyi, E.B.; Hohlfeld, D.; Bechtold, T. Towards efficient design optimization of a miniaturized thermoelectric generator for electrically active implants via model order reduction and submodeling technique. *Int. J. Numer. Methods Biomed. Eng.* **2020**, *36*, e3311. [CrossRef]
49. Yuan, C.; Hu, S.; Bechtold, T. Stable Compact Modeling of Piezoelectric Energy Harvester Devices. *COMPEL* **2020**, *39*, 868. [CrossRef]
50. Yuan, C.; Hu, S.; Castagnotto, A.; Lohmann, B.; Bechtold, T. Implicit Schur Complement for Model Order Reduction of Second Order Piezoelectric Energy Harvester Model. In *MATHMOD 2018 Extended Abstract Volume*; Volume ARGESIM Report 55; ARGESIM: Vienna, Austria, 2018; pp. 77–78, ISBN 978-3-901608-91-9.
51. Schütz, A.; Maeter, S.; Bechtold, T. System-Level Modelling and Simulation of a Multiphysical Kick and Catch Actuator System. *Actuators* **2021**, *10*, 279. [CrossRef]
52. Schütz, A.; Farny, M.; Olbrich, M.; Hoffmann, M.; Ament, C.; Bechtold, T. Model Order Reduction of a Nonlinear Electromechanical Beam Actuator by Clustering Nonlinearities. In *ACTUATOR 2022*; GMM-Fachbericht, VDE VERLAG: Berlin, Germany; Offenbach, Germany, 2022; pp. 354–357.
53. Pierquin, A.; Henneron, T.; Clenet, S.; Brisset, S. Model-Order Reduction of Magnetoquasi-Static Problems Based on POD and Arnoldi-Based Krylov Methods. *IEEE Trans. Magn.* **2015**, *51*, 1–4. [CrossRef]
54. Bonotto, M.; Cenedese, A.; Bettini, P. Krylov Subspace Methods for Model Order Reduction in Computational Electromagnetics. *IFAC-PapersOnLine* **2017**, *50*, 6355–6360. [CrossRef]
55. Chang, H.; Zhang, Y.; Xie, J.; Zhou, Z.; Yuan, W. Integrated Behavior Simulation and Verification for a MEMS Vibratory Gyroscope Using Parametric Model Order Reduction. *J. Microelectromech. Syst.* **2010**, *19*, 282–293. [CrossRef]
56. Han, J.S.; Rudnyi, E.B.; Korvink, J.G. Efficient optimization of transient dynamic problems in MEMS devices using model order reduction. *J. Micromech. Microeng.* **2005**, *15*, 822–832. [CrossRef]
57. Liang, Y.C.; Lin, W.Z.; Lee, H.P.; Lim, S.P.; Lee, K.H.; Sun, H. Proper Orthogonal Decomposition and Its Applications—Part II: Model Reduction for Mems Dynamical Analysis. *J. Sound Vib.* **2002**, *256*, 515–532. [CrossRef]
58. Binion, D.; Chen, X. A Krylov enhanced proper orthogonal decomposition method for frequency domain model reduction. *Eng. Comput.* **2017**, *34*, 285–306. [CrossRef]
59. Corigliano, A.; Dossi, M.; Mariani, S. Domain decomposition and model order reduction methods applied to the simulation of multi-physics problems in MEMS. *Comput. Struct.* **2013**, *122*, 113–127. [CrossRef]
60. Gobat, G.; Opreni, A.; Fresca, S.; Manzoni, A.; Frangi, A. Reduced order modeling of nonlinear microstructures through Proper Orthogonal Decomposition. *Mech. Syst. Signal Process.* **2022**, *171*, 108864. [CrossRef]
61. Benner, P.; Gugercin, S.; Willcox, K. A Survey of Projection-Based Model Reduction Methods for Parametric Dynamical Systems. *SIAM Rev.* **2015**, *57*, 483–531. [CrossRef]
62. Benner, P.; Ohlberger, M.; Patera, A.; Rozza, G.; Urban, K. (Eds.) *Model Reduction of Parametrized Systems*; MS&A; Springer International Publishing: Cham, Switerland, 2017.
63. Rudnyi, E.B.; Moosmann, C.; Greiner, A.; Bechtold, T.; Korvink, J.G. Parameter Preserving Model Reduction for MEMS System-level Simulation and Design. In *MATHMOD*; Vienna University of Technology: Vienna, Austria, 2006.
64. Moosmann, C. ParaMOR–Model Order Reductionfor Parameterized MEMS Applications. Ph.D. Thesis, Albert-Ludwigs-Universität Freiburg, Freiburg, Germany, 2007.
65. Panzer, H.; Mohring, J.; Eid, R.; Lohmann, B. Parametric Model Order Reduction by Matrix Interpolation. *Automatisierungstechnik* **2010**, *58*, 958. [CrossRef]
66. Bond, B.; Daniel, L. Parameterized model order reduction of nonlinear dynamical systems. In Proceedings of the ICCAD-2005, IEEE/ACM International Conference on Computer-Aided Design, San Jose, CA, USA, 6–10 November 2005; pp. 487–494.
67. Baur, U.; Benner, P.; Greiner, A.; Korvink, J.G.; Lienemann, J.; Moosmann, C. Parameter preserving model order reduction for MEMS applications. *Math. Comput. Model. Dyn. Syst.* **2011**, *17*, 297–317. [CrossRef]
68. Feng, L.; Benner, P.; Korvink, J.G. Subspace recycling accelerates the parametric macro-modeling of MEMS. *Int. J. Numer. Methods Eng.* **2013**, *94*, 84–110. [CrossRef]
69. del Tin, L. Reduced-Order Modelling, Circuit-Level Design and SOI Fabrication of Microelectromechanical Resonators. Ph.D. Thesis, Università di Bologna, Bologna, Italy, 2007.
70. Rewieński, M.; White, J. A trajectory piecewise-linear approach to model order reduction and fast simulation of nonlinear circuits and micromachined devices. *IEEE Trans. Comput.-Aided Des. Integr. Circuits Syst.* **2003**, *22*, 155–170. [CrossRef]
71. Yang, Y.J.; Shen, K.Y. Nonlinear heat-transfer macromodeling for MEMS thermal devices. *J. Micromech. Microeng.* **2005**, *15*, 408–418. [CrossRef]
72. Liu, Y.; Chang, H.; Yuan, W. A Global Maximum Error Controller-Based Method for Linearization Point Selection in Trajectory Piecewise-Linear Model Order Reduction. *IEEE Trans. Comput.-Aided Des. Integr. Circuits Syst.* **2014**, *33*, 1100–1104.
73. Liu, Y.; Wang, X. A Two-Step Global Maximum Error Controller-Based TPWL MOR with POD Basis Vectors and Its Applications to MEMS. *Math. Probl. Eng.* **2017**, *2017*, 5014235. [CrossRef]
74. Albunni, M.N. Model Order Reduction of Moving Nonlinear Electromagnetic Devices. Ph.D. Thesis, Technischen Universität München, Munich, Germany, 2010.

75. Chen, J.; Kang, S.M.; Zou, J.; Liu, C.; Schutt-Aine, J.E. Reduced-Order Modeling of Weakly Nonlinear MEMS Devices with Taylor-Series Expansion and Arnoldi Approach. *J. Microelectrmech. Syst.* **2004**, *13*, 441–451. [CrossRef]

76. Mignolet, M.P.; Przekop, A.; Rizzi, S.A.; Spottswood, S.M. A review of indirect/non-intrusive reduced order modeling of nonlinear geometric structures. *J. Sound Vib.* **2013**, *332*, 2437–2460. [CrossRef]

77. Chaturantabut, S.; Sorensen, D.C. Nonlinear Model Reduction via Discrete Empirical Interpolation. *SIAM J. Sci. Comput.* **2010**, *32*, 2737–2764. [CrossRef]

78. Roy, A.; Nabi, M. Modeling of MEMS Electrothermal Microgripper employing POD-DEIM and POD method. *Microelectron. Reliab.* **2021**, *125*, 114338. [CrossRef]

79. Hochman, A.; Bond, B.N.; White, J. A stabilized discrete empirical interpolation method for model reduction of electrical, thermal, and microelectromechanical systems. In Proceedings of the 48th Design Automation Conference, San Diego, CA, USA, 5–10 June 2011; Stok, L., Dutt, N., Hassoun, S., Eds.; ACM: New York, NY, USA, 2011; pp. 540–545.

80. Farhat, C.; Avery, P.; Chapman, T.; Cortial, J. Dimensional reduction of nonlinear finite element dynamic models with finite rotations and energy-based mesh sampling and weighting for computational efficiency. *Int. J. Numer. Methods Eng.* **2014**, *98*, 625–662. [CrossRef]

81. Farhat, C.; Chapman, T.; Avery, P. Structure-preserving, stability, and accuracy properties of the energy-conserving sampling and weighting method for the hyper reduction of nonlinear finite element dynamic models. *Int. J. Numer. Methods Eng.* **2015**, *102*, 1077–1110. [CrossRef]

82. Chapman, T. Nonlinear Model Order Reduction for Structural Systems with Contact. Ph.D. Thesis, Stanford University, Stanford, CA, USA, 2019.

83. Gao, H.; Wang, J.X.; Zahr, M.J. Non-intrusive model reduction of large-scale, nonlinear dynamical systems using deep learning. *Phys. D Nonlinear Phenom.* **2020**, *412*, 132614. [CrossRef]

84. Cicci, L.; Fresca, S.; Manzoni, A. Deep-HyROMnet: A deep learning-based operator approximation for hyper-reduction of nonlinear parametrized PDEs. *J. Sci. Comput.* **2022**, *93*, 57. [CrossRef]

85. Simpson, T.W.; Poplinski, J.D.; Koch, P.N.; Allen, J.K. Metamodels for Computer-based Engineering Design: Survey and recommendations. *Eng. Comput.* **2001**, *17*, 129–150. [CrossRef]

86. Schrag, G.; Wachutka, G. System-Level Modeling of MEMS Using Generalized Kirchhoffian Networks-Basic Principles. In *System-Level Modeling of MEMS*; Bechtold, T., Schrag, G., Feng, L., Eds.; Advanced Micro and Nanosystems; Wiley-VCH Verlag GmbH & Co. KGaA: Weinheim, Germany, 2013; pp. 19–51.

87. Lenk, A.; Ballas, R.G.; Werthschützky, R.; Pfeifer, G. *Electromechanical Systems in Microtechnology and Mechatronics: Electrical, Mechanical and Acoustic Networks, Their Interactions and Applications*; Microtechnology and MEMS; Springer: Berlin, Germany, 2010.

88. Bosetti, G.; Manz, J.; Dehé, A.; Krumbein, U.; Schrag, G. Modeling and physical analysis of an out-of-plane capacitive MEMS transducer with dynamically coupled electrodes. *Microsyst. Technol.* **2019**, *3*, 81. [CrossRef]

89. Bosetti, G.; Schrag, G. Efficient Modeling of Acoustic Channels – Towards Tailored Frequency Response of Airborne Ultrasonic MEMS Transducers. In Proceedings of the 2022 23rd International Conference on Thermal, Mechanical and Multi-Physics Simulation and Experiments in Microelectronics and Microsystems (EuroSimE), Graz, Austria, 25–27 April 2022; pp. 1–5.

90. Gruber, A.; Gunzburger, M.; Ju, L.; Wang, Z. A comparison of neural network architectures for data-driven reduced-order modeling. *Comput. Methods Appl. Mech. Eng.* **2022**, *393*, 114764. [CrossRef]

91. Peherstorfer, B.; Willcox, K. Data-driven operator inference for nonintrusive projection-based model reduction. *Comput. Methods Appl. Mech. Eng.* **2016**, *306*, 196–215. [CrossRef]

92. Benner, P.; Goyal, P.; Kramer, B.; Peherstorfer, B.; Willcox, K. Operator inference for non-intrusive model reduction of systems with non-polynomial nonlinear terms. *Comput. Methods Appl. Mech. Eng.* **2020**, *372*, 113433. [CrossRef]

93. Qian, E.; Kramer, B.; Peherstorfer, B.; Willcox, K. Lift & Learn: Physics-informed machine learning for large-scale nonlinear dynamical systems. *Phys. D Nonlinear Phenom.* **2020**, *406*, 132401.

94. Antoulas, A.C. *Approximation of Large-Scale Dynamical Systems*; Advances in Design and Control; Society for Industrial and Applied Mathematics: Philadelphia, PA, USA, 2005.

95. Ryckelynck, D. A priori hyperreduction method: An adaptive approach. *J. Comput. Phys.* **2005**, *202*, 346–366. [CrossRef]

96. Rewieński, M. A Trajectory Piecewise-Linear Approach to Model Order Reduction of Nonlinear Dynamical Systems. Ph.D. Thesis, Technical University of Gdansk, Gdansk, Poland, 2003.

97. Schütz, A.; Bechtold, T. Performance Comparison for Stable Compact Modelling of Piezoelectric Microactuator. In Proceedings of the 2022 23rd International Conference on Thermal, Mechanical and Multi-Physics Simulation and Experiments in Microelectronics and Microsystems (EuroSimE), Graz, Austria, 25–27 April 2022; pp. 1–8.

98. Rudnyi, E.B.; Lienemann, J.; Greiner, A.; Korvink, J.G. *mor4ansys: Generating Compact Models Directly from ANSYS Models*; Routledge: Oxfordshire, UK, 2004.

99. Harris, C.R.; Millman, K.J.; van der Walt, S.J.; Gommers, R.; Virtanen, P.; Cournapeau, D.; Wieser, E.; Taylor, J.; Berg, S.; Smith, N.J.; et al. Array programming with NumPy. *Nature* **2020**, *585*, 357–362. [CrossRef] [PubMed]

100. Virtanen, P.; Gommers, R.; Oliphant, T.E.; Haberland, M.; Reddy, T.; Cournapeau, D.; Burovski, E.; Peterson, P.; Weckesser, W.; Bright, J.; et al. SciPy 1.0: Fundamental algorithms for scientific computing in Python. *Nat. Methods* **2020**, *17*, 261–272. [CrossRef] [PubMed]

101. Kudryavtsev, M.; Rudnyi, E.B.; Korvink, J.G.; Hohlfeld, D.; Bechtold, T. Computationally efficient and stable order reduction methods for a large-scale model of MEMS piezoelectric energy harvester. *Microelectron. Reliab.* **2015**, *55*, 747–757. [CrossRef]
102. Hu, S.; Yuan, C.; Castagnotto, A.; Lohmann, B.; Bouhedma, S.; Hohlfeld, D.; Bechtold, T. Stable reduced order modeling of piezoelectric energy harvesting modules using implicit Schur complement. *Microelectron. Reliab.* **2018**, *85*, 148–155. [CrossRef]
103. Farny, M.; Hoffmann, M. Kick & Catch: Elektrostatisches Rotieren einer Kugel. In *MikroSystemTechnik Kongress*; VDE Verlag GmbH: Stuttgart-Ludwigsburg, Germany, 2021; pp. 274–277.

Article

Bistable Actuation Based on Antagonistic Buckling SMA Beams

Xi Chen [1], Lars Bumke [2], Eckhard Quandt [2] and Manfred Kohl [1,*]

[1] Institute of Microstructure Technology, Karlsruhe Institute of Technology, 76344 Eggenstein-Leopoldshafen, Germany
[2] Inorganic Functional Materials, Institute for Materials Science, Kiel University, 24143 Kiel, Germany
* Correspondence: manfred.kohl@kit.edu

Abstract: Novel miniature-scale bistable actuators are developed, which consist of two antagonistically coupled buckling shape memory alloy (SMA) beams. Two SMA films are designed as buckling SMA beams, whose memory shapes are adjusted to have opposing buckling states. Coupling the SMA beams in their center leads to a compact bistable actuator, which exhibits a bi-directional snap-through motion by selectively heating the SMA beams. Fabrication involves magnetron sputtering of SMA films, subsequent micromachining by lithography, and systems integration. The stationary force–displacement characteristics of monostable actuators consisting of single buckling SMA beams and bistable actuators are characterized with respect to their geometrical parameters. The dynamic performance of bistable actuation is investigated by selectively heating the SMA beams via direct mechanical contact to a low-temperature heat source in the range of 130–190 °C. The bistable actuation is characterized by a large stroke up to 3.65 mm corresponding to more than 30% of the SMA beam length. Operation frequencies are in the order of 1 Hz depending on geometrical parameters and heat source temperature. The bistable actuation at low-temperature differences provides a route for waste heat recovery.

Keywords: bistability; bistable actuation; shape memory actuator; antagonistic coupling; waste heat recovery

Citation: Chen, X.; Bumke, L.; Quandt, E.; Kohl, M. Bistable Actuation Based on Antagonistic Buckling SMA Beams. *Actuators* **2023**, *12*, 422. https://doi.org/10.3390/act12110422

Academic Editor: Wei Min Huang

Received: 2 October 2023
Revised: 6 November 2023
Accepted: 9 November 2023
Published: 11 November 2023

1. Introduction

Bistable mechanical systems exhibit the unique property of two stable equilibrium states characterized by local minima of potential energy [1]. The stable positions are retained without power supply and, thus, power consumption is only required upon switching between the positions. When subjected to specific stimuli or loading conditions, a snap-through buckling motion can occur resulting in a rapid transition between the stable states. Bistable microactuators are basic components in engineering applications, such as micro-electro-mechanical systems (MEMS) [2–4], microfluidics [5], and constitute key building blocks in emerging digital mechanical microsystems based on multistable microactuation [6].

In the past, bistable mechanisms have been developed consisting of compliant beams and plates, e.g., [7,8]. In this case, bistable actuation relies on the design of the movable compliant structures and the method of external loading. In particular, an external force needs to be applied to trigger the snap-through of the bistable structures. Considerable effort has been devoted to the design and fabrication of bistable MEMS devices including microvalves [9], micro-switches and relays [10–12], as well as fiber-optic switches [13]. In recent years, the development of smart materials such as shape memory [14,15] and piezoelectric materials [16,17] has led to the development of bistable mechanisms that do no longer rely on an external load but utilize their intrinsic transducer properties to induce bistability. Thus, novel smart bistable actuators have been developed making use of an extended range of stimuli including electrical [18,19], magnetic [20,21], thermal [22], and coupled fields [5,23].

Among the thermally responsive actuators, shape memory alloy (SMA) actuators are of special interest as they can convert small amounts of heat into mechanical work output in a narrow temperature interval. Depending on the SMA material, shape recovery can be induced well below 200 °C, enabling the conversion of low temperature waste heat. Due to the high power/weight ratio and ease-of-micromachining, film-basedSMA actuators have been widely implemented in microsystems, e.g., [24,25], and in robotics, e.g., [26]. The drawback of an external triggering force in conventional bistable systems can be avoided by the design of buckling SMA actuators with a predefined shape that is memorized by thermo-mechanical treatment (memory shape) and the use of the intrinsic thermally induced shape-recovery force [27]. As the one-way shape memory effect can only cause one-way snap-through behavior, a reset mechanism is required. Although the two-way shape memory effect could be implemented to switch between the stable states, the recovery stress upon cooling is much lower compared to the one-way shape memory effect and, thus, the switching force might be too low in most cases.

In the following, we present the design, fabrication, and characterization of novel miniature-scale bistable actuators consisting of two antagonistically coupled buckling SMA beams. In order to enable flexible downscaling, the fabrication technology is based on SMA films that are deposited by magnetron sputtering and micromachined by lithography. Thermal actuation is performed by direct mechanical contact between the SMA beams and a low-temperature heat source. The stationary and dynamic performance of the actuators are investigated to assess the influence of geometrical parameters and heat source temperature.

2. Material Characterization

The base material for this study was $Ti_{53.9}Ni_{30.4}Cu_{15.7}$ films of 15 μm thickness fabricated by magnetron sputtering and subsequent crystallization by rapid thermal annealing at 700 °C for 15 min [28,29]. The phase transformation between austenite and martensite state of the films was investigated by differential scanning calorimetry (DSC). Therefore, the temperature of a SMA test sample was ramped at a constant cooling and heating rate of 10 °C/min. Figure 1a shows a DSC measurement indicating that the phase transformation occurred in the temperature range between 30 and 70 °C. The tangential method was used to determine the critical phase transformation temperatures, i.e., martensite start and finish temperatures; M_s = 41.9 °C and M_f = 31.9 °C, as well as the austenite start and finish temperatures; A_s = 58.6 °C and A_f = 70.3 °C, respectively. In a martensitic state at a temperature below M_f, the SMA film could be deformed easily by mechanical loading. When the SMA film was heated above A_f, the remanent strain after loading could be reset and, thus, the SMA film restored its initial memory shape.

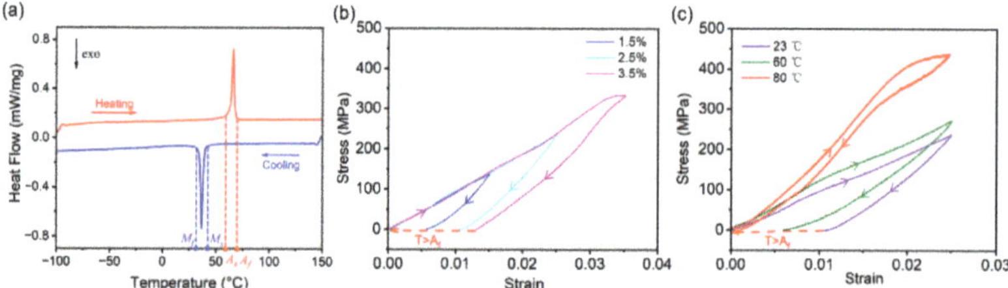

Figure 1. (**a**) Differential scanning calorimetry (DSC) measurement of a TiNiCu film of 15 μm thickness; (**b**) stress–strain characteristics of a TiNiCu tensile test specimen at different maximum strain in the range of from 1.5 to 3.5% at a low strain rate of 10^{-3}/s at room temperature; (**c**) stress–strain characteristics of a TiNiCu tensile test specimen at different temperatures in the range of from 23 to 80 °C at a strain rate of 10^{-3}/s.

A tensile test set-up was used to investigate the stress–strain characteristics of TiNiCu test specimens with the dimensions length l × width w of 15 × 2 mm^2. Therefore, the strain was ramped step-wise and stress was monitored within a sufficiently long time interval to allow for quasi-stationary conditions. A temperature chamber was utilized for control of the ambient temperature. Figure 1b shows the stress–strain characteristics at room temperature (23 °C) at a strain rate of 10^{-3} s^{-1} for a maximum strain up to 3.5%. As the SMA film was in a martensitic state, the stress–strain characteristics show the typical nonlinear quasi-plastic behavior, reflecting the accommodation of martensite variants into the external load. After reaching the strain limit and subsequent load release, a remanent strain occurred depending on the maximum strain, which could be reset by heating above A_f. Figure 1c shows the corresponding stress–strain characteristics at elevated temperatures. When the specimen was heated from ambient temperature to 60 °C, partial phase transformation occurred resulting in an initial increase in the slope of the stress–strain curve and a reduction in remanent strain. At 80 °C, the strain was fully reset upon unloading and the length of the sample returned to the initial state due to the reversible-stress-induced martensitic phase transition. Taking the difference in stress between martensite (23 °C) and austenite (80 °C) at 2% strain, a maximum shape recovery stress of about 150 MPa was determined. At larger strain, stress-induced martensite causes a stress plateau (Figure 1c) and, thus, the shape recovery stress saturates and does not increase further.

3. Actuator Design and Fabrication

Buckling beam structures are an effective way to design bistable devices, but they require an external force to overcome the energy barrier between the two stable states. Here, we design buckling SMA beams with a predefined memory shape to utilize the intrinsic thermally induced shape recovery force instead of an external force. Figure 2 shows the layout of the bistable SMA actuator, which is sketched in its two equilibrium positions. The major components were two SMA beams of TiNiCu, whose memory shapes were adjusted to be opposing buckling states, a spacer separating the SMA beams, and two heat sources located above and below the SMA beams. When SMA beam 1 came into contact with heat source 1 (Figure 2a), it was heated selectively and transformed from martensite to an austenite state. Due to the shape memory effect, SMA beam 1 returned to the opposing buckling state and, therefore, pushed SMA beam 2 towards heat source 2. This motion was supported by the snap-through motion towards the second equilibrium position. In this case, SMA beam 2 came into contact with heat source 2 (Figure 2b) and was heated above its A_f temperature, while SMA beam 1 cooled back down below its M_f temperature. Consequently, the actuator was reset to its initial state, where the actuation cycle started again. Alternately heating the SMA beams resulted in an oscillatory snap-through motion.

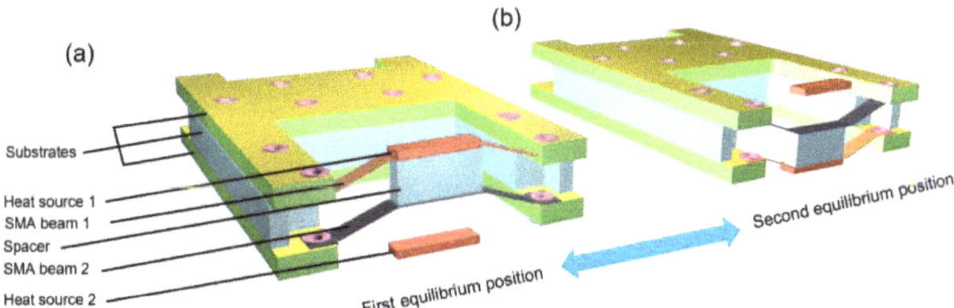

Figure 2. Layout of the bistable SMA actuator. At sufficiently high temperature of the heat sources 1 and 2 an oscillatory motion occurs between the two equilibrium positions shown in (**a**,**b**).

Figure 3 illustrates the shape setting and assembly process of the bistable actuator. Flat SMA beams were placed in customized molds and deformed in opposite bending directions as shown in Figure 3a. Subsequent heat treatment was performed in a tube furnace in constraint condition for 30 min at 465 °C to obtain pairs of SMA beams with opposing buckling memory shapes. By setting the geometrical parameters of the mold, the heat treatment resulted in a pre-deformed shape that matched the specific mold with spacer length s and pre-deflection h. Figure 3b shows a photo of an SMA beam before and after the heat treatment. The SMA beams were fabricated by micromachining the magnetron-sputtered TiNiCu films using optical lithography. The width of the SMA beams was 1 mm and the distance between the two clamping sides of each SMA beam was 10 mm. The spacer height was adjusted to 3 mm, which enabled sufficient thermal isolation between the SMA beams.

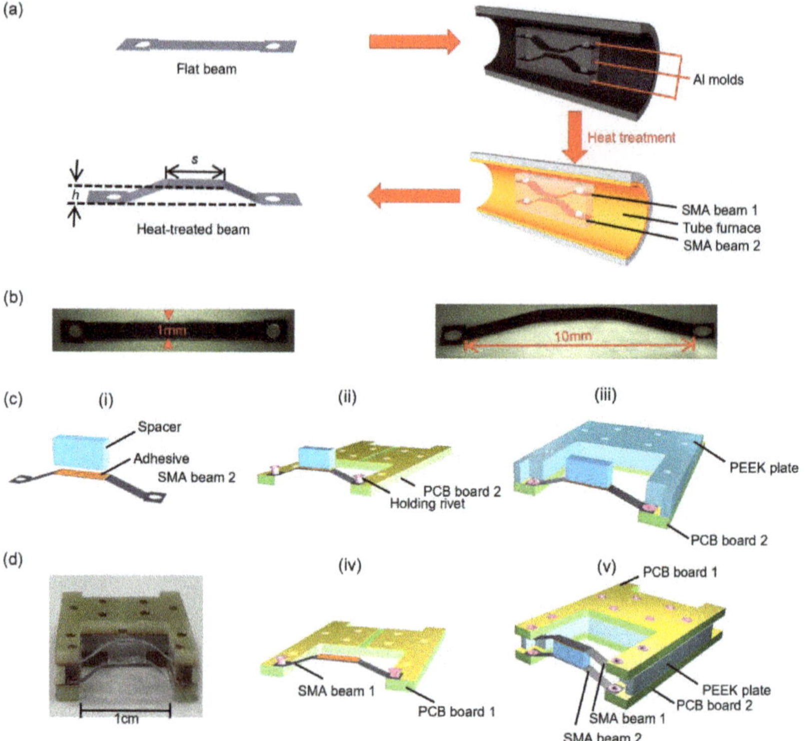

Figure 3. Shape-setting and assembly process of the bistable actuator. (**a**) Schematic of heat treatment in a vacuum tube furnace; (**b**) photos of the SMA beam before and after heat treatment; (**c**) schematic assembly process; (**d**) photo of the bistable SMA actuator. Heat sources 1 and 2 are not shown for clarity. Legend: *s*—spacer length, *h*—pre-deflection.

Bistable actuators were fabricated by coupling the pairs of SMA beams in their center by a spacer. As illustrated in Figure 3c, the assembly process started with bonding the spacer on SMA beam 2 (i). Then, both SMA beam 2 and the attached spacer were then mounted on milled PCB board 2 and fixed by rivets (ii). Subsequently, a PEEK plate was positioned on PCB board 2 (iii), while SMA beam 1 was mounted on PCB board 1 (iv). The final assembly is sketched in (v). The PEEK plate between the two PCBs was designed to prevent short circuits and to reduce heat transfer between the SMA beams. A photo of the bistable SMA actuator is shown in Figure 3d without heat sources.

4. Stationary Force–Displacement Characteristics

4.1. Monostable SMA Actuators

A series of monostable SMA actuators consisting of a single SMA beam with buckling memory shape were investigated for different spacer lengths s and pre-deflection h. Therefore, the center of the SMA beam was deflected in out-of-plane direction under quasi-static loading conditions using a tensile test system to characterize the force–displacement characteristics in a martensite and an austenite state. The single SMA beam shows monostable behavior as it adopts one stable deflection after heating. Figure 4a illustrates the measurement setup. A spacer of length s was attached at the beam center, which complied with the interface of the load cell of the tensile test system. In a martensitic state, the measurement of force F_M started in the undeflected state of the SMA beam center at displacement $z = 0$. Therefore, F_M denotes the mechanical loading force of the tensile testing machine required to deform the SMA beam at room temperature in martensitic condition. This deformation was partly elastic and partly quasi-plastic due to reorientation of martensite variants. The shape recovery force F_A was determined by the mechanical loading force required to deform the SMA beam in austenitic condition after Joule heating to austenitic state. By comparing the resistive and ambient heating conditions shown in Figure S1 in the Supplementary Information, we conclude that an electrical power of >430 mW is required to increase the average temperature above A_f temperature.

In this case, the measurement started in fully deflected state at maximum displacement of the SMA beam center, which depends on the pre-deflection h. The forces F_M and F_A were determined whilst increasing and decreasing displacement using displacement-control under quasi-stationary conditions, respectively.

Typical force–displacement characteristics of a monostable SMA actuator are shown in Figure 4b for the case of spacer length s and pre-deflection h of 4 and 1mm, respectively. In a martensitic state, the force F_M pointed in a negative z-direction acting against the force of the load cell (positive direction). It initially increased strongly with increasing deflection until a maximum was reached. When further increasing the deflection, the force decreased up to a minimum and then increased again. After release of the applied deflection load near the maximum displacement, the SMA beam stayed quasi-plastically deformed due to the self-accommodation behavior of the martensite variants. Figure 4b also shows that the course of shape recovery force F_A in the case of the memory shape corresponds to the deflected state at 2 mm displacement. Therefore, F_A was zero at 2 mm and pointed in a positive z-direction acting against the force of the load cell (negative z-direction). F_A increased during decreasing displacement and exhibited a large maximum followed by a shallow minimum. For all displacements, F_A was larger compared to F_M, in this case indicating that any deflection of the SMA beam in a martensitic state could be reset by an SMA beam in an austenitic state. In particular, the sum of positive force F_A and negative force F_M ($F_A - F_M$) stayed positive for all displacements except for the maximum displacement close to the memory shape. This result has important consequences for coupled SMA beam systems, as it indicates that selective heating of one of the two SMA beams always causes a sufficiently high shape recovery force at any deflection to induce a snap-through motion back to the corresponding memory shape. In particular, bistable snap-through should occur between the two stable deflection states when heating the two SMA beams alternately. Figure 4b also indicates that two coupled SMA beams may only show bistability for those geometrical parameters, for which the force minimum in an austenitic state is larger compared to the force maximum in a martensitic state.

Figure 4c shows the effect of spacer length s on the stationary force–displacement characteristics of monostable SMA actuators for a pre-deflection h of 1 mm. In a martensitic state, the maximum force increases from 67 to 98 mN when increasing the spacer length s from 3 to 5 mm. At the same time, the force minimum also increases from 11 to 22 mN. Again, the SMA beams stay quasi-plastically deformed after the release of the deflection load near the maximum displacement. In an austenitic state, the maximum force increases from 129 to 184 mN with s increasing from 3 to 5 mm, while the force minimum F_A rises from 80 to 106 mN. These trends of maximum and minimum force values are summarized in Figure 4d. When increasing spacer length s, the difference in force minimum in an austenitic state $F_A(\text{min})$ and the force maximum in a martensitic state $F_M(\text{max})$ decrease and eventually becomes negative between 5 and 6 mm. At a spacer length s of 6 mm, $F_M(\text{max})$ exceeds $F_A(\text{min})$, indicating that the shape recovery force is no longer sufficient to induce a snap-through motion back to the memory shape and, thus, the coupled SMA beam system is not bistable.

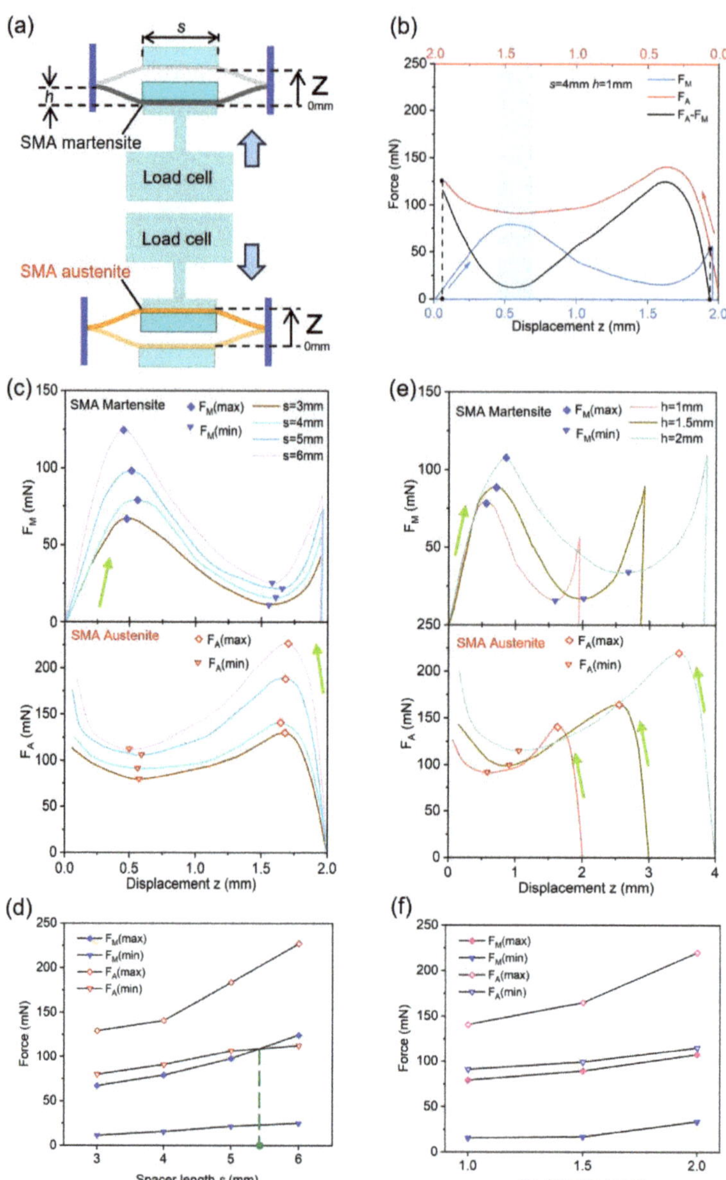

Figure 4. (**a**) Schematic of force–displacement measurement of monostable SMA actuators. (**b**) Force–displacement characteristics of a monostable SMA actuator with an SMA beam in a martensitic state F_M (blue) and in an austenitic state F_A (red) with the memory shape corresponding to the deflected state at 2 mm displacement. The combined force F_A–F_M is shown in black. (**c**) Force–displacement characteristics for different spacer lengths s at a pre-deflection h of 1 mm. (**d**) Summary of maximum/minimum forces in martensitic and an austenitic state $F_M(\mathrm{max})/F_M(\mathrm{min})$ and $F_A(\mathrm{max})/F_A(\mathrm{min})$, respectively, determined from (**c**). (**e**) Force–displacement characteristics for different pre-deflections h at a spacer length s of 4 mm. (**f**) Summary of maximum/minimum forces in a martensitic and an austenitic state $F_M(\mathrm{max})/F_M(\mathrm{min})$ and $F_A(\mathrm{max})/F_A(\mathrm{min})$, respectively, determined from (**e**). The arrows in (**b**–**e**) indicate the direction of loading by the load cell.

Figure 4e shows the effect of pre-deflection h on the stationary force–displacement characteristics of the monostable SMA actuators for a spacer length s of 4 mm. In a martensitic state, the maximum force $F_M(\text{max})$ increases from 79.1 to 107.7 mN when increasing pre-deflection h from 1 to 2 mm, while the minimum force in an austenitic state $F_A(\text{min})$ increases from 91.1 to 115.1 mN. Thus, $F_M(\text{max})$ remains below $F_A(\text{min})$ in the investigated parameter range, as summarized in Figure 4f, indicating that the coupled SMA beam system displays bistable performance in all cases.

The effects of the parameters s and h on the maximum and minimum forces of single SMA beams show a similar trend and, thus, add up. Figure S2 summarizes the effects of all parameter combinations of s and h in the Supplementary Information.

4.2. Bistable SMA Actuators

In the following, we investigate the performance of a series of bistable actuators for different spacer lengths s and pre-deflections h. The center of the coupled SMA beams is deflected in the out-of-plane direction under quasi-static conditions using a tensile test system. Force–displacement characteristics were determined by selectively heating one SMA beam and measuring the resulting net-force F_n at discrete displacement steps by controlling the travel distance of the load cell. The measurement setup shown in Figure 5a illustrates a case in which SMA beams 1 and 2 are in a martensitic and an austenitic state, respectively. In this case, F_n points in a positive direction in the first equilibrium position at zero displacement. Similarly, F_n points in a negative direction in the second equilibrium position at maximum displacement, when the states of SMA beams 1 and 2 are reversed.

The upper part of Figure 5b shows the force–displacement characteristics of bistable SMA actuators for the case illustrated in (a) for different spacer length s at a pre-deflection h of 1 mm. For small-enough spacer lengths s in the range of 3 to 5 mm, F_n decreases and passes through a minimum. Then, it strongly increases to a maximum and finally decreases to zero at the second equilibrium position of the actuator. The minimal net force decreases with increasing s and eventually becomes zero for spacer lengths s exceeding 5 mm. In particular, for $s = 6$ mm, F_n decreases to zero after a small displacement of 0.3 mm, indicating that the actuator cannot reach the second equilibrium position. Evidently, the actuator is no longer bistable when the spacer length s becomes too large. This performance is in line with the course of the combined forces of the monostable actuator F_A–F_M displayed in Figure 4b and the trend of maximum and minimum force values summarized in Figure 4d. We conclude that the shape recovery force becomes too low to induce a snap-through motion back to the memory shape and, thus, the coupled SMA beam system looses bistability at large spacer lengths. A similar performance is observed in the lower part of Figure 5b, in which the states of SMA beams 1 and 2 are reversed. In this case, the actuator starts at the second equilibrium position, whereby F_n drives the actuator towards the first equilibrium position at zero displacement.

Figure 5c shows the effect of pre-deflection h on the stationary force–displacement characteristics of the bistable SMA actuators for a spacer length s of 4 mm. Again, the upper part displays the case wherein SMA beams 1 and 2 are in martensitic and an austenitic state, respectively. When increasing h from 1 to 2 mm, the maximum displacement increases to 3.6 mm. This large bistable actuation stroke corresponds to more than 30% of the SMA beam length. F_n decreases but stays positive, indicating that the actuators display bistable performance in all cases. A similar performance is observed during actuation in the reverse direction in the lower part of Figure 5c. This performance is in line with the course of the combined forces of the monostable actuator F_A-F_M displayed in Figure 4e,f, showing that the shape recovery forces dominate the forces in a martensitic state at all displacements. The effect of all parameter combinations of s and h on minimum net-forces $F_n(\text{min})$ is presented in Figure 6, providing an overview of their combined effect on bistability.

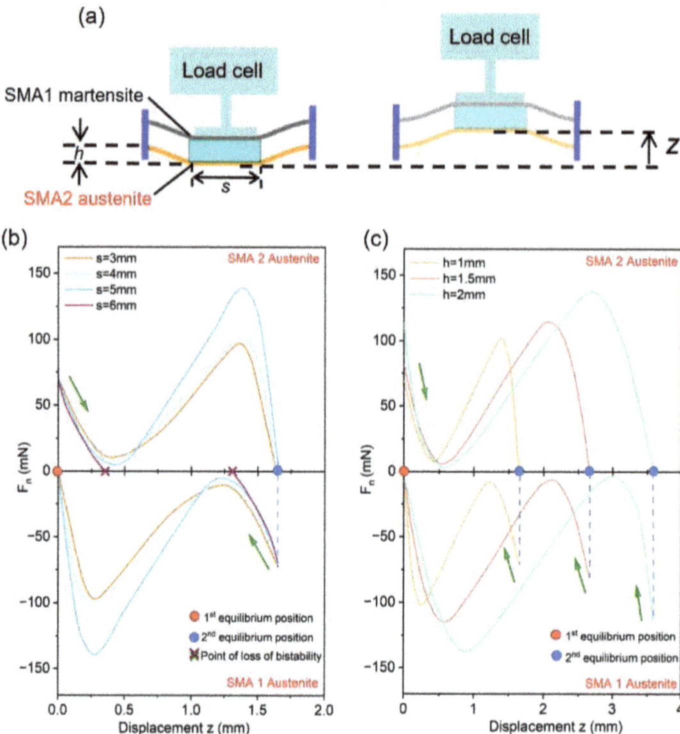

Figure 5. (**a**) Schematic of force–displacement measurement of bistable SMA actuators for the case of selectively heating SMA beam 2. (**b**) Force–displacement characteristics for different spacer length *s* at a pre-deflection *h* of 1 mm. The upper panel shows the case of selectively heating SMA beam 2 to an austenitic state keeping SMA beam 1 in a martensitic state, and in the lower panel vice versa. (**c**) Force–displacement characteristics for different pre-deflections *h* at a spacer length *s* of 4 mm. The upper panel shows the case of selectively heating SMA beam 2 to an austenitic state keeping SMA beam 1 in a martensitic state, and in the lower panel vice versa. The arrows in (**b**,**c**) indicate the direction of loading by the load cell.

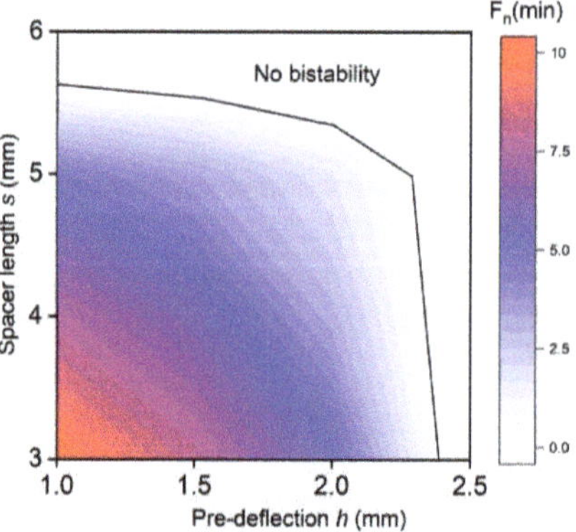

Figure 6. The combined effects of spacer length *s* and pre-deflection *h* on the minimum net-force F_n(min) of the bistable SMA actuator. The limit of bistability is reached when F_n(min) becomes zero, as indicated by the solid line.

5. Dynamic Actuation Performance

In addition to direct Joule heating of the SMA beams, thermal actuation via direct mechanical contact of the SMA beams to a low-temperature heat source is an attractive alternative. In the following, we investigate the performance of a bistable SMA actuator, which is driven by additional heat sources above and below the SMA beams, as sketched in Figure 2. The spacer height in between the SMA beams of 3 mm enables sufficient thermal isolation. This has been verified by electrical resistance measurements of both beams, in which the electrical heating power correlates with the average temperature of the beams, as shown in the new Figure S1 in the Supplementary Information. Figure 7a presents the typical time-resolved switching performance of a bistable SMA actuator with spacer length s and pre-deflection h of 5 and 1 mm, respectively. The time-resolved displacement was determined using a laser displacement sensor. The heat sources were set to 160 °C using temperature sensors attached to the heat sources and closed-loop feedback control. In this case, an oscillatory snap-through motion of about 1 Hz occurred, reflecting the periodic heating and cooling of the two antagonistic SMA beams. The time constants required for heating τ_h and for switching between equilibrium positions τ_{sw} were determined to be 0.21 and 0.36 s, respectively.

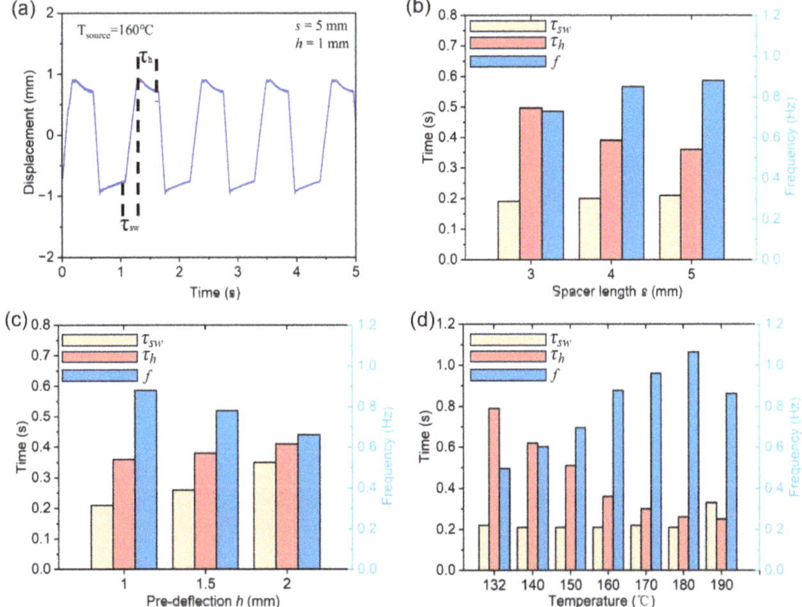

Figure 7. (**a**) Typical time-resolved switching performance of a bistable SMA actuator with a spacer length s of 5 mm and pre-deflection h of 1 mm at a heat source temperature of 160 °C. (**b**) Switching time τ_{sw}, heating time τ_h, and frequency for different spacer lengths s at $h = 1$ mm. (**c**) Switching time τ_{sw}, heating time τ_h, and frequency for different pre-deflections h at $s = 5$ mm. (**d**) Switching time τ_{sw}, heating time τ_h, and frequency for different heat source temperatures at $s = 5$ mm and $h = 1$ mm.

The dynamic performance depended on the spacer length s, pre-deflection h, and heat source temperatures. When the spacer length s was increased from 3 to 5 mm, the switching time τ_{sw} remained almost unchanged, while the heating time τ_h decreased from 0.5 to 0.36 s, as shown in Figure 7b. The increasing spacer length resulted in a larger contact area between the SMA beam and the heat source, which allowed for enhanced heat transfer and, thus, reduced the time required for phase transformation from martensite to austenite. Therefore, the frequency of the oscillatory motion increased by about 20%.

Figure 7c illustrates the time-resolved switching performance of the bistable actuator, when increasing the pre-deflection h from 1 to 2 mm at a spacer length s of 4 mm and temperature of heat source of 160 °C. As h increased from 1 to 2 mm, the switching time τ_{sw} increased from 0.21 to 0.35 s. The slight increase in heating time from 0.39 to 0.41 s was due to the change in beam length. The oscillation frequency decreased from 0.88 to 0.66 Hz, while the actuation stroke increased by a factor of 2. Despite the larger stroke, the switching speed of the SMA beams increased due to the larger net force F_n. The higher switching speed ensured that there was only a minor decrease in frequency. Figure 7d summarizes the effect of increasing the heat source temperature on the dynamic response of the bistable SMA actuator. When the heat source temperature increased from 132 to 180 °C, the switching time τ_{sw} remained almost constant, while the heating time decreased from 0.79 to 0.26 s. The higher temperature of the heat source enhanced the heat transfer between SMA beam and heat source, while the contact area and contact force remained about the same. The reduced heating time to induce the phase transformation caused an increase in the frequency of the oscillatory motion by about a factor of 2. The optimal temperature range is 160–180 °C. Below 160 °C, heating times increase substantially until no oscillation is observed below 132 °C. Above 180 °C, the oscillation frequency decreases again due to overheating caused by limited heat transfer. Consequently, continuous oscillation is lost above 190 °C.

6. Conclusions

We present the design, fabrication, and performance of miniature-scale bistable SMA actuators which consist of two antagonistically coupled buckling SMA beams. The SMA beams were fabricated by magnetron sputtering, lithography, and a dedicated heat treatment to set their memory shapes to adopt opposing buckling states. Coupling the SMA beams in their center leads to compact bistable actuators, which exhibit a bi-directional snap-through motion between two equilibrium positions by selectively heating the SMA beams.

The stationary performance of the actuators is strongly affected by the spacer length and pre-deflection of the SMA beams. The stationary force–displacement characteristics of monostable SMA actuators consisting of a single SMA beam with buckling memory shape reveal minimal shape recovery forces in an austenitic state $F_A(\min)$ and maximal forces in a martensitic state $F_M(\max)$, which pose important constraints on the design of bistable coupled SMA beam systems. Bistability is only achieved if the $F_A(\min)$ of one SMA beam exceeds the $F_M(\max)$ of the antagonistic SMA beam at any displacement to enable a snap-through motion in between the corresponding memory shapes. This result is confirmed by the stationary force–displacement characteristics of the bistable SMA actuators showing bistable switching in between two equilibrium positions only for spacer lengths up to 5 mm, where the net force of the coupled SMA beams stays positive throughout the displacement range. We demonstrate that this condition is no longer fulfilled at a larger spacer length of 6 mm, where the net force decreases to zero after a small displacement and, thus, the actuator cannot reach the second equilibrium position. Increasing the pre-deflection of the SMA beams from 1 to 2 mm gives rise to an enhanced actuation stroke by a factor of 2, while the bistable performance is retained. The presented bistable SMA actuator can be operated thermally through direct heat transfer at mechanical contact between the SMA beams and a heat source, which is an effective means to convert low-temperature waste heat below 200 °C into mechanical energy. We present dynamic measurements of thermally driven bistable SMA actuators operating at frequencies of up to 1.1 Hz and actuation strokes of up to 3.65 mm, corresponding to more than 30% of the SMA beam length. This performance provides an interesting route for the design of novel thermal bistable actuators that are driven by waste-heat at low-temperature differences.

Supplementary Materials: The following supporting information can be downloaded at: https://www.mdpi.com/article/10.3390/act12110422/s1, Figure S1: Normalized electrical resistance of the monostable SMA actuator determined under Joule heating conditions (yellow). For comparison, the normalized electrical resistance of the SMA film material was determined under ambient heating conditions (blue) in a thermostat. Normalization was performed using the expression $(R - R_{min})/(R_{max} - R_{min})$. Comparing both measurements indicates that an electrical power of >430 mW was required to increase the average temperature above A_f temperature; Figure S2: The combined effects of spacer length s and pre-deflection h on the minimum and maximum forces of monostable SMA actuators in martensite (F_M(min), F_M(max)) (a) and austenite condition (F_A(min), F_A(max)) (b) as well as on the corresponding force difference F_A(min) $- F_M$(max). Both parameters s and h affect the forces in a similar way and, thus, both effects add up.

Author Contributions: X.C. designed, fabricated the bistable SMA actuator, performed the experiments, analyzed the data, and wrote the original draft. L.B. and E.Q. designed, fabricated, and optimized TiNiCu films. M.K. conceived the concept, designed the bistable SMA actuator, supervised the experiments, and wrote the manuscript. All authors have read and agreed to the published version of the manuscript.

Funding: X.C. received financial support from the China Scholarship Council (CSC201908140104). L.B. and E.Q. acknowledge funding by the German Science Foundation (DFG) through project (413288478).

Data Availability Statement: The data presented in this study are available on request from the corresponding author. The data are not publicly available, because they are used in ongoing research.

Acknowledgments: We gratefully acknowledge the expertise of Christof Megnin in interconnection technology.

Conflicts of Interest: The authors declare no conflict of interest.

References

1. Qiu, J.; Lang, J.H.; Slocum, A.H. A Curved-Beam Bistable Mechanism. *J. Microelectromech. Syst.* **2004**, *13*, 137–146. [CrossRef]
2. Dorfmeister, M.; Schneider, M.; Schmid, U. Static and Dynamic Performance of Bistable MEMS Membranes. *Sens. Actuators A Phys.* **2018**, *282*, 259–268. [CrossRef]
3. Wang, D.A.; Chen, J.H.; Pham, H.T. A Constant-Force Bistable Micromechanism. *Sens. Actuators A Phys.* **2013**, *189*, 481–487. [CrossRef]
4. Hussein, H.; Younis, M.I. Analytical Study of the Snap-Through and Bistability of Beams with Arbitrarily Initial Shape. *J. Mech. Robot.* **2020**, *12*, 041001. [CrossRef]
5. Megnin, C.; Barth, J.; Kohl, M. A Bistable SMA Microvalve for 3/2-Way Control. *Sens. Actuators A Phys.* **2012**, *188*, 285–291. [CrossRef]
6. Kohl, M.; Seelecke, S.; Wallrabe, U. (Eds.) . *Cooperative Microactuator Systems*; MDPI: Basel, Switzerland, 2023.
7. Howell, L.L.; Rao, S.S.; Midha, A. Reliability-Based Optimal Design of a Bistable Compliant Mechanism. *J. Mech. Design* **1994**, *116*, 1115–1121. [CrossRef]
8. Tsay, J.; Su, L.Q.; Sung, C.K. Design of a Linear Micro-Feeding System Featuring Bistable Mechanisms. *J. Micromech. Microeng.* **2005**, *15*, 63–70. [CrossRef]
9. Yang, B.; Wang, B.; Schomburg, W.K. A Thermopneumatically Actuated Bistable Microvalve. *J. Micromech. Microeng.* **2010**, *20*, 095024. [CrossRef]
10. Taher, M.; Saif, A. On a Tunable Bistable MEMS-Theory and Experiment. *J. Microelectromech. Syst.* **2000**, *9*, 157.
11. Baker, M.S.; Howell, L.L. On-Chip Actuation of an in-Plane Compliant Bistable Micromechanism. *J. Microelectromech. Syst.* **2002**, *11*, 566–573. [CrossRef]
12. Masters, N.D.; Howell, L.L. A Self-Retracting Fully Compliant Bistable Micromechanism. *J. Microelectromech. Syst.* **2003**, *12*, 273–280. [CrossRef]
13. Pieri, F.; Piotto, M. A Micromachined Bistable 1 x 2 Switch for Optical Fibers. *Microelectron. Eng.* **2000**, *53*, 561–564. [CrossRef]
14. Chen, T.; Shea, K. An Autonomous Programmable Actuator and Shape Reconfigurable Structures Using Bistability and Shape Memory Polymers. *3D Print Addit. Manuf.* **2018**, *5*, 91–101. [CrossRef]
15. Jeong, H.Y.; Lee, E.; Ha, S.; Kim, N.; Jun, Y.C. Multistable Thermal Actuators Via Multimaterial 4D Printing. *Adv. Mater. Technol.* **2019**, *4*, 1800495. [CrossRef]
16. Arrieta, A.F.; Van Gemmeren, V.; Anderson, A.J.; Weaver, P.M. Dynamics and Control of Twisting Bi-Stable Structures. *Smart Mater. Struct.* **2018**, *27*, 025006. [CrossRef]
17. Giddings, P.F.; Kim, H.A.; Salo, A.I.T.; Bowen, C.R. Modelling of Piezoelectrically Actuated Bistable Composites. *Mater. Lett.* **2011**, *65*, 1261–1263. [CrossRef]

18. Hamouche, W.; Maurini, C.; Vidoli, S.; Vincenti, A. Multi-Parameter Actuation of a Neutrally Stable Shell: A Flexible Gear-Less Motor. *Proc. R. Soc. A Math. Phys. Eng. Sci.* **2017**, *473*, 20170364. [CrossRef]
19. Shao, H.; Wei, S.; Jiang, X.; Holmes, D.P.; Ghosh, T.K. Bioinspired Electrically Activated Soft Bistable Actuators. *Adv. Funct. Mater.* **2018**, *28*, 1802999. [CrossRef]
20. Loukaides, E.G.; Smoukov, S.K.; Seffen, K.A. Magnetic Actuation and Transition Shapes of a Bistable Spherical Cap. *Int. J. Smart Nano Mater.* **2014**, *5*, 270–282. [CrossRef]
21. Hou, X.; Liu, Y.; Wan, G.; Xu, Z.; Wen, C.; Yu, H.; Zhang, J.X.J.; Li, J.; Chen, Z. Magneto-Sensitive Bistable Soft Actuators: Experiments, Simulations, and Applications. *Appl. Phys. Lett.* **2018**, *113*, 221902. [CrossRef]
22. Hussein, H.; Khan, F.; Younis, M.I. A Monolithic Tunable Symmetric Bistable Mechanism. *Smart Mater. Struct.* **2020**, *29*, 075033. [CrossRef]
23. Kim, H.A.; Betts, D.N.; Salo, A.I.T.; Bowen, C.R. Shape Memory Alloy-Piezoelectric Active Structures for Reversible Actuation of Bistable Composites. *AIAA J.* **2010**, *48*, 1265–1268. [CrossRef]
24. Kohl, M. *Shape Memory Microactuators*; Springer: Berlin/Heidelberg, Germany, 2010; ISBN 3-540-20635-3.
25. Kohl, M.; Ossmer, H.; Gueltig, M.; Megnin, C. SMA Foils for MEMS: From Material Properties to the Engineering of Microdevices. *Shape Mem. Superelasticity* **2018**, *4*, 127–142. [CrossRef]
26. Firouzeh, A.; Paik, J. Robogami: A Fully Integrated Low-Profile Robotic Origami. *J. Mech. Robot.* **2015**, *7*, 021009. [CrossRef]
27. Chen, X.; Bumke, L.; Quandt, E.; Kohl, M. A Thermal Energy Harvester Based on Bistable SMA Microactuation. In Proceedings of the ACTUATOR 2022, Mannheim, Germany, 29–30 June 2022.
28. Gu, H.; Bumke, L.; Chluba, C.; Quandt, E.; James, R.D. Phase Engineering and Supercompatibility of Shape Memory Alloys. *Mater. Today* **2018**, *21*, 265–277. [CrossRef]
29. De Miranda, R.L.; Zamponi, C.; Quandt, E. Micropatterned Freestanding Superelastic TiNi Films. *Adv. Eng. Mater.* **2013**, *15*, 66–69. [CrossRef]

Article

Resonant Self-Actuation Based on Bistable Microswitching

Joel Joseph [1], **Makoto Ohtsuka** [2], **Hiroyuki Miki** [3] **and Manfred Kohl** [1,*]

1   Institute of Microstructure Technology, Karlsruhe Institute of Technology (KIT), Postfach 3640, D-76021 Karlsruhe, Germany; joel.joseph@kit.edu
2   Institute of Multidisciplinary Research for Advanced Materials, Tohoku University, Sendai 980-8577, Japan; makoto.ohtsuka.d7@tohoku.ac.jp
3   Faculty of Science and Engineering, Ishinomaki Senshu University, Miyagi 986-8580, Japan; hiroyuki.miki.k6@isenshu-u.ac.jp
*   Correspondence: manfred.kohl@kit.edu

Abstract: We present the design, simulation, and characterization of a magnetic shape-memory alloy (MSMA) film actuator that transitions from bistable switching to resonant self-actuation when subjected to a stationary heat source. The actuator design comprises two Ni-Mn-Ga films of 10 µm thickness integrated at the front on either side of an elastic cantilever that moves freely between two heatable miniature permanent magnets and, thus, forms a bistable microswitch. Switching between the two states is induced by selectively heating the MSMA films above their Curie temperature Tc. When continuously heating the permanent magnets above Tc, the MSMA film actuator exhibits an oscillatory motion in between the magnets with large oscillation stroke in the frequency range of 50–60 Hz due to resonant self-actuation. A lumped-element model (LEM) is introduced to describe the coupled thermo-magnetic and magneto-mechanical performance of the actuator. We demonstrate that this performance can be used for the thermomagnetic energy generation of low-grade waste heat (T < 150 °C) with a high power output per footprint in the order of 2.3 µW/cm^2.

Keywords: bistable actuator; thermomagnetic generator; energy harvesting; energy conversion; heat transfer; ferromagnetism; Heusler alloys; magnetic shape-memory alloys, Ni-Mn-Ga, thin films

check for updates

Citation: Joseph, J.; Ohtsuka, M.; Miki, H.; Kohl, M. Resonant Self-Actuation Based on Bistable Microswitching. *Actuators* 2023, 12, 245. https://doi.org/10.3390/act12060245

Academic Editor: Jose Luis Sanchez-Rojas

Received: 17 May 2023
Revised: 7 June 2023
Accepted: 9 June 2023
Published: 13 June 2023

1. Introduction

Magnetic shape-memory alloys (MSMAs) are multifunctional materials that offer both thermal shape-memory, magnetic properties, as well as various multiferroic coupling effects [1]. The unique features of MSMAs are due to pronounced magnetic ordering and large magnetocrystalline anisotropy, resulting in an energetically preferred orientation of magnetic moments. In addition, MSMAs undergo a first-order martensitic phase transformation, which involves a large abrupt change in both the structural and magnetic properties. The phase transformation proceeds in a reversible manner, enabling a thermal shape-memory effect as well as superelasticity. The presence of coupled ferromagnetic and metastable ferroelastic (martensitic) domains in MSMA materials gives rise to multiferroic coupling effects, such as magnetic field-induced reorientation and magnetic shape-memory effect [2]. A prominent example is the Heusler alloy Ni-Mn-Ga [2–4], which is ferromagnetic at room temperature. Metamagnetic SMAs, such as the quaternary Heusler alloys Ni–Mn–X–Y (X = Ga, In, Sn; Y = Co, Fe, Al), are non-magnetic or antiferromagnetic in the martensitic state at low temperatures and undergo an abrupt change of magnetization with narrow hysteresis at the first-order transformation to the austenitic state [5].

Due to these properties, MSMAs exhibit different modes of combined sensing and actuation capabilities [6,7]. As a consequence, MSMA actuators may be designed as a single piece of material but may still perform various functions. This option is particularly interesting for applications in small dimensions, where the effect size, multifunctionality, and ease of fabrication are important factors. Previous approaches to the development of miniature-scale MSMA actuators include the thinning of bulk single crystals to foil specimens [8,9], the deposition of MSMA films [10–12], as well as 3D printing [13]. So far, several actuation mechanisms have been exploited at miniature scales. MSMA linear actuators have been developed using the magnetic field-induced reorientation effect in single crystalline Ni–Mn–Ga [14–17]. A bidirectional microactuator has been developed consisting of a polycrystalline Ni–Mn–Ga double-beam cantilever, which generates antagonistic bending forces near a miniature permanent magnet due to an electrical current-induced transition between ferromagnetic martensite and paramagnetic austenite. Thus, pulsed heating causes fast oscillatory motion of the cantilever in microscanner applications [18]. The abrupt change of magnetization in the first-order phase transformation in metamagnetic SMA films has been considered in [19,20] to enable thermally induced switching of magnetic forces.

Another field of application of MSMA actuation is thermomagnetic generation (TMG) at miniature scales [21–25]. Film-based TMGs have been developed using metamagnetic [22] and ferromagnetic SMA films [23–25]. The MSMA film actuator was integrated at the front of a freely movable cantilever near a heatable permanent magnet serving at the same time as the magnetic field and heat source. In this case, due to the large surface-to-volume ratio, rapid heat transfer is achieved through direct mechanical contact between the MSMA film and heat source, enabling a short duration of the thermal duty cycle in the order of the mechanical eigenoscillation. When heating the permanent magnet above the Curie temperature Tc, this performance gives rise to resonant self-actuation, which is characterized by a large cantilever deflection and a high oscillation frequency [24,25]. Recent demonstrators based on Heusler alloy films of Ni-Mn-Ga showed oscillation frequencies in the order of 100 Hz. By integrating a pick-up coil at the cantilever front, electrical power densities of up to 120 mW/cm^3 have been generated [24,25].

In this investigation, we extend this approach by integrating two MSMA film actuators on opposite sides of the cantilever front and using two heatable permanent magnets above and below the cantilever to form a bistable actuation system. Bistable systems are of special interest both for actuation, as energy is only consumed in case of switching, and for energy harvesting, as the power output could be enhanced by the additional bistable forces. A lumped-element model (LEM) is presented to determine the magneto-mechanical performance of the new actuator design as a function of the MSMA film temperature, which complements the experimental study of the dynamic mechanical actuator performance.

We demonstrate that the MSMA film actuator shows bistable switching between two stable end positions due to thermally induced switching of magnetic latching forces, which transitions to resonant self-actuation in between the end positions when continuously heating the permanent magnets above Tc. We report on large oscillation strokes in the order of 20% of the cantilever length at oscillation frequencies of 50–60 Hz, which allows for thermomagnetic generation with a power per footprint of up to 2.3 μW/cm^2.

2. Material Properties

Sputtering is the most favorable deposition method to obtain thick films on a large scale, as is required for MEMS technology [6,12]. The resulting film structure depends on various parameters including the substrate, deposition temperature, sputtering power, and annealing conditions. Consequently, the as-prepared films can exhibit a polycrystalline structure [26–28] or an epitaxial relation to the substrate [10,29]. A process has been developed for the fabrication of freestanding polycrystalline Ni–Mn–Ga films showing 10 M martensite at room temperature [26,30]. Here, the MSMA films are fabricated by magnetron sputtering with the target composition of Ni$_{49.5}$Mn$_{28}$Ga$_{22.5}$. Film deposition is performed on polyvinyl alcohol (PVA) substrates that can be dissolved after sputtering. A sputtering power of 200 W is used and the sputtering time is adjusted to obtain a homogenous film thickness of 10 μm. Argon gas flow is maintained at 230 mm^3 s^{-1}, and the substrate temperature is kept at 50 °C. Under these conditions, the as-deposited films are amorphous and require heat treatment to adjust the crystal structure and phase transformation properties. The effect of heat treatment on the performance of polycrystalline Ni–Mn–Ga films is described in [26]. The final composition after heat treatment of 10 h at 800 °C is determined by the inductive plasma method to be $Ni_{51.4}Mn_{28.3}Ga_{20.3}$.

Our investigations showed that the sputtering power can be used as a means to control the chemical composition of the films. Therefore, the Ni and Mn contents of the films decrease with increasing sputtering power for both targets, whereas the Ga content increases. However, little change in composition is observed for different heat treatment temperatures between 650 and 1000 °C and heat treatment times between 1 h and 10 h. A temperature-dependent measurement of the electrical resistance is shown in Figure 1. The measurement reveals the typical features of a first-order martensitic phase transformation, namely, a jump-like drop upon heating and a hysteresis upon subsequent cooling. The finish temperatures of the martensitic and austenitic transformations, M_f/A_f, are determined to be −20/25 °C. The measurement also exhibits a characteristic kink at 98 °C due to the ferromagnetic transition. Typical temperature-dependent magnetization characteristics are shown in the Appendix A (Figure A1). When heating the Ni-Mn-Ga films above T_c, the magnetization drops to zero, which will be used in the following to control the magnetic attraction forces.

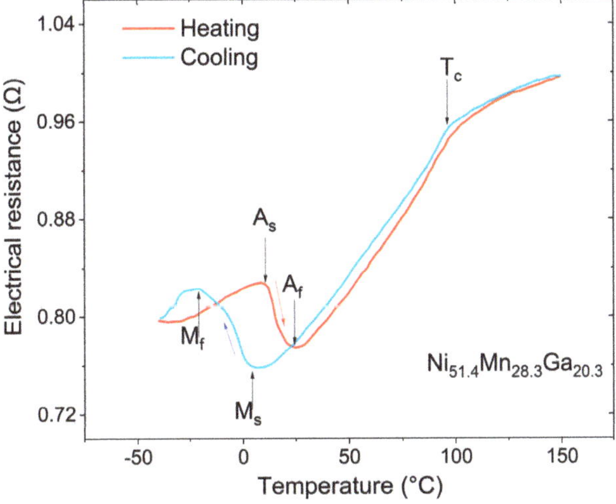

Figure 1. Four-wire electrical resistance measurements of the Ni-Mn-Ga films fabricated by magnetron sputtering. Legend: $A_{s/f}$—austenite start/finish temperature, $M_{s/f}$—martensite start/finish temperature, T_c—Curie temperature.

3. Layout and Operation Principle

The bistable actuator comprises two MSMA films that are mounted at the tip of a free-standing cantilever on either side, as shown in Figure 2a. Furthermore, two permanent magnets are arranged above and below the cantilever, providing magnetic fields to magnetize the MSMA films and magnetic field gradients to induce magnetic attraction forces on the MSMA films. These magnetic forces can be varied in a wide range by increasing/decreasing the temperature of the MSMA films above/below their Curie temperature T_c. The magnets are placed at an angle on both sides of the cantilever to achieve maximum contact. The coupling of the elastic and magnetic forces, F_{el} and F_m, in the system is illustrated schematically in Figure 2b. This coupling gives rise to bistable performance, i.e., the system can adopt two stable deflection states, whereby external energy is required to switch between these states. When providing sufficient external energy in the form of mechanical vibration, for instance, the cantilever tip would oscillate between the two deflection states. Here, we consider the supply of thermal energy.

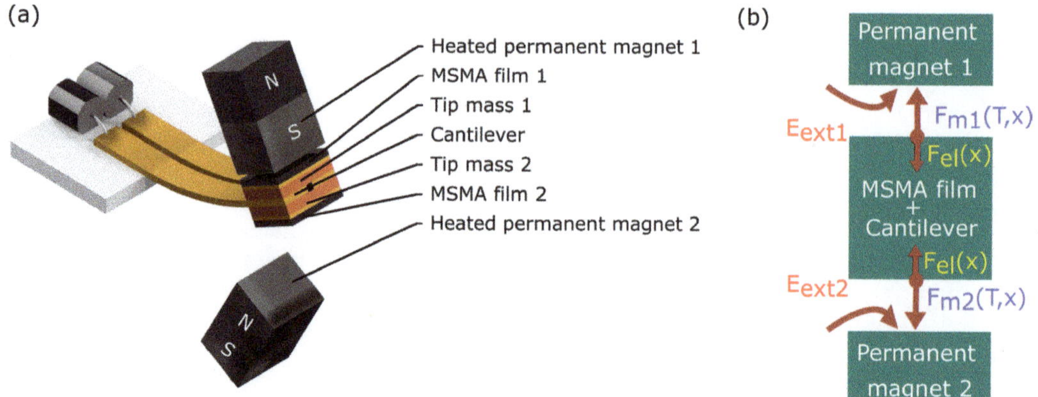

Figure 2. (**a**) Schematic of the bistable MSMA film actuator consisting of a free-standing cantilever, a tip mass, and an MSMA film on both sides of the cantilever allowing for two stable end positions; (**b**) schematic of the film actuator system and the elastic and magnetic forces, F_{el} and F_m, that act on the freely movable cantilever causing bistable performance. Switching between the two stable deflection states requires external energy E_{ext}.

One operation mode is bistable switching by the selective heating of the MSMA films. At room temperature, the MSMA films are in a ferromagnetic state and, therefore, the cantilever is attracted to either of the two permanent magnets. Any asymmetry results in a preferred initial deflection state. Selective heating of the MSMA film being in contact with its corresponding magnet above T_c causes a reduction in magnetization and a decrease in the magnetic attraction force F_m until it drops below the restoring force of the cantilever F_{el}. In this case, the contact is released and the cantilever deflects back towards its initial straight position. Due to inertial forces, the cantilever overshoots and is attracted to the opposite permanent magnet. Due to symmetry, resetting occurs when selectively heating the opposite MSMA film.

The second operation mode is a thermally induced oscillation between the two deflection states. In this case, both MSMA films are heated alternately in a periodic manner, which is easily accomplished by continuously heating the permanent magnets and maintaining their temperature above T_c. Then, the heating of the MSMA films occurs by direct contact with the heated magnets. Periodic heating and cooling of the MSMA films cause an oscillatory motion of the cantilever, as long as the thermal energy supply is maintained.

The cantilever is fabricated by laser micromachining of a CuZn film of 20 µm thickness. The inertial forces in the system are adjusted by providing additional tip masses on either side of the cantilever. The two MSMA films of Ni-Mn-Ga are attached on top of the respective tip masses, allowing for thermal decoupling from the cantilever beam. A photo of the final device is presented in the Appendix A.

4. Lumped Element Modelling of Performance

A lumped element model (LEM) in Matlab SIMSCAPE 2022b is introduced to describe the coupled magneto-thermo-mechanical coupling within the device and to optimize the performance parameters. If a small lateral displacement of the cantilever is neglected, the bending motion of the cantilever can be assumed to be one-dimensional. Thus, a spring mass damper system can be used to represent the dynamic motion of the microactuator. At the miniature scale, the gravitational force on the system can be neglected as well. Thus, the effective net force F_{net} acting on the movable masses mi of the cantilever tip is described by the force balance as follows:

$$F_{net} = c\dot{x} + kx + \sum_{i=1}^{2} m_i \ddot{x}_i + F_{mi} \tag{1}$$

whereby mi is approximated by

$$m_i = \frac{33}{140} m_{cant} + m_{fi} \tag{2}$$

The parameters m_{cant} and m_{fi} represent the masses of the cantilever and the magnetic SMA films, respectively. c and k denote the damping coefficient and the spring constant of the structure, respectively. The attraction force between the magnetic SMA films and the permanent magnets is denoted by F_{mi}, which is a function of both position and temperature.

$$F_{mi} = V_m M_{Ti} \frac{\partial B_{xi}}{\partial x_i} \tag{3}$$

Thereby, V_m denotes the volume of the MSMA films that are magnetized by the applied field of the permanent magnet. M_{Ti} is the magnetization of the films as a function of temperature and magnetic field, which is mapped experimentally as a function of the position of the two films. The position dependencies are used to determine the magnetic field gradients $\frac{\partial B_{xi}}{\partial x_i}$ analytically [31]. Since there are two magnets, the magnetic fields and field gradients are superimposed to calculate the resulting magnetic field at a given position.

The heat transfer at contact and the thermal dissipation during actuation are modeled by using thermal equivalent circuits. A Cauer model is utilized for representing different components of the device as a combination of thermal resistance and thermal capacitance [32]. Figure 3 shows a schematic of the thermal network of the bistable magnetic SMA microactuator device. The overall impedance of the system governs its heat intake and dissipation. This is given by Equation (4), where Z_{th} is the overall thermal impedance of the device, $R_{th(i)}$ is the thermal resistance, and $C_{th(i)}$ ($i = 1, 2, 3 \ldots n$) is the thermal capacitance of the individual components, respectively. In order to obtain the thermal gradient within the individual components, each component is represented by a combination of several thermal resistances and capacitances with a resolution proportional to the number of resistances and capacitances representing one component. Depending on the dimensions of each component and thermal properties, the heat flow and dynamic temperature change are obtained [32].

$$Z_{th} = \cfrac{1}{C_{th(1)} + \cfrac{1}{R_{th(1)} + \cfrac{1}{C_{th(2)} + \cfrac{1}{R_{th(2)} + \cfrac{1}{\ldots C_{th(n)} + \frac{1}{R_{th(n)}}}}}} \tag{4}$$

Coupling both the thermal circuit model and the mechanical spring damper model allows for the time-resolved simulation of the thermal performance, as well as the dynamic actuation of the bistable device. In order to compare different multistable nonlinear systems and quantify their stability, a model-based description of the involved energies is appropriate. This allows the determination of available stable states at the energy minima and the energy barrier that needs to be overcome in order to switch between the states.

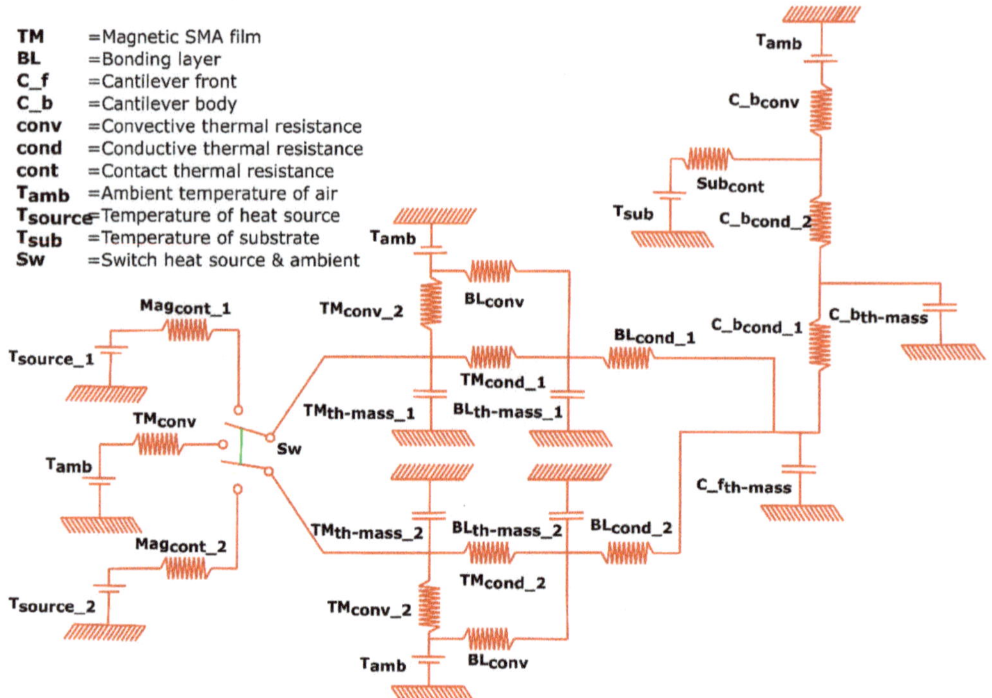

Figure 3. Thermal lumped element model of the bistable MSMA film actuator. Both MSMA films reject most of the heat transferred during contact with the heat source into the cantilever by conduction. Thus, the cantilever acts as the main heat sink. A minor fraction of the heat is dissipated by convection to the ambient air during actuation.

5. Resonant Self-Actuation

When continuously heating the permanent magnets above T_c, the cantilever undergoes an oscillatory motion, which is governed by the interplay of magnetic attraction, elasticity, inertia, and damping forces. In particular, the dynamics of heat transfer have an important influence on the oscillation performance. If heat transfer from the heated magnet to the MSMA film is slower than the duration of the cantilever eigen oscillations, the cantilever will rest at the magnet and possibly perform damped eigen oscillations until the magnetic attraction force decreases enough after a sufficient heating time of the MSMA film to retract from the magnet. On the other hand, if the cooling of the MSMA film is too slow to be attracted to the magnet surface, it will undergo chaotic oscillations without touching the magnet until the attraction force of one of the magnets is sufficiently large. In the optimum case, however, matching the time constants of heat transfer and mechanical oscillation will lead to resonant self-actuation [33]. In the present actuator design, the dimensions of the MSMA films are adjusted for a sufficiently large surface-to-volume ratio, and the eigenfrequency of the moving mass is tuned to meet the matching conditions. Once the coupled thermo-magneto-mechanical cycles are balanced, resonant self-actuation occurs, which is characterized by a continuous periodic oscillation with large strokes and high frequencies. Figure 4a shows the simulated time-resolved deflection of the cantilever tip in resonant self-actuation mode. The corresponding time-resolved changes in the MSMA film temperatures are shown in Figure 4b.

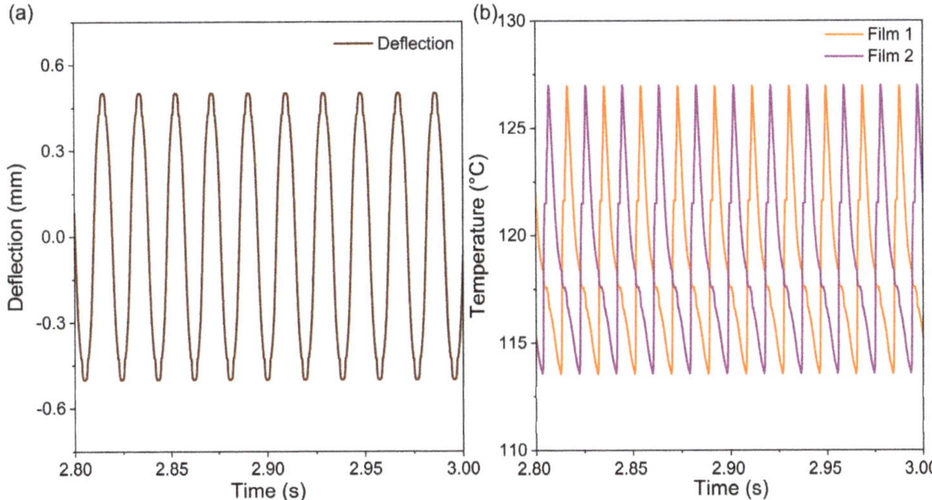

Figure 4. (**a**) LEM simulation of the time-resolved deflection of the cantilever tip of the bistable MSMA film actuator in resonant self-actuation mode; (**b**) time-resolved temperature change of the MSMA films 1 and 2 during resonant self-actuation.

Due to the symmetry of the bistable MSMA film actuator, the cantilever tip performs equal deflections in positive and negative directions. As the MSMA films are heated alternately while in contact with the permanent magnets, they undergo separate thermal cycles with a phase shift of 180 degrees for each mechanical cycle. The experimental results on the time-dependent performance of the bistable MSMA film actuator are presented in Section 7. Due to the constraints of the experimental setup, time-resolved measurements of the actuator deflections were not possible. Therefore, the simulation model was validated using temperature-dependent electrical measurements (see Section 7).

6. Transition from Bistable Switching to Resonant Self-Actuation

Once the permanent magnets are continuously heated above T_c, the bistable MSMA film actuator undergoes a transition from bistable switching to resonant self-actuation. Next, this transition is analyzed via LEM simulations of the evolution of magnetic attraction forces and the corresponding potential energies.

The magnetic attraction forces acting on the MSMA films 1 and 2 (F_{m1} and F_{m2}) result in an effective force <F> acting on the cantilever tip. Figure 5a shows the simulated change in the effective magnetic attraction force <F> versus cantilever deflection during the initial oscillation cycles before the MSMA film actuator reaches stationary operation conditions. In the first actuation cycle, the microactuator switches from the initial stable state at -0.5 mm to the second stable state at 0.5 mm, where it rests until selective heating of the MSMA film at contact with the magnet causes the microactuator to switch back to the initial position in the second actuation cycle. Within the first 16 cycles, the system transitions from the initial bistable switching mode to the resonant self-actuation mode. Thereby, the effective force maxima <F_{max}> at the permanent magnets decrease from ±6.5 mN to about ±2.5 mN. This reduction is caused by an increase in the average temperatures of the MSMA films until they reach stationary values. Figure 5b shows the magnetic attraction forces F_{m1} and F_{m2}, as well as the resulting effective magnetic force <F> under stationary operation conditions.

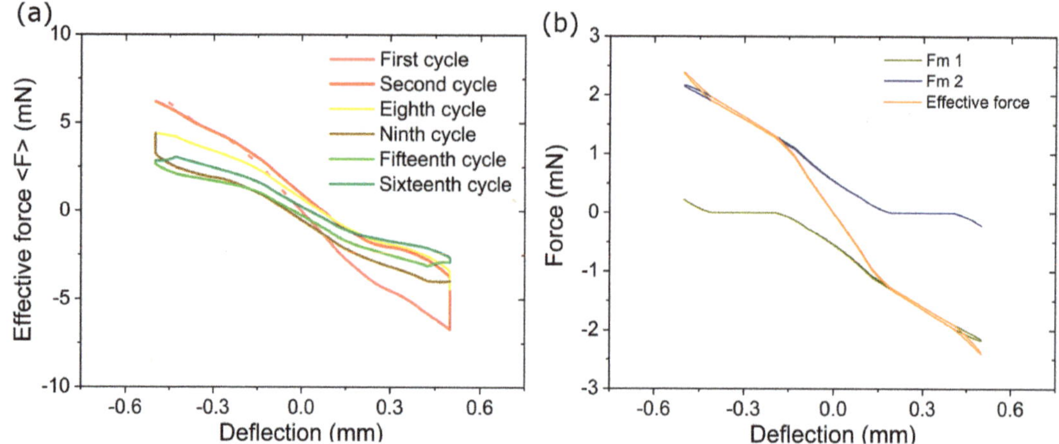

Figure 5. (a) LEM simulation of the effective magnetic attraction force <*F*> on the cantilever tip of the bistable MSMA film actuator versus deflection of the cantilever tip during the initial phase of operation for the selected actuation cycles as indicated. Stationary operation conditions are reached after about 16 actuation cycles; **(b)** simulated course of magnetic attraction forces F_{m1}, F_{m2} and the resulting effective magnetic force <*F*> in stationary condition.

Figure 6 shows the corresponding simulated changes in the potential energies, whereby the total potential energy comprises the contributions of the elastic potential energy of the cantilever and the magnetic potential energies of the two magnetic subsystems. Figure 6a depicts the change in the total potential energy versus deflection of the cantilever tip during the initial oscillation cycles before the MSMA film actuator reaches stationary operation conditions. The dotted line in Figure 6a indicates the symmetry of the total potential energy in the initial state. In the simulation, the initial position is at zero deflection in the center between the permanent magnets, where the potential energy shows a pronounced maximum of about 10 µJ. In addition, two potential minima occur at the surfaces of the two permanent magnets, which correspond to the two stable deflection states of the actuator. Without a continuous supply of thermal energy, the MSMA film actuator would deflect towards one of the two deflection states and rest. The continuous supply of thermal energy during the contact between the MSMA films and their respective permanent magnets strongly affects the potential energies. Within the first 16 cycles, the maximum potential energy decreases considerably to less than 3 µJ. Simultaneously, two pronounced potential energy minima evolve on either side of the actuator center, whereby they become increasingly pronounced and propagate towards the actuator center until they become stationary at about ±2 µm. This reduction is caused by an increase in the average temperatures of the MSMA films, which affects the magnetic forces (Figure 5) alike.

Figure 6b shows the potential energies of both magnetic subsystems, the elastic energy, and the resulting total energy versus deflection once stationary conditions are reached. In this case, the total energy exhibits two energy minima well apart from the maximum possible deflection states and a small energy barrier in the actuator center, which is sufficiently low to enable continuous oscillation. Continuous oscillations occur as long as the kinetic energy of the moving mass exceeds the barrier height. Therefore, a critical upper limit of the elastic potential energy exists, above which the energy barrier becomes too high for resonant self-actuation.

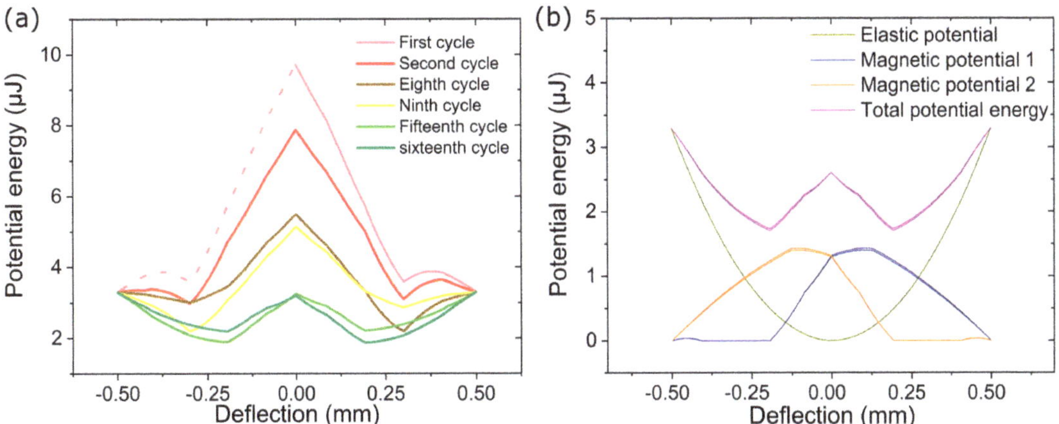

Figure 6. (**a**) LEM simulation of the total potential energy versus deflection of the cantilever tip of the bistable MSMA film actuator during the initial phase of operation for selected actuation cycles as indicated. Stationary operation conditions are reached after about 16 actuation cycles; (**b**) simulated course of potential energies of both magnetic subsystems, the elastic energy, and the resulting total energy in stationary condition after the sixteenth cycle.

7. Bistable Thermal Energy Harvesting Based on Resonant Self-Actuation

The bistable MSMA film actuator is highly attractive for thermal energy harvesting, as the operation mode of resonant self-actuation enables an optimum conversion of thermal energy into kinetic energy of mechanical eigenoscillation. By integrating two pick-up coils, 1 and 2, next to the MSMA films 1 and 2, respectively, instead of the tip masses, a symmetric energy harvester design is obtained. The core of the pick-up coil is made of copper to allow thermal conduction between the MSMA film and the cantilever. The outer sides of the coils are made of brass. The dimensions of the Cu core and brass layers are tailored for optimum heat transfer, following the guidelines given in [33]. To simulate the energy harvesting performance, the LEM is extended to include the electromagnetic coupling of the movable pick-up coil following the approach in [34]. In this case, the time-dependent performance of the bistable MSMA film actuator can be determined via LEM simulations and electrical measurements, which allows the validation of the simulation model.

Figure 7a,b show experimental and simulated electrical performance characteristics of the bistable thermal energy harvester, respectively, for a heat-source temperature of 150 °C. The output current is determined by connecting the two pick-up coils in parallel. The output occurs in pairs of large and small peaks due to the changing direction of the magnetic field gradient in the actuation center. The simulation model reproduces the current amplitude and frequency of the bistable MSMA film actuator very well. Further experimental and simulation results on the electrical current output at higher heat source temperatures of 160 and 170 °C are presented in the Appendix A (Figure A2), which further validate the LEM simulations. The achieved average power of a single device is 0.23 μW at a heat source temperature of 150 °C and a load resistance of 1000 Ohms, which corresponds to a power per footprint of about 2.3 μW/cm².

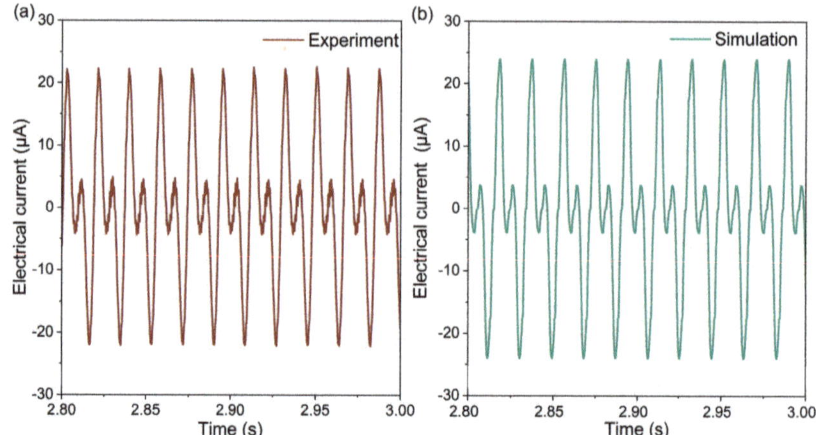

Figure 7. Performance of the bistable thermal energy harvester for a heat source temperature of 150 °C. (**a**) Experimentally determined electrical current in the pick-up coils connected in parallel; (**b**) corresponding LEM simulation.

8. Conclusions

A bistable magnetic shape-memory alloy (MSMA) film actuator is presented consisting of two Ni-Mn-Ga films at the front, on either side of a freestanding cantilever located between two heatable miniature permanent magnets. By selective heating of the MSMA films, the actuator operates as a bistable switch, whereby the cantilever can be switched between two deflection states at the two magnets. Continuous heating of the permanent magnets above their Curie temperature T_c gives rise to a thermally induced oscillation between the two deflection states. We demonstrate that the cantilever oscillation can be optimized by matching the time constants of the heat transfer and mechanical oscillation to achieve resonant self-actuation. A lumped-element model (LEM) is presented to determine the coupled thermal and magneto-mechanical performance of the bistable MSMA film actuator. Using LEM simulations, we investigate the transition from bistable switching to the resonant self-actuation mode. At low MSMA film temperatures, the magnetic potential of the system is higher compared to the elastic potential, and thus, the system rests in a stable deflection state at one of the two magnets. Once the MSMA film temperatures exceed T_c during contact with the permanent magnets, the actuator starts oscillating between the heated magnets. Therefore, the average temperature of the MSMA film increases, resulting in a decrease in magnetization, and a corresponding decrease in the magnetic attraction force and in the magnetic potential energy. Under stationary conditions, the total energy of the system exhibits two energy minima separated by a small energy barrier in the actuator center, which is low enough to enable continuous oscillation. In resonant self-actuation mode, the large oscillation stroke in the order of 20% of the cantilever length and high oscillation frequency at 50–60 Hz enable thermal power generation of 2.3 μW/cm^2 at a heat source temperature of 150 °C.

9. Patents

The authors hold a patent related to this work.

Author Contributions: Conceptualization, M.K.; design, M.K. and J.J.; fabrication, J.J., H.M. and M.O.; software and model validation, J.J.; investigation, J.J., H.M. and M.O.; writing—original draft preparation, J.J.; writing—review and editing, M.K. and J.J.; visualization, J.J.; supervision, project administration, and funding acquisition, M.K. All authors have read and agreed to the published version of the manuscript.

Funding: This work was funded by the German Science Foundation (DFG) through the project "Thervest II" and partly supported by the Core-to-Core Program A "Advanced Research Networks" of the Japanese Science Foundation (JSPS).

Data Availability Statement: The data that support the findings of this study are available from the corresponding authors upon reasonable request.

Acknowledgments: This work was partly carried out with the support of the Karlsruhe Nano Micro Facility (KNMFi, www.knmf.kit.edu).

Conflicts of Interest: The authors declare no conflict of interest.

Appendix A

Appendix A.1. Simulation Parameters

Table A1. Summary of LEM simulation parameters. The parameters depend on the MSMA material and the detailed operation conditions of the TMG device. Table A1 also shows specific values used for LEM simulation of a TMG demonstrator device using Ni-Mn-Ga films of 10 μm thickness with a Curie temperature T_c of 371 K.

Parameter	TMG Device Based on a Ni-Mn-Ga Film	Reference
Length of each MSMA film	2 mm	This work
Width of each MSMA film	2 mm	This work
Thickness of each MSMA film	10 μm	This work
Density of each MSMA material	8020 kg/m^3	[35]
Length of the cantilever beam	5 mm	This work
Width of the cantilever beam	2 mm	This work
Thickness of the cantilever beam	20 μm	This work
Density of cantilever material	8500 kg/m^3	[36]
Length of bonding layer	2 mm	This work
Width of bonding layer	2 mm	This work
Thickness of bonding layer	10 μm	This work
Density of bonding layer	1250 kg/m^3	[37]
Young's modulus of cantilever material (E)	1×10^{11} N/m^2	[38]
Contact stiffness beam-magnet (k_{cont})	1×10^4 N/m	This work
Structural damping (c)	1.12×10^{-5} Ns/m	This work
Impact damping (c_{cont})	0.1 Ns/m	This work
Thermal conductivity of MSMA material	23.2 W/mK	[39]
Specific heat capacity of MSMA material	490 J/kgK	[39]
Thermal conductivity of cantilever material	109 W/mK	[40]
Specific heat capacity of cantilever material	400 J/kgK	[36]
Thermal conductivity of bonding layer	0.225 W/mK	[37]
Specific heat capacity of bonding layer	2100 J/kgK	[37]
Area of thermal contact	4 mm^2	This work
Max. Conductive heat transfer coefficient (Kc_{ond})	8400 W/m^{2}K	This work
Max. Convective heat transfer coefficient (K_{conv})	120 W/m^{2}K	This work
Remanent magnetic field	1.07 T	This work
Number of turns of coil	800	This work
Area of coil	1.96×10^{-6} m^2	This work
Electrical resistance of coil (internal resistance)	250 Ω	This work
Electrical load resistance	400 Ω	This work
Operation temperature	423 K	This work

Appendix A.2. Experimental Performance Characteristics

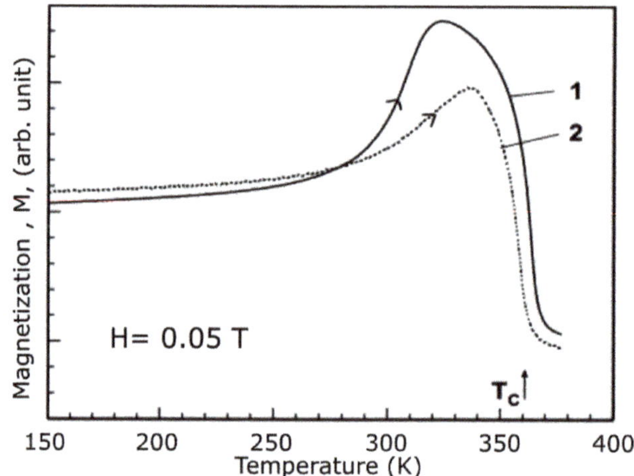

Figure A1. Typical low-field magnetization versus temperature curve of the investigated Ni-Mn-Ga films with two different thicknesses of 5 μm (**1**) and 0.4 μm (**2**) M, adapted from [41].

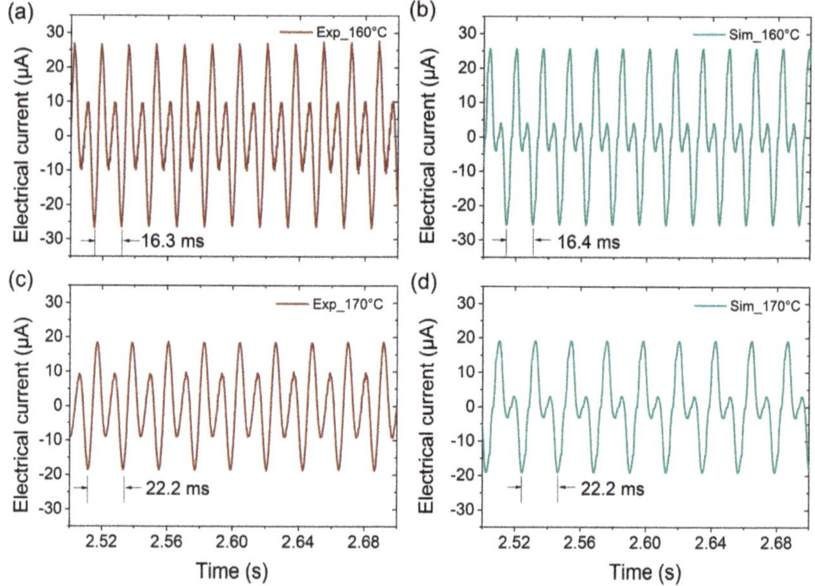

Figure A2. Performance of the bistable thermal energy harvester for heat source temperatures of 160 and 170 °C. (**a,c**) experimentally determined electrical current in the pick-up coils connected in parallel, (**b,d**) are the corresponding LEM simulations. The simulation model reproduces the amplitude and frequency of the bistable MSMA film actuator very well, while the small peaks due to the changing direction of the magnetic field gradient are less pronounced.

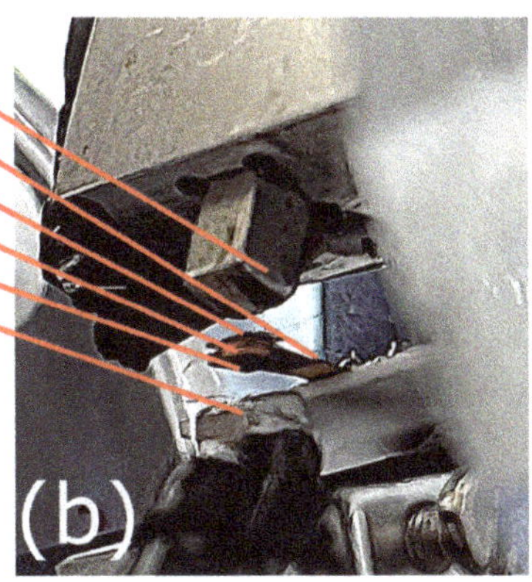

Figure A3. Experimental setup with the bistable MSMA film actuator shown in front view (**a**) and side view (**b**). It consists of a free-standing cantilever, two pick-up coils, and MSMA films 1 and 2 on both sides of the cantilever.

References

1. Heczko, O.; Seiner, H.; Fähler, S. Coupling between Ferromagnetic and Ferroelastic Transitions and Ordering in Heusler Alloys Produces New Multifunctionality. *MRS Bull.* **2022**, *47*, 618–627. [CrossRef]
2. Ullakko, K.; Huang, J.K.; Kantner, C.; O'Handley, R.C.; Kokorin, V.V. Large Magnetic-field-induced Strains in Ni 2 MnGa Single Crystals. *Appl. Phys. Lett.* **1996**, *69*, 1966–1968. [CrossRef]
3. Murray, S.J.; Marioni, M.A.; Kukla, A.M.; Robinson, J.; O'Handley, R.C.; Allen, S.M. Large Field Induced Strain in Single Crystalline Ni–Mn–Ga Ferromagnetic Shape Memory Alloy. *J. Appl. Phys.* **2000**, *87*, 5774–5776. [CrossRef]
4. Sozinov, A.; Likhachev, A.A.; Lanska, N.; Ullakko, K. Giant Magnetic-Field-Induced Strain in NiMnGa Seven-Layered Martensitic Phase. *Appl. Phys. Lett.* **2002**, *80*, 1746–1748. [CrossRef]
5. Kainuma, R.; Oikawa, K.; Ito, W.; Sutou, Y.; Kanomata, T.; Ishida, K. Metamagnetic Shape Memory Effect in NiMn-Based Heusler-Type Alloys. *J. Mater. Chem.* **2008**, *18*, 1837. [CrossRef]
6. Kohl, M.; Reddy, Y.S.; Khelfaoui, F.; Krevet, B.; Backen, A.; Fähler, S.; Eichhorn, T.; Jakob, G.; Mecklenburg, A. Recent Progress in FSMA Microactuator Developments. *Mater. Sci. Forum* **2009**, *635*, 145–154. [CrossRef]
7. Kohl, M.; Gueltig, M.; Pinneker, V.; Yin, R.; Wendler, F.; Krevet, B. Magnetic Shape Memory Microactuators. *Micromachines* **2014**, *5*, 1135–1160. [CrossRef]
8. Khelfaoui, F.; Kohl, M.; Szabo, V.; Mecklenburg, A.; Schneider, R. Development of Single Crystalline Ni-Mn-Ga Foil Microactuators. In *ICOMAT*; John Wiley & Sons, Inc.: Hoboken, NJ, USA, 2013; pp. 215–222.
9. Heczko, O.; Soroka, A.; Hannula, S.-P. Magnetic Shape Memory Effect in Thin Foils. *Appl. Phys. Lett.* **2008**, *93*, 022503. [CrossRef]
10. Dong, J.W.; Chen, L.C.; Xie, J.Q.; Müller, T.A.R.; Carr, D.M.; Palmstrøm, C.J.; McKernan, S.; Pan, Q.; James, R.D. Epitaxial Growth of Ferromagnetic Ni2MnGa on GaAs(001) Using NiGa Interlayers. *J. Appl. Phys.* **2000**, *88*, 7357–7359. [CrossRef]
11. Jakob, G.; Elmers, H.J. Epitaxial Films of the Magnetic Shape Memory Material. *J. Magn. Magn. Mater.* **2007**, *310*, 2779–2781. [CrossRef]
12. Backen, A.; Yeduru, S.R.; Diestel, A.; Schultz, L.; Kohl, M.; Fähler, S. Epitaxial Ni-Mn-Ga Films for Magnetic Shape Memory Alloy Microactuators. *Adv. Eng. Mater.* **2012**, *14*, 696–709. [CrossRef]
13. Mostafaei, A.; Rodriguez De Vecchis, P.; Stevens, E.L.; Chmielus, M. Sintering Regimes and Resulting Microstructure and Properties of Binder Jet 3D Printed Ni-Mn-Ga Magnetic Shape Memory Alloys. *Acta Mater.* **2018**, *154*, 355–364. [CrossRef]
14. Kohl, M.; Krevet, B.; Yeduru, S.R.; Ezer, Y.; Sozinov, A. A Novel Foil Actuator Using the Magnetic Shape Memory Effect. *Smart Mater. Struct.* **2011**, *20*, 094009. [CrossRef]
15. Kabla, M.; Ben-David, E.; Shilo, D. A Novel Shape Memory Alloy Microactuator for Large In-Plane Strokes and Forces. *Smart Mater. Struct.* **2016**, *25*, 075020. [CrossRef]
16. Musiienko, D.; Saren, A.; Ullakko, K. Magnetic Shape Memory Effect in Single Crystalline Ni-Mn-Ga Foil Thinned down to 1 µm. *Scr. Mater.* **2017**, *139*, 152–154. [CrossRef]
17. Musiienko, D.; Straka, L.; Klimša, L.; Saren, A.; Sozinov, A.; Heczko, O.; Ullakko, K. Giant Magnetic-Field-Induced Strain in Ni-Mn-Ga Micropillars. *Scr. Mater.* **2018**, *150*, 173–176. [CrossRef]
18. Kohl, M.; Brugger, D.; Ohtsuka, M.; Takagi, T. A Novel Actuation Mechanism on the Basis of Ferromagnetic SMA Thin Films. *Sens. Actuators A Phys.* **2004**, *114*, 445–450. [CrossRef]
19. Kainuma, R.; Imano, Y.; Ito, W.; Sutou, Y.; Morito, H.; Okamoto, S.; Kitakami, O.; Oikawa, K.; Fujita, A.; Kanomata, T.; et al. Magnetic-Field-Induced Shape Recovery by Reverse Phase Transformation. *Nature* **2006**, *439*, 957–960. [CrossRef] [PubMed]

20. Srivastava, V.; Chen, X.; James, R.D. Hysteresis and Unusual Magnetic Properties in the Singular Heusler Alloy Ni45Co5Mn40Sn10. *Appl. Phys. Lett.* **2010**, *97*, 014101. [CrossRef]
21. Srivastava, V.; Song, Y.; Bhatti, K.; James, R.D. The Direct Conversion of Heat to Electricity Using Multiferroic Alloys. *Adv. Energy Mater.* **2011**, *1*, 97–104. [CrossRef]
22. Post, A.; Knight, C.; Kisi, E. Thermomagnetic Energy Harvesting with First Order Phase Change Materials. *J. Appl. Phys.* **2013**, *114*, 033915. [CrossRef]
23. Gueltig, M.; Ossmer, H.; Ohtsuka, M.; Miki, H.; Tsuchiya, K.; Takagi, T.; Kohl, M. High Frequency Thermal Energy Harvesting Using Magnetic Shape Memory Films. *Adv. Energy Mater.* **2014**, *4*, 1400751. [CrossRef]
24. Gueltig, M.; Wendler, F.; Ossmer, H.; Ohtsuka, M.; Miki, H.; Takagi, T.; Kohl, M. High-Performance Thermomagnetic Generators Based on Heusler Alloy Films. *Adv. Energy Mater.* **2017**, *7*, 1601879. [CrossRef]
25. Joseph, J.; Ohtsuka, M.; Miki, H.; Kohl, M. Upscaling of Thermomagnetic Generators Based on Heusler Alloy Films. *Joule* **2020**, *4*, 2718–2732. [CrossRef]
26. Ohtsuka, M.; Itagaki, K. Effect of Heat Treatment on Properties of Ni-Mn-Ga Films Prepared by a Sputtering Method. *Int. J. Appl. Electromagn. Mech.* **2001**, *12*, 49–59. [CrossRef]
27. Castaño, F.J.; Nelson-Cheeseman, B.; O'Handley, R.C.; Ross, C.A.; Redondo, C.; Castaño, F. Structure and Thermomagnetic Properties of Polycrystalline Ni–Mn–Ga Thin Films. *J. Appl. Phys.* **2003**, *93*, 8492–8494. [CrossRef]
28. Hakola, A.; Heczko, O.; Jaakkola, A.; Kajava, T.; Ullakko, K. Ni–Mn–Ga Films on Si, GaAs and Ni–Mn–Ga Single Crystals by Pulsed Laser Deposition. *Appl. Surf. Sci.* **2004**, *238*, 155–158. [CrossRef]
29. Backen, A.; Yeduru, S.R.; Kohl, M.; Baunack, S.; Diestel, A.; Holzapfel, B.; Schultz, L.; Fähler, S. Comparing Properties of Substrate-Constrained and Freestanding Epitaxial Ni–Mn–Ga Films. *Acta Mater.* **2010**, *58*, 3415–3421. [CrossRef]
30. Suzuki, M.; Ohtsuka, M.; Suzuki, T.; Matsumoto, M.; Miki, H. Fabrication and Characterization of Sputtered Ni2MnGa Thin Films. *Mater. Trans. JIM* **1999**, *40*, 1174–1177. [CrossRef]
31. Camacho, J.M.; Sosa, V. Alternative Method to Calculate the Magnetic Field of Permanent Magnets with Azimuthal Symmetry. *Rev. Mex. Fis. E* **2013**, *59*, 8–17.
32. Schütze, T. *AN2008-03: Thermal Equivalent Circuit Models. Application Note. V1.0*; Infineon Technologies AG: Neubiberg, Germany, 2008.
33. Joseph, J.; Ohtsuka, M.; Miki, H.; Kohl, M. Thermal Processes of Miniature Thermomagnetic Generators in Resonant Self-Actuation Mode. *iScience* **2022**, *25*, 104569. [CrossRef]
34. Joseph, J.; Ohtsuka, M.; Miki, H.; Kohl, M. Lumped Element Model for Thermomagnetic Generators Based on Magnetic SMA Films. *Materials* **2021**, *14*, 1234. [CrossRef] [PubMed]
35. Tickle, R.; James, R.D. Magnetic and Magnetomechanical Properties of Ni2MnGa. *J. Magn. Magn. Mater.* **1999**, *195*, 627–638. [CrossRef]
36. Beiss, P. Non-Ferrous Materials. In *Powder Metallurgy Data*; Beiss, P., Ruthardt, R., Warlimont, H., Eds.; Landolt-Börnstein—Group VIII Advanced Materials and Technologies; Springer: Berlin/Heidelberg, Germany, 2002; Volume 2A2, pp. 194–204; ISBN 3-540-42961-1.
37. Assmus, W.; Brühne, S.; Charra, F.; Chiarotti, G.; Fischer, C.; Fuchs, G.; Goodwin, F.; Gota-Goldman, S.; Guruswamy, S.; Gurzadyan, G.; et al. *Springer Handbook of Condensed Matter and Materials Data*; Martienssen, W., Warlimont, H., Eds.; Springer: Berlin/Heidelberg, Germany, 2005; ISBN 978-3-540-44376-6.
38. Gere, J.M.; Timoshenko, S. *Mechanics of Materials*, 4th ed.; PWS: Boston, MA, USA, 1997; ISBN 9780534934293.
39. Söderberg, O.; Aaltio, I.; Ge, Y.; Heczko, O.; Hannula, S.P. Ni-Mn-Ga Multifunctional Compounds. *Mater. Sci. Eng. A* **2008**, *481–482*, 80–85. [CrossRef]
40. Young, H.; Freedman, R.A. *Sears and Zemansky's University Physics with Modern Physics*, 14th ed.; Pearson Education Limited: Harlow, UK, 2016; ISBN 978-1-292-10032-6.
41. Chernenko, V.A.; Ohtsuka, M.; Kohl, M.; Khovailo, V.V.; Takagi, T. Transformation Behavior of Ni-Mn-Ga Thin Films. *Smart Mater. Struct.* **2005**, *14*, S245–S252. [CrossRef]

Article

Variational Reduced-Order Modeling of Thermomechanical Shape Memory Alloy Based Cooperative Bistable Microactuators

Muhammad Babar Shamim , Marian Hörsting and Stephan Wulfinghoff *

Institute for Materials Science, Computational Materials Science, Kiel University, 24143 Kiel, Germany
* Correspondence: swu@tf.uni-kiel.de

Abstract: This article presents the formulation and application of a reduced-order thermomechanical finite strain shape memory alloy (SMA)-based microactuator model for switching devices under thermal loading by Joule heating. The formulation is cast in the generalized standard material framework with an extension for thermomechanics. The proper orthogonal decomposition (POD) is utilized for capturing a reduced basis from a precomputed finite element method (FEM) full-scale model. The modal coefficients are computed by optimization of the underlying incremental thermomechanical potential, and the weak form for the mechanical and thermal problem is formulated in reduced-order format. The reduced-order model (ROM) is compared with the FEM model, and the exemplary mean absolute percentage errors for the displacement and temperature are 0.973% and 0.089%, respectively, with a speedup factor of 9.56 for a single SMA-based actuator. The ROM presented is tested for single and cooperative beam-like actuators. Furthermore, cross-coupling effects and the bistability phenomenon of the microactuators are investigated.

Keywords: microactuators; shape memory alloy; thermomechanics; reduced-order modeling; proper orthogonal decomposition

Citation: Shamim, M.B.; Hörsting, M.; Wulfinghoff, S. Variational Reduced-Order Modeling of Thermomechanical Shape Memory Alloy Based Cooperative Bistable Microactuators. *Actuators* **2023**, *12*, 36. https://doi.org/10.3390/act12010036

Academic Editor: Micky Rakotondrabe

Received: 1 December 2022
Revised: 4 January 2023
Accepted: 5 January 2023
Published: 10 January 2023

1. Introduction

Since the discovery of shape memory alloys (SMAs), they have been used in various applications. For example, as orthodontic appliances, in mobile phones, valves, space robotics and microactuators [1,2]. Their frequent use in these applications stems from their unique properties, namely the super-elasticity and the shape memory effect [3–6]. SMA-based microactuators are in increased demand [7,8] because they have a high mechanical strength and work density, which results in lightweight structures that can repeatedly generate large forces with small device sizes [9,10]. SMA-based actuators have the disadvantage of low actuation speed caused by the latency of thermal cooling, which is strongly size-dependent. Song et al. [11] proposed a smart soft composite actuator that is capable of large deformations with fast bending actuation. Stachiv et al. [12] proposed an approach to use SMA in combination with an elastic substrate as tunable active layer. Samal et al. [13] investigated the bending stiffness and radius of curvature of polymethyl methacrylate (PMMA) and NiTi SMA composites upon application of an external thermal load. The discussed actuator in this paper is based on a concept explored by Winzek et al. [14]. They investigated a combination of SMA and polymer that leads to the actuator behaving in a bistable manner.

These SMA-based actuators usually undergo transformation due to a thermal loading during their operation. Therefore, it is necessary to understand the complex, nonlinear thermomechanical behavior of the SMA-based actuator, which makes the numerical implementation challenging [15].

The shape memory and super-elastic effect occur in SMAs due to a phase transformation between austenite and martensite, which is induced by thermomechanical loadings [9,16–18]. SMA models can be developed based on a continuum thermodynamics framework. In this work, we take into account the thermomechanical potential developed by Yang et al. [19] and assume a potential including internal variables [5,17,20]. Sielenkämper et al. [21] extended the SMA potential formulation from Sedlak et al. [22] to the finite strain case, which was developed for small-strain SMA modeling. It captures the super-elasticity and shape memory effect, as well as the volume change due to phase transformation, which is found in some SMAs [23]. Frost et al. [24] also developed a thermomechanical model for SMAs within the framework of generalized standard materials that could predict the martensite transformation and its localization. Solomou et al. [25] proposed a coupled thermomechanical SMA-based actuator model that is capable of simulating heat transfer and convection effects in the actuator. There, the constitutive equations for SMAs from Lagoudas' model [10] are used to predict the coupled thermomechanical response of an SMA beam.

It is often time-consuming to perform an experimental investigation for the design optimization of new components [26]. For nonlinear multiphysics problems, such optimizations are usually perceived by numerical simulations [27,28], which might have millions of degrees of freedom (DOF), and can thus be computationally expensive [29] and also require a continuously evolving discretized model for nonlinear problems [30–32]. Performing simulations using an FEM model sometimes becomes unfeasible when performing hundreds of simulations for different, varying parameters such as different loading conditions, different values of Young's modulus and Poisson's ratio to design a new structure [26,33]. Therefore, we need a reduced-order model to reduce the computational cost with respect to solving parametrized nonlinear partial differential equations (PDEs) [32]. Different model order reduction techniques that perform well in reducing the computational time while preserving model accuracy have been utilized. Model order reduction (MOR) is a class of techniques to approximate a higher dimensional system by a low-dimensional model [26]. POD has been utilized by many researchers as a basis for MOR techniques [34–36].

Reduced-order modeling for a compact FEM model could speed up the investigation process significantly while keeping a certain level of model accuracy. Vettermann et al. [37] developed a strategy for coupled thermomechanical problems with nonlinear boundary conditions for faster simulations of machine tools. Ummunakwe et al. [38] developed a reduced-order thermomechanical model for packaged chips using Krylov subspace methods with excellent model accuracy.

JinXiu et al. [39] used POD to analyze transient heat conduction problems and proposed a fast reduced-order model that could interpolate and extrapolate the field variable at an unknown time. Jia et al. [40] proposed a reduced-order thermal model for a multifin field effect transistor, which provides accurate solutions. Bikora et al. [41] developed a low-order model to predict the thermal deformation in reticles for extreme ultraviolet lithography, which is based on a large-scale thermomechanical FEM model. Hernandez et al. [42] worked on a thermomechanical reduced-order model for machine tools that predicts the thermal response in a frequency range and also predicts the coupling between the mechanical and thermal responses. Das et al. [43] proposed a thermomechanical reduced-order model using POD to simulate the thermomechanical processes that occur during the fabrication of photovoltaic cells.

The objective of this paper is to develop a fast and reliable reduced-order model for the FEM model developed by Sielenkämper et al. [21]. The novelty of this ROM is the application to the coupled thermomechanical potential, which was first considered by Yang et al. [19]. First, samples of snapshots are generated from high-fidelity FEM solutions for different parameter sets. This so-called offline computation is computationally expensive. Subsequently, a reduced basis that could best represent the full snapshot matrix is obtained from the snapshots through statistical tools. Then, the most significant modes contained in the reduced basis are used to predict the overall behavior of the nonlinear SMA actuator model.

The article is organized as follows: The second section provides, in compact form, all necessary equations for the thermomechanical SMA-based model. Then, a reduced-order model is presented. Section 3 presents the application of the ROM to different examples. Finally, Section 4 concludes the paper, along with suggestions for future work.

We used bold letters for first- and second-order tensors, as well as tensor-valued functions such as P, and light face letters for scalars and scalar valued functions such as T.

2. Modeling of the Thermomechanical SMA-Based Actuator

This section briefly presents the mathematical formulation for the SMA-based actuator model.

2.1. Continuum Model

Within a continuum body with reference configuration $\mathcal{B}_0$, the displacement vector u of a material point with reference position vector X at time t is defined as the difference between the current and reference position vectors x and X, respectively ($u(X, t) = x(X, t) - X$). The deformation gradient F describes the mapping of infinitesimal line elements from the reference to the current configuration and can be expressed as

$$F = \text{Grad}(x(X, t)) \tag{1}$$

In the reference configuration, the linear momentum balance for a body $\mathcal{B}_0$ can be written as

$$\text{Div}P + \rho_0 b = 0, \tag{2}$$

where b is the body force, ρ_0 is the reference mass density and P is the first Piola–Kirchoff stress tensor. The boundary conditions are $u = \bar{u}$ on $\partial\mathcal{B}_{0u}$ and $\hat{t} = PN$ on $\partial\mathcal{B}_{0t}$, with the given traction vector $\hat{t}$ and the external normal N in the reference configuration.

We consider the free energy density (with respect to the reference configuration) $\psi = \psi(C, T, Z)$, which depends on the right Cauchy–Green tensor $C = F^T F$, the absolute temperature T and an array of internal variables Z. Inserting this function into the Clausius–Planck inequality and using Coleman and Noll's reasoning, we can write

$$P = \frac{\partial\psi}{\partial F}; \quad s = -\frac{\partial\psi}{\partial T}, \tag{3}$$

where s denotes the entropy density per unit reference volume.

We assume a scalar dissipation potential for the isothermal case (the nonisothermal case is discussed below, see Equation (7)) $\phi(\dot{Z}, Z)$ to depend on internal variables Z and the rate of the internal variables $\dot{Z}$. For general standard dissipative solids, the rates of the internal variables are identified by solving Biot's equation

$$\frac{\partial\psi}{\partial Z} + \frac{\partial\phi}{\partial\dot{Z}} \ni 0, \tag{4}$$

where $\partial\phi/\partial\dot{Z}$ in general denotes a subdifferential. We obtain the dissipation density as

$$\mathcal{D} = \frac{\partial\phi(\dot{Z}, Z)}{\partial\dot{Z}} \cdot \dot{Z}. \tag{5}$$

Then, the balance of energy with Fourier's law, for simplicity taken as $Q = -\kappa\text{Grad}T$, where κ is the heat conductivity and Q is the heat flux vector in the reference configuration, can be represented as

$$T\dot{s} = -\text{Div}(\kappa Q) + w + \mathcal{D}, \tag{6}$$

where w represents the heat source density. The Dirichlet boundary condition for the thermal part is $T = \bar{T}$ on $\partial\mathcal{B}_{0T}$. The Robin-type boundary condition is given as $Q \cdot N = \alpha(T - T_a)$ on $\partial\mathcal{B}_{0Q}$. Here, α is the heat transfer coefficient and T_a is the ambient temperature. Neumann-type boundary conditions are not considered.

Consider the following rate potential for the nonisothermal case in its time-discrete format for the material modeling of the SMA (compare Yang et al. [19]).

$$\psi(\mathbf{C}, T, \mathbf{Z}) - (e_n - T s_n) + \Delta t \phi\left(\frac{T}{T_n}\mathring{\mathbf{Z}}, \mathbf{Z}\right), \tag{7}$$

where e_n and s_n are the internal energy density and entropy density at time t_n, respectively. $\mathring{\mathbf{Z}}$ is a numerical approximation of the time derivative of $\mathbf{Z}$. This means that any $\mathring{\mathbf{Z}}$ satisfying $\lim_{\Delta t \to 0} \mathring{\mathbf{Z}} = \dot{\mathbf{Z}}$ is admissible. A typical example is $\mathring{\mathbf{Z}} = \Delta \mathbf{Z}/\Delta t$. Additionally, $\Delta t \phi$ is denoted as ϕ_Δ and is assumed homogeneous of degree one in its first variable in the following. Note that the major difference of the dissipation potential in the nonisothermal case compared with the isothermal case is the factor T/T_n in the first argument.

The temperature and the displacement vector are the unknown state variables. Fourier's heat conduction law is assumed along with dissipative terms. Both state variables are coupled through the following time-discrete potential

$$\pi_\Delta = \psi - (e_n - T s_n) + \phi_\Delta - \kappa\frac{\Delta t}{2T_n}\|\mathrm{Grad}T\|^2 + \kappa\frac{\Delta t T}{T_n^2}\|\mathrm{Grad}T_n\|^2, \tag{8}$$

where the last two terms in Equation (8) represent the heat conduction in the body. The integral form of the incremental potential is expressed as follows:

$$\Pi_\Delta = \int_{\mathcal{B}_0} \pi_\Delta \mathrm{d}V - \int_{\partial\mathcal{B}_{0t}} \hat{\mathbf{t}} \cdot \mathbf{u}\mathrm{d}S - \int_{\mathcal{B}_0} \rho_0 \mathbf{b} \cdot \mathbf{u}\mathrm{d}V - \int_{\partial\mathcal{B}_{0Q}} \frac{\alpha\Delta t}{2T_n}(T - T_n)^2\mathrm{d}S + \int_{\mathcal{B}_0} \Delta t\frac{T}{T_n}w\mathrm{d}V. \tag{9}$$

Now, the state variables and internal variables are obtained by solving the following saddle point problem

$$\inf_{\mathbf{u}\in\kappa_u} \sup_{T\in\kappa_T} \inf_{\mathbf{Z}} \Pi_\Delta. \tag{10}$$

Here, $\kappa_u = \{\mathbf{u} : \mathbf{u} = \bar{\mathbf{u}} \text{ on } \partial\mathcal{B}_{0u}\}$ is the set of admissible displacement values fulfilling the Dirichlet boundary conditions on the boundary $\partial\mathcal{B}_{0u}$, and $\kappa_T = \{T : T = \bar{T} \text{ on } \partial\mathcal{B}_{0T}\}$ is the set of admissible temperatures that fulfill the Dirichlet boundary conditions on the boundary $\partial\mathcal{B}_{0T}$.

The weak form of the quasi-static linear momentum balance can be obtained by applying a variation of the potential with respect to the displacement:

$$\delta_u\Pi_\Delta = \int_{\mathcal{B}_0} \mathbf{P} : \delta\mathbf{F}\,\mathrm{d}V - \int_{\partial\mathcal{B}_{0t}} \hat{\mathbf{t}} \cdot \delta\mathbf{u}\,\mathrm{d}S - \int_{\mathcal{B}_0} \rho_0\mathbf{b} \cdot \delta\mathbf{u}\mathrm{d}V, \tag{11}$$

where $\delta\mathbf{F}$ and $\delta\mathbf{u}$ are the variations of the deformation gradient and displacement vector, respectively. In the same way, the weak form of the energy balance can be obtained by the variation with the temperature T as follows:

$$\delta_T\Pi_\Delta = \int_{\mathcal{B}_0} \left(\left(\frac{\partial\psi}{\partial T} + s_n + \frac{1}{T_n}\phi_\Delta(\mathring{\mathbf{Z}}, \mathbf{Z})\right)\delta T - \frac{\kappa\Delta t}{T_n}\mathrm{Grad}(T) \cdot \mathrm{Grad}(\delta T) \right.$$

$$\left. + \Delta t\frac{\delta T}{T_n^2}\kappa\|\mathrm{Grad}(T_n)\|^2 + \frac{\Delta t}{T_n}w \right)\delta T\mathrm{d}V - \int_{\mathcal{B}_{0Q}} \alpha\frac{\Delta t}{T_n}(T - T_n)\delta T\mathrm{d}S = 0. \tag{12}$$

It can be easily verified that Equations (11) and (12) yield Equations (2) and (6), as well as the corresponding boundary conditions in the limit $\Delta t \to 0$, which proves the consistency of the time-discrete potential with the time-continuous theory.

Furthermore, the variation of the potential with respect to the internal variables yields the nonisothermal form of Biot's equation, which represents the stationarity condition with respect to the internal variables:

$$\delta_{\mathbf{Z}}\Pi_\Delta = \frac{\partial\psi}{\partial\mathbf{Z}} + \Delta t\frac{\partial\phi}{\partial\mathbf{Z}} \ni \mathbf{0}. \tag{13}$$

2.2. Reduced-Order Modeling

After solving a fully resolved FEM model for a wide range of certain parameter values, snapshots (i.e., the model state variable vectors) are stored in a snapshot matrix. After applying POD on the snapshot matrix, a reduced basis that can best represent the solution in terms of a few displacement "modes" $\boldsymbol{\psi}_i(\boldsymbol{X})$ $(i = 1, \ldots N_u)$ and temperature modes $\chi_i(\boldsymbol{X})$ $(m = 1, \ldots N_T)$ is obtained. The reduced solution for the displacement vector with the mode coefficient $xi(t)$ can be expressed using this reduced basis as

$$\tilde{\boldsymbol{u}}(\boldsymbol{X}, t) = \sum_{i=1}^{N_u} \xi_i(t) \boldsymbol{\psi}_i(\boldsymbol{X}); \quad \mathrm{Grad}\tilde{\boldsymbol{u}}(\boldsymbol{X}, t) = \sum_{i=1}^{N_u} \xi_i \mathrm{Grad}\boldsymbol{\psi}_i(\boldsymbol{X}). \tag{14}$$

Likewise, the reduced solution for the temperature can be expressed via the modes as:

$$\tilde{T}(\boldsymbol{X}, t) = \sum_{m=1}^{N_T} \mu_m(t) \chi_m(\boldsymbol{X}); \quad \mathrm{Grad}\tilde{T}(\boldsymbol{X}, t) = \sum_{m=1}^{N_T} \mu_m(t) \mathrm{Grad}\chi_m(\boldsymbol{X}), \tag{15}$$

where the subscript N_u and N_T represent the number of modes selected for displacement and temperature calculation, respectively. Furthermore, $\tilde{\boldsymbol{u}}(\boldsymbol{X}, t)$ and $\tilde{T}(\boldsymbol{X}, t)$ are the reduced approximations of the displacement vector and temperature that depend on the position vector $\boldsymbol{X}$ and time t.

Now, the aim is to obtain a weak formulation of the thermomechanical model but with a reduced number of dimensions, which therefore can be solved faster. The major novelty in this work is to combine the MOR-scheme described above with the incremental thermomechanical potential (Equation (9)). This enables a further performance enhancement, as the existence of a potential automatically ensures a symmetric tangent, which in turn enables the use of more efficient linear equation solvers compared with the nonsymmetric case. Furthermore, the potential-based formulation may be useful for the formulation of advanced nonlinear equation solvers and when mathematically analyzing the reduced-order model, which is beyond the scope of the work at hand. The weak form for the linear momentum balance in reduced-order format is given as

$$\delta_\xi \Pi_\Delta = \sum_{i=1}^{N_u} \delta\xi_i \underbrace{\left[\int_{\mathcal{B}_0} \boldsymbol{\tau} : \mathbf{g}_i^s \mathrm{d}V - \int_{\partial\mathcal{B}_{0t}} \hat{\boldsymbol{t}} \cdot \boldsymbol{\psi}_i \mathrm{d}S - \int_{\mathcal{B}_0} \rho_0 \boldsymbol{b} \cdot \boldsymbol{\psi}_i \mathrm{d}V \right]}_{R_i^u}, \tag{16}$$

where $\boldsymbol{\tau} = \boldsymbol{F}\boldsymbol{S}\boldsymbol{F}^T$ is the Kirchhoff stress tensor, $\boldsymbol{S} = \boldsymbol{F}^{-1}\boldsymbol{P}$ is the second Piola–Kirchhoff stress tensor and $\mathbf{g}_i^s$ is the symmetric part of $\boldsymbol{G}_i\boldsymbol{F}^{-1}$, with $\boldsymbol{G}_i = \mathrm{Grad}\boldsymbol{\psi}_i(\boldsymbol{X})$.

Likewise, the weak form of the energy balance can be expressed as

$$\delta_T \Pi_\Delta = \int_{\mathcal{B}_0} \left(\frac{\partial\pi_\Delta}{\partial T} \delta T + \frac{\partial\pi_\Delta}{\partial\mathrm{Grad}T} \cdot \mathrm{Grad}\delta T \right) \mathrm{d}V - \int_{\partial\mathcal{B}_{0Q}} \frac{\partial\pi_\Delta^s}{\partial T} \delta T \mathrm{d}S + \int_{\mathcal{B}_0} \frac{\Delta t}{T_n} w \delta T \mathrm{d}V. \tag{17}$$

After rearranging and inserting Equation (15), it reads

$$\delta_T \Pi_\Delta = \sum_{m=1}^{N_T} \delta\mu_m \underbrace{\left[\int_{\mathcal{B}_0} \left(\left(\frac{\partial\pi_\Delta}{\partial T} + \frac{\Delta t}{T_n} w \right) \chi_m + \frac{\partial\pi_\Delta}{\partial\mathrm{Grad}T} \cdot \boldsymbol{W}_m \right) \mathrm{d}V - \int_{\partial\mathcal{B}_{0Q}} \frac{\partial\pi_\Delta^s}{\partial T} \chi_m \mathrm{d}S \right]}_{R_m^T}, \tag{18}$$

where $\boldsymbol{W}_m = \mathrm{Grad}\chi_m(\boldsymbol{X})$ and $\pi_\Delta^s = \alpha\Delta t(T - T_n)^2/(2T_n)$.

The square bracket terms in Equations (16) and (18) are the residuals for the mechanical and thermal subproblems, respectively. Now, we can linearize both subproblems with respect to the DOFs ξ_i and μ_m to obtain the tangent moduli required for the global Newton scheme. Subsequently, the global residual vector and global tangent moduli are expressed as

$$\mathbf{R} = \begin{bmatrix} \mathbf{R}^u \\ \mathbf{R}^T \end{bmatrix}; \quad \mathbf{K} = \begin{bmatrix} \mathbf{K}^{uu} & \mathbf{K}^{uT} \\ \mathbf{K}^{Tu} & \mathbf{K}^{TT} \end{bmatrix}, \tag{19}$$

where $\mathbf{K}$ is symmetric as a result of the potential-based formulation.

3. Results and Discussion

This section presents the results obtained from MOR. The constitutive models (in particular the SMA model) are taken from Sielenkämper et al. [21]. For the material properties of SMA and PMMA, see Table 1 and Section 4.3.2, respectively, in [21]. The FEM model is solved using the finite element program FEAP [44], and the snapshots for the displacement and the temperature are stored. These snapshots are then used in the MOR scheme, which is written in FORTRAN. The visualization of the results is carried out in Paraview [45].

A bistable microactuator beam is investigated, which is made up of three layers: an SMA NiTiHf layer, a molybdenum layer, and a PMMA layer, as shown in Figure 1. To achieve bistability, the polymer layer plays a very important role. On the one hand, if the polymer layer is too thick, it will make the actuation speed slower with a higher energy consumption. On the other hand, if it is too thin, it will not be able to hold its shape. At $t = 0$, we assume the initial conditions to be given by zero displacements and room temperature at 20 °C. At the left face, Dirichlet boundary conditions are applied for the displacement, and for the temperature, where a constant value of 20 °C is considered. Initially, thermal eigenstrains are introduced, as described in [21], such that the actuator would be stress-free at 500 °C (annealing temperature). Joule heating is cyclically realized by a heat source applied to the molybdenum and SMA material in the beam. A sine function is used to model the magnitude of the heat source (see Equation (A1) in Appendix A).

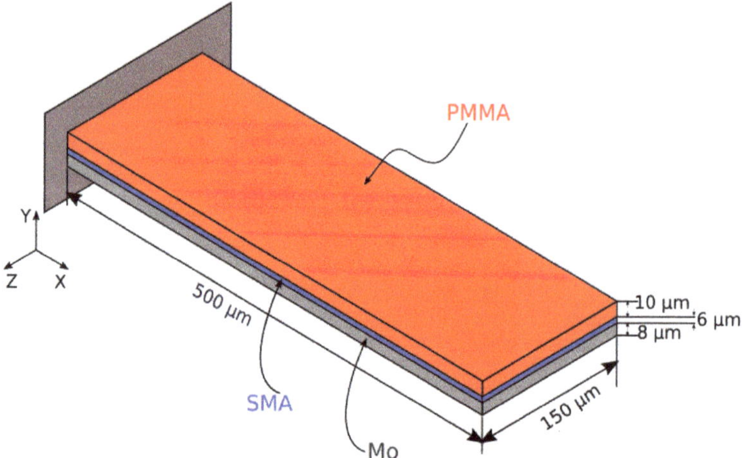

Figure 1. Dimensions of the actuator assembly, which is clamped at the left side. The red layer is polymer, blue is SMA and gray is molybdenum.

The heat convection is realized by applying a Robin-type boundary condition to the top and bottom faces of the actuator. The heat convection term at the remaining air enclosed surfaces is neglected. The bistability actuation principle of the presented actuator depends upon the difference in the coefficients of thermal expansion (CTE) between the layers, the shape memory effect and the difference in volume between the austenite and martensite phases. To achieve this bistability in the actuator, a heating cycle is applied as follows. From the annealing temperature (500 °C), the actuator cools down and the polymer stiffens when the glass transition temperature T_g is reached. Upon further cooling, austenite transforms to unoriented (twinned) martensite. Then, the actuator is heated (this is where the simulation actually starts) just until the glass transition temperature T_g of the polymer is reached, which softens the PMMA layer and causes martensite to twin, and thus the actuator to relax. Then, cooling down will lock the position of the actuator in place due to the polymer stiffening below T_g, which makes the actuator take its first stable state at room temperature. Again, heating the actuator above the austinite finish temperature will bend it up due to SMA phase transformation, and then cooling to room temperature will lock this position, because the polymer hardens before the martensite finish temperature, obtaining the second stable state at room temperature. The tip displacement difference between the two stable states is called the bistable stroke, as shown in Figure 7 (see Section 3.3). A typical aim is to maximize the bistable stroke while keeping the actuator small, which comes with a low power consumption [21]. We utilized trilinear elements with reduced integration and hourglass stabilization for all materials for the FEM computations. We refer to [21] for more information on the fully resolved FEM model.

3.1. Single SMA Actuator

We consider here a single actuator for one heat load cycle with appropriate boundary conditions for reduced-order modeling and simulation. For this example, the training data are obtained for different values of Young's modulus and Poisson's ratio, replacing the molybdenum layer by solving the full-order model as shown in Table 1. No external force is applied for all the examples, only Joule heating is performed. The snapshots are collected for one heat load cycle. One cycle in which the maximum temperature succeeds the austenite finish temperature (A_f) is enough to observe the actuation behavior of our bistable actuator. We acquire 150 snapshots that represent the solution obtained through Equations (16) and (18), respectively. We determine solutions for a wide range of parameter values, as mentioned in Table 1. In total, we collect 1350 snapshots for different material properties. The larger our snapshot matrix, the more accurately we can determine the optimal reduced basis.

POD [34–36] is utilized separately to obtain the reduced bases for the displacement and temperature. The model is trained in an offline computation phase. During the online phase, we predict the actuator behavior using any value in the considered parameter range. We minimize the thermomechanical potential (Equation (9)) with respect to the mechanical degrees of freedom to obtain the residual equation for the displacement and, simultaneously, we maximize the same potential with respect to thermal degrees of freedom to obtain the residual equation for the temperature. Then, we obtain the modes which are the global Galerkin ansatz functions for the state variables (displacement and temperature) separately. In this sense, displacements and temperature are comparable and of equal "value" or importance. Moreover, the snapshots calculated from FEM contain nonvanishing Dirichlet boundary conditions for temperature (20 °C), which is technically realized by a homogeneous temperature mode in addition to the other modes, which are normalized by subtracting from all nodes of a given mode the temperature at the clamping position (at $x = 0$), such that the modes are zero at that location. In the reduced-order model, temperature Dirichlet boundary conditions are realized, for simplicity, through a penalty approach in the form of a Robin-type boundary condition with very high heat transfer coefficient. The material model developed in [21] was fitted to experimental tensile tests performed at different temperatures and tensile loads. The FEM model predicts reasonable results, and the dependency on the temperature is also captured accurately. Please see Figure 2 in Curtis et al. [46]. In that paper, further experiments with bimorph beam structures similar to the one in this manuscript are also discussed.

Table 1. Parameter data set for training the reduced-order model for beam type trimorph actuator.

Parameter Set	Set 1	Set 2	Set 3
E (MPa)	300	300	300
ν	0.31	0.40	0.49
	Set 4	**Set 5**	**Set 6**
E (MPa)	325	325	325
ν	0.31	0.40	0.49
	Set 7	**Set 8**	**Set 9**
E (MPa)	350	350	350
ν	0.31	0.40	0.49

We use the following percentage error measure (E_{per}) between FEM and MOR

$$E_{per} = \frac{1}{N_{Nd}} \sum_{i=1}^{N_{Nd}} \frac{|u_i^{MOR} - u_i^{FEM}|}{|u_i^{FEM}|} \times 100\%, \tag{20}$$

where u^{MOR} is the displacement solution calculated using MOR, u^{FEM} is the displacement solution calculated for FEM and N_{Nd} is the total number of nodes taken for the error calculation. The error measure used for the temperature is defined in analogy. In addition to this global error measure, we use the local measures specified below.

Figures 2 and 3 illustrate the results for FEM and MOR for an "untrained" parameter set, i.e., using values for E and ν, which were not used when computing (see caption of Figure 2) the snapshot matrix. These displacement and temperature graphs are obtained at the left tip of the beam actuator, as marked in the Figure 2 with green circles. We consider 10 and 34 modes for the displacement computations based on the frequency graph of the modes that were obtained after POD computations, as shown in Figure 4. The number of temperature modes taken is equal to the number of displacement modes.

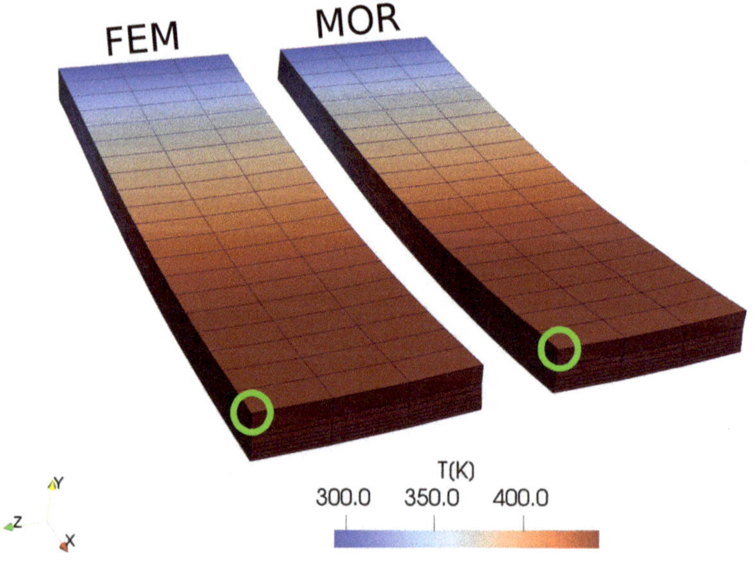

Figure 2. FEM and MOR result comparison for temperature for an "untrained" parameter set with 10 modes ($E = 331$ MPa and $\nu = 0.35$).

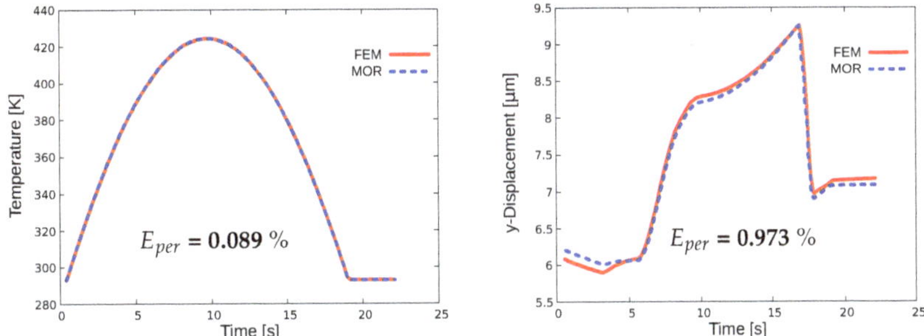

Figure 3. FEM and MOR comparison for displacement and temperature for one heat cycle (**left**). Temperature graph with 10 modes. Displacement graph with 34 modes (**right**).

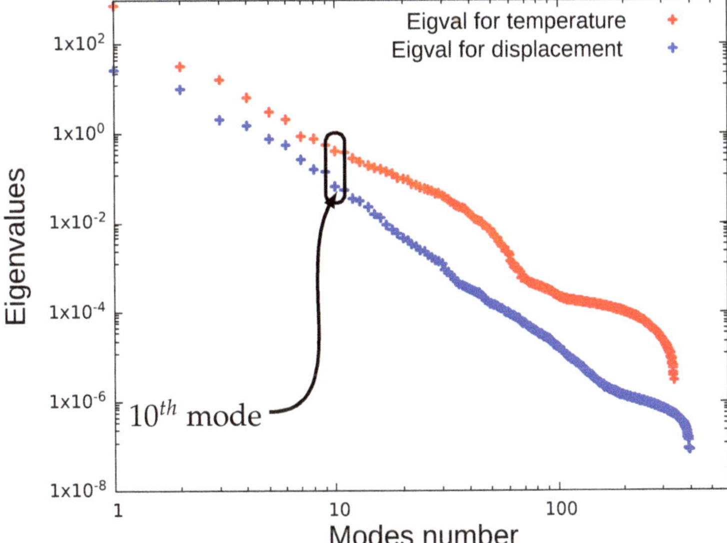

Figure 4. Eigenvalue graph for modes obtained after POD computation for the displacement and temperature.

The total time to run the full-scale model for one complete heat cycle and for one parameter value is ~132 s. Considering the three sets of values for E and ν, this takes ~1188 s in total to train the MOR against these parameters in an offline stage. After training the MOR against these parameters, the online computations are carried out for an untrained set of parameters. These simulations are carried out on an Intel® Core™ i7-8850H CPU @ 2.60 GHz with 32 GB RAM. The speedup factors obtained for this example are summarized in Table 2.

Table 2. Speedup factor for the model considered for one parameter set and one heat load cycle.

CPU Time FEM	Displacement Modes	CPU Time MOR	Speedup	Displacement Error
132 s	10	4.5 s	29.33	2.489%
132 s	34	14.2 s	9.29	0.973%

We also carried out the MOR computations for different untrained sets of parameter values ($E = 320$ MPa and $v = 0.38$, $E = 330$ MPa and $v = 0.33$), and the displacement error value is similar for these calculations ($E_{per} = 0.98\%$, $E_{per} = 0.971\%$).

3.2. Cross-Coupling between Actuators

The purpose of this analysis is to comprehend the cross-sensitivities and coupling effects if two actuators cooperate, as well as to investigate the performance of the MOR for a more complicated geometry. In confined spaces, actuating one microactuator that is part of an array might have an influence on the other actuators for cooperative multistable operation. These are unwanted effects that cause a loss of precision in the actuation. For example, if there is a huge cross-coupling between the actuators, activating one actuator through Joule heating will cause the neighboring actuators to actuate, which restricts the design space of the actuator. These cross-coupling effects need to be understood properly [47].

Two actuators are connected at the base to a molybdenum block. The bottom and back-face of the molybdenum base are fixed. A zero Dirichlet boundary condition is applied for the displacement at the fixed locations, and we have a nonzero Dirichlet boundary condition at the back-face of the molybdenum block for temperature, which is 20 °C, and the heat load is applied only at the left actuator. We investigate cross-coupling at different distances d between the actuators using our developed ROM after training the model against different size scales of the actuators. For this case, the snapshots are obtained for the parameters shown in Table 3. The predictions are given for the untrained case of dimensions, which are $500 \times 150 \times 24$ µm. Figure 5 illustrates the simulation results for the cross-coupling effect due to thermal load application for FEM and MOR.

Table 3. Parameter training data set for cross-coupling effect investigation.

Scale	Dimensions of One Actuator $L \times W \times T$	d_1	d_2	d_3
mm	$5 \times 1.5 \times 0.34$	1	0.5	0.1
µm	$5 \times 1.5 \times 0.34$	1	0.5	0.1
nm	$500 \times 150 \times 3.4$	100	50	10

Figure 6 compares the FEM and MOR results for cross coupling sensitivities. The MOR results generated for this graph are for 52 modes of displacement. The number of temperature modes is taken to be equal to the number of displacement modes. These graphs are obtained for the point at the right tip of the first actuator and the point at the base of second actuator, as illustrated in Figure 5 with green circles. For this analysis, actuating the left actuator has a minor influence on the neighboring actuator, i.e., the sensitivity is not very high. When the temperature reaches 440 K at the left actuator, at a distance of 100 micrometers between the actuators (see Figure 6c), the maximum rise in temperature is 20 K, which is measured at the base of the second actuator, where the maximum temperature is found. This increase in temperature is not enough to initiate the actuation of the second actuator, because the temperature increase in the second actuator has not yet reached T_g of the PMMA and the phase transformation temperature of SMA. The heat is transferred via conduction. The base that connects the two actuators plays an important role as, if we use any material that has a higher conductivity, we may see cross-coupling effects between the actuators.

For the displacement, the error for a small number of modes is large (see Table 4). By enriching the basis with more modes, the error decreases. Table 4 summarizes this finding, including the speedup factor for this example.

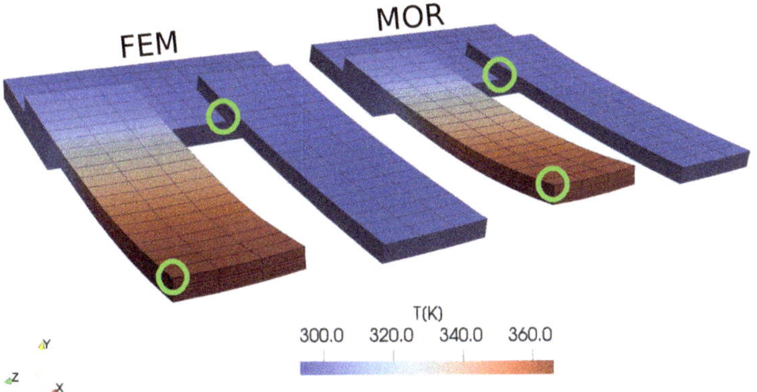

Figure 5. FEM and MOR temperature results for cross-coupling effect investigation for an untrained actuator geometry ($500 \times 150 \times 24\,\mu m$).

Table 4. Speedup factors for the model considered for one parameter set and one heat load cycle at $d = 100\,\mu m$.

CPU Time FEM	Displacement Modes	CPU Time MOR	Speedup	Displacement Error
202 s	10	12.2 s	16.5	6.54%
202 s	52	42 s	4.8	1.29%

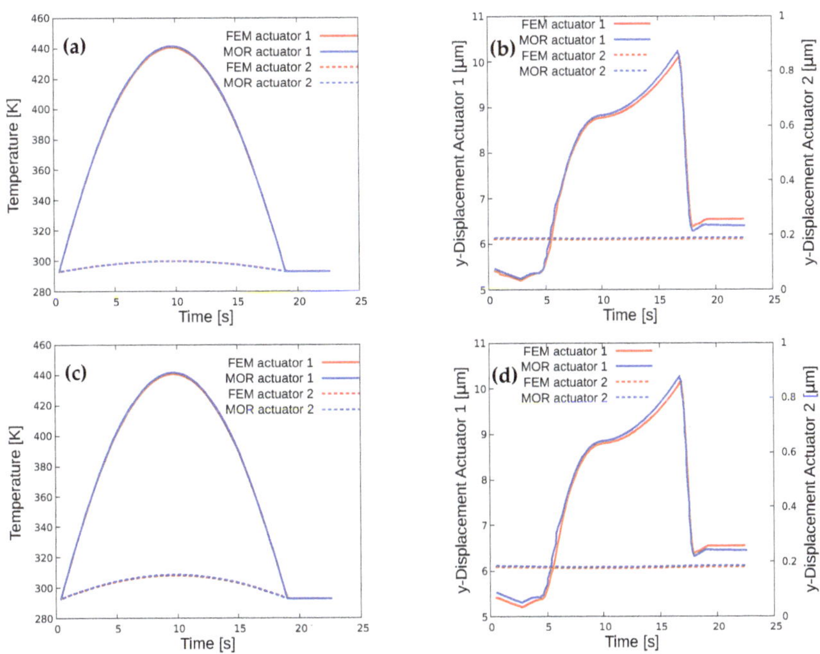

Figure 6. Temperature and displacement graphs for different distances between actuators for the investigation of cross-coupling effects: (**a**) Temperature prediction at $d = 1000\,\mu m$. (**b**) Displacement prediction at $d = 1000\,\mu m$. (**c**) Temperature prediction at $d = 100\,\mu m$. (**d**) Displacement prediction at $d = 100\,\mu m$.

3.3. Bistability

The term bistability describes the actuator's ability to be able to hold two stable states at room temperature, which is useful for energy efficient switching of microdevices [48]. We investigate the trimorph NiTiHf/Mo/PMMA-based microactuator's bistable stroke capabilities using the proposed reduced-order model. The resulting stable states at room temperature are shown in Figure 7. This phenomenon was already discussed in Section 3. Please see, respectively, Sections 4.3 and 3.4 in [21,46] for further details about the bistability phenomenon.

For the bistability performance investigation, we apply two heat load cycles for observing the two stable states of the actuator. We examine the actuator dimensions of $10 \times 5 \times 0.19$ mm^3 with 10, 20 and 160 µm thicknesses for NiTiHf, Mo and PMMA, respectively.

Figure 7 illustrates the bistability and shows the comparison of FEM and MOR results for displacement and temperature. The first heat cycle heats the actuator just above the polymers glass transition temperature T_g and the second cycle just above the austenite finish temperature of the SMA. The two stable positions at room temperature obtained with the SMA-based actuator are depicted in Figure 7. Here, for the displacement and temperature, using 10 POD modes, the FEM and MOR results agree well. The speedup factor acquired for this analysis is 58.29, with 1061 s and 18.2 s for the FEM and MOR computations, respectively.

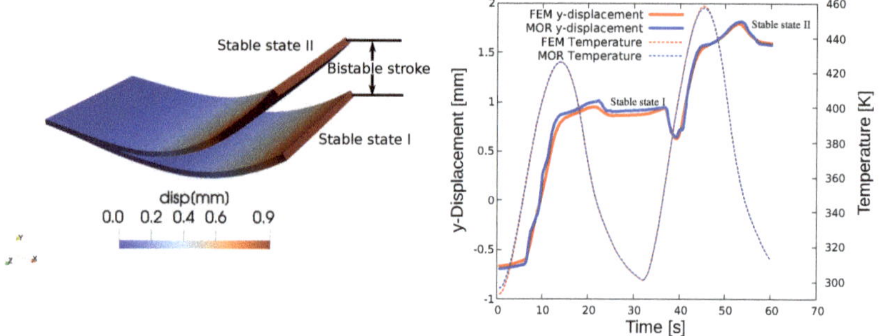

Figure 7. MOR results for displacement magnitude at two different stable states (**left**): Bistable positions for the selected actuator with FEM and MOR comparison (**right**).

4. Conclusions

A reduced-order model for a thermomechanical finite strain shape memory alloy actuator is presented in this paper. Using POD, the reduced basis is obtained. The weak forms of the mechanical and thermal problem are formulated in the reduced-order format. As a major novelty, the reduced-order model is derived from an incremental thermomechanical potential, which ensures a symmetric tangent, and thus allows for efficient solvers and may enable a mathematical model analysis and the employment of enhanced solution methods. The ROM is tested for a single actuator example, cross-coupling effects for two actuators and bistability performance. It was shown from the numerical examples that the model can predict the properties of a parameterized actuator model with controlled accuracy and with less computational effort. The error can be reduced further if the snapshot matrix is enriched with more offline computations.

The ROM developed in this paper still needs further improvement in predicting new parameter values more accurately in less computational time. Therefore, we intend to perform hyper-reduction, in which instead of using each integration point, a reduced set of integration points is evaluated to reduce the computational time.

Author Contributions: Conceptualization, M.B.S. and S.W.; methodology, M.B.S.; software, M.B.S., M.H. and S.W.; validation, M.B.S.; formal analysis, M.H. and S.W.; investigation, M.B.S.; resources, M.B.S. and S.W.; writing—original draft preparation, M.B.S.; writing—review and editing, M.H. and S.W.; visualization, M.B.S.; supervision, S.W.; project administration, S.W.; funding acquisition, S.W. All authors have read and agreed to the published version of the manuscript.

Funding: We gratefully acknowledge the financial support of subproject "A3" Cooperative Actuator Systems for Nanomechanics and Nanophotonics: Coupled Simulation of the Priority Programme SPP 2206 by the German Research Foundation (DFG) (Grant WU847/3-1 and WU847/3-2).

Institutional Review Board Statement: Not applicable.

Informed Consent Statement: Not applicable.

Data Availability Statement: Not applicable.

Conflicts of Interest: The authors of this article declared no conflict of interest with respect to the content, research and publication of this paper.

Appendix A

For a geometry in millimeter scale as in for single actuator example 1, we realized the heating of the actuator through the following sine function:

$$w = 3107 \left| \sin\left(0.1 \left(\frac{t}{s} + 6 \right) \right) - 0.6 \right| \frac{\Delta t}{T_n} \qquad \mathrm{mWmm}^{-3}, \tag{A1}$$

where t is the simulation time in seconds, Δt is the time step size and T_n is the temperature at t_n. For a geometry in micrometer scale as in example 2, we realized the heating of the actuator through the following sine function:

$$w = 273460 \left| \sin\left(0.1 \left(\frac{t}{s} + 6 \right) \right) - 0.6 \right| \frac{\Delta t}{T_n} \qquad \mathrm{mWmm}^{-3}, \tag{A2}$$

In example 3, two heat loading cycles are applied, one just above the T_g and the second cycle above the austenite finish temperature to visualize the bistability phenomenon

$$w_u = 807 \left| \sin\left(0.1 \left(\frac{t}{s} + 6 \right) \right) - 0.6 \right| \frac{\Delta t}{T_n} \qquad \mathrm{mWmm}^{-3}, \tag{A3}$$

$$w_l = 1491 \left| \sin\left(0.1 \left(\frac{t}{s} + 6 \right) \right) - 0.6 \right| \frac{\Delta t}{T_n} \qquad \mathrm{mWmm}^{-3}, \tag{A4}$$

where w_u and w_l are the upper and lower heat load cycles.

References

1. Chaudhari, R.; Vora, J.J.; Parikh, D.M. A review on applications of nitinol shape memory alloy. *Recent Adv. Mech. Infrastruct.* **2021**, 123–132. [CrossRef]
2. Jani, J.M.; Leary, M.; Subic, A.; Gibson, M.A. A review of shape memory alloy research, applications and opportunities. *Mater. Des.-(1980–2015)* **2014**, *56*, 1078–1113. [CrossRef]
3. Serry, M.Y.; Moussa, W.A.; Raboud, D.W. Finite-element modeling of shape memory alloy components in smart structures, part II: Application on shape-memory-alloy-embedded smart composite for self-damage control. In Proceedings of the International Conference on MEMS, NANO and Smart Systems, Banff, AB, Canada, 23–23 July 2003; pp. 423–429.
4. Shibly, H.; Söffker, D. Mathematical models of shape memory alloy behavior for online and fast prediction of the hysteretic behavior. *Nonlinear Dyn.* **2010**, *62*, 53–66. [CrossRef]
5. Liang, C.; Rogers, C.A. One-dimensional thermomechanical constitutive relations for shape memory materials. *J. Intell. Mater. Syst. Struct.* **1997**, *8*, 285–302. [CrossRef]
6. Oka, S.; Saito, S.; Onodera, R. Mathematical Model of Shape Memory Alloy Actuator for Resistance Control System. In Proceedings of the 2021 International Conference on Advanced Mechatronic Systems (ICAMechS), Tokyo, Japan, 9–12 December 2021; pp. 7–11.
7. Huang, W. On the selection of shape memory alloys for actuators. *Mater. Des.* **2002**, *23*, 11–19. [CrossRef]
8. AbuZaiter, A.; Nafea, M.; Mohd Faudzi, A.A.; Kazi, S.; Mohamed Ali, M.S. Thermomechanical behavior of bulk NiTi shape-memory-alloy microactuators based on bimorph actuation. *Microsyst. Technol.* **2016**, *22*, 2125–2131. [CrossRef]

9. Terriault, P.; Brailovski, V. Modeling of shape memory alloy actuators using Likhachev's formulation. *J. Intell. Mater. Syst. Struct.* **2011**, *22*, 353–368. [CrossRef]

10. Lagoudas, D.C.; Miller, D.A.; Rong, L.; Kumar, P.K. Thermomechanical fatigue of shape memory alloys. Smart Materials and Structures. 2009, 18, 085021. [CrossRef]

11. Song, S.H.; Lee, J.Y.; Rodrigue, H.; Choi, I.S.; Kang, Y.J.; Ahn, S.H. 35 Hz shape memory alloy actuator with bending-twisting mode. *Sci. Rep.* **2016**, *6*, 1–3. [CrossRef] [PubMed]

12. Stachiv, I.; Gan, L. Hybrid shape memory alloy-based nanomechanical resonators for ultrathin film elastic properties determination and heavy mass spectrometry. *Materials* **2019**, *12*, 3593. [CrossRef] [PubMed]

13. Samal, S.; Kosjakova, O.; Vokoun, D.; Stachiv, I. Shape Memory Behaviour of PMMA-Coated NiTi Alloy under Thermal Cycle. *Polymers* **2022**, *14*, 2932. [CrossRef] [PubMed]

14. Winzek, B.; Schmitz, S.; Rumpf, H.; Sterzl, T.; Hassdorf, R.; Thienhaus, S.; Feydt, J.; Moske, M.; Quandt, E. Recent developments in shape memory thin film technology. *Mater. Sci. Eng. A* **2004**, *378*, 40–46. [CrossRef]

15. Machairas, T.T.; Solomou, A.G.; Karakalas, A.A.; Saravanos, D.A. Effect of shape memory alloy actuator geometric non-linearity and thermomechanical coupling on the response of morphing structures. *J. Intell. Mater. Syst. Struct.* **2019**, *30*, 2166–2185. [CrossRef]

16. Chang, B.C.; Shaw, J.A.; Iadicola, M.A. Thermodynamics of shape memory alloy wire: Modeling, experiments, and application. *Contin. Mech. Thermodyn.* **2006**, *18*, 83–118. [CrossRef]

17. Roh, J.H.; Han, J.H.; Lee, I. Nonlinear finite element simulation of shape adaptive structures with SMA strip actuator. *J. Intell. Mater. Syst. Struct.* **2006**, *17*, 1007–1022. [CrossRef]

18. Popov, P.; Lagoudas, D.C. A 3-D constitutive model for shape memory alloys incorporating pseudoelasticity and detwinning of self-accommodated martensite. *Int. J. Plast.* **2007**, *23*, 1679–1720. [CrossRef]

19. Yang, Q.; Stainier, L.; Ortiz, M. A variational formulation of the coupled thermo-mechanical boundary-value problem for general dissipative solids. *J. Mech. Phys. Solids* **2006**, *54*, 401–424. [CrossRef]

20. Saleeb, A.F.; Dhakal, B.; Hosseini, M.S.; Padula, S.A., II. Large scale simulation of NiTi helical spring actuators under repeated thermomechanical cycles. *Smart Mater. Struct.* **2013**, *22*, 094006. [CrossRef]

21. Sielenkämper, M.; Wulfinghoff, S. A thermomechanical finite strain shape memory alloy model and its application to bistable actuators. *Acta Mech.* **2022**, *233*, 3059–3094. [CrossRef]

22. Sedlak, P.; Frost, M.; Benešová, B.; Zineb, T.B.; Šittner, P. Thermomechanical model for NiTi-based shape memory alloys including R-phase and material anisotropy under multi-axial loadings. *Int. J. Plast.* **2012**, *39*, 132–151. [CrossRef]

23. Potapov, P.L.; Shelyakov, A.V.; Gulyaev, A.A.; Svistunov, E.L.; Matveeva, N.M.; Hodgson, D. Effect of Hf on the structure of Ni-Ti martensitic alloys. *Mater. Lett.* **1997**, *32*, 247–250. [CrossRef]

24. Frost, M.; Benešová, B.; Seiner, H.; Kružík, M.; Šittner, P.; Sedlák, P. Thermomechanical model for NiTi-based shape memory alloys covering macroscopic localization of martensitic transformation. *Int. J. Solids Struct.* **2021**, *221*, 117–129. [CrossRef]

25. Solomou, A.G.; Machairas, T.T.; Saravanos, D.A. A coupled thermomechanical beam finite element for the simulation of shape memory alloy actuators. *J. Intell. Mater. Syst. Struct.* **2014**, *25*, 890–907. [CrossRef]

26. Shah, N.V.; Girfoglio, M.; Quintela, P.; Rozza, G.; Lengomin, A.; Ballarin, F.; Barral, P. Finite element based Model Order Reduction for parametrized one-way coupled steady state linear thermo-mechanical problems. *Finite Elem. Anal. Des.* **2022**, *212*, 103837. [CrossRef]

27. Chemisky, Y.; Duval, A.; Patoor, E.; Zineb, T.B. Constitutive model for shape memory alloys including phase transformation, martensitic reorientation and twins accommodation. *Mech. Mater.* **2011**, *43*, 361–376. [CrossRef]

28. Merzouki, T.; Duval, A.; Zineb, T.B. Finite element analysis of a shape memory alloy actuator for a micropump. *Simul. Model. Pract. Theory* **2012**, *27*, 112–126. [CrossRef]

29. Hickey, D.; Hoffait, S.; Rothkegel, J.; Kerschen, G.; Brüls, O. Model Order Reduction Techniques for Thermomechanical Systems with Nonlinear Radiative Heat Transfer Using Proper Order Decomposition. Available online: https://www.semanticscholar.org/paper/Model-order-reduction-techniques-for-systems-with-Hickey-Hoffait/09ff384c345b140035ac20271f4ec9fbf07ac503 (accessed on 10 October 2022).

30. Hickey, D.; Masset, L.; Kerschen, G.; Brüls, O. Proper orthogonal decomposition for nonlinear radiative heat transfer problems. In Proceedings of the International Design Engineering Technical Conferences and Computers and Information in Engineering Conference, Washington, DC, USA, 28–31 August 2011; Volume 54785, pp. 407–418.

31. Binion, D.; Chen, X. A Krylov enhanced proper orthogonal decomposition method for efficient nonlinear model reduction. *Finite Elem. Anal. Des.* **2011**, *47*, 728–738. [CrossRef]

32. Choi, Y.; Carlberg, K. Space-time least-squares Petrov–Galerkin projection for nonlinear model reduction. *SIAM J. Sci. Comput.* **2019**, *41*, A26–A58. [CrossRef]

33. Guo, M.; Hesthaven, J.S. Reduced order modeling for nonlinear structural analysis using Gaussian process regression. *Comput. Methods Appl. Mech. Eng.* **2018**, *341*, 807–826. [CrossRef]

34. Lin, W.Z.; Lee, K.H.; Lim, S.P.; Liang, Y.C. Proper orthogonal decomposition and component mode synthesis in macromodel generation for the dynamic simulation of a complex MEMS device. *J. Micromech. Microeng.* **2003**, *13*, 646. [CrossRef]

35. Kerschen, G.; Golinval, J.C.; Vakakis, A.F.; Bergman, L.A. The method of proper orthogonal decomposition for dynamical characterization and order reduction of mechanical systems: An overview. *Nonlinear Dyn.* **2005**, *41*, 147–169. [CrossRef]

36. Friderikos, O.; Olive, M.; Baranger, E.; Sagris, D.; David, C. A Space-Time POD Basis Interpolation on Grassmann Manifolds for Parametric Simulations of Rigid-Viscoplastic FEM. *MATEC WEB Conf.* **2020**, *318*, 01043. [CrossRef]
37. Vettermann, J.; Steinert, A.; Brecher, C.; Benner, P.; Saak, J. Compact thermo-mechanical models for the fast simulation of machine tools with nonlinear component behavior. *at-Automatisierungstechnik* **2022**, *70*, 692–704. [CrossRef]
38. Umunnakwe, C.B.; Zawra, I.; Yuan, C.; Rudnyi, E.B.; Hohlfeld, D.; Niessner, M.; Bechtold, T. Model Order Reduction of a Thermo-Mechanical Packaged Chip Model for automotive MOSFET applications. In Proceedings of the 2022 23rd International Conference on Thermal, Mechanical and Multi-Physics Simulation and Experiments in Microelectronics and Microsystems (EuroSimE), St Julian, Malta, 25–27 April 2022; pp. 1–5.
39. Hu, J.; Zhang, B.; Gao, X. Reduced order model analysis method via proper orthogonal decomposition for transient heat conduction. *Sci. Sin. Phys. Mech. Astron.* **2015**, *45*, 14602. [CrossRef]
40. Jia, W.; Helenbrook, B.T.; Cheng, M.C. A reduced order thermal model with application to multi-fin field effect transistor structure. In Proceedings of the Fourteenth Intersociety Conference on Thermal and Thermomechanical Phenomena in Electronic Systems (ITherm), Orlando, FL, USA, 27–30 May 2014; pp. 1–8.
41. Bikcora, C.; Weiland, S.; Coene, W.M. Thermal deformation prediction in reticles for extreme ultraviolet lithography based on a measurement-dependent low-order model. *IEEE Trans. Semicond. Manuf.* **2014**, *27*, 104–117. [CrossRef]
42. Hernández-Becerro, P.; Spescha, D.; Wegener, K. Model order reduction of thermo-mechanical models with parametric convective boundary conditions: Focus on machine tools. *Comput. Mech.* **2021**, *67*, 167–184. [CrossRef]
43. Das, A.; Khoury, A.; Divo, E.; Huayamave, V.; Ceballos, A.; Eaglin, R.; Kassab, A.; Payne, A.; Yelundur, V.; Seigneur, H. Real-time thermomechanical modeling of PV cell fabrication via a pod-trained RBF interpolation network. *Comput. Model. Eng. Sci.* **2020**, *122*, 757–777. [CrossRef]
44. Taylor, R.L.; Govindjee, S. *FEAP-A Finite Element Analysis Program, Programmer Manual: v8. 6.* University of California: Berkeley, CA, USA, 2022. Available online: http://projects.ce.berkeley.edu/feap/ (accessed on 25 October 2022).
45. Ahrens, J.; Geveci, B.; Law, C. Paraview: An end-user tool for large data visualization. *Vis. Handb.* **2005**, *717*.
46. Curtis, S.M.; Sielenkämper, M.; Arivanandhan, G.; Dengiz, D.; Li, Z.; Jetter, J.; Hanke, L.; Bumke, L.; Quandt, E.; Wulfinghoff, S.; et al. TiNiHf/SiO$_2$/Si shape memory film composites for bi-directional micro actuation. *Int. J. Smart Nano Mater.* **2022**, *13*, 1–22. [CrossRef]
47. Habineza, D.; Zouari, M.; Le Gorrec, Y.; Rakotondrabe, M. Multivariable compensation of hysteresis, creep, badly damped vibration, and cross couplings in multiaxes piezoelectric actuators. *IEEE Trans. Autom. Sci. Eng.* **2017**, *15*, 1639–1653. [CrossRef]
48. Barth, J.; Krevet, B.; Kohl, M. A bistable shape memory microswitch with high energy density. *Smart Mater. Struct.* **2010**, *19*, 094004. [CrossRef]

Communication

Dielectric Elastomer Cooperative Microactuator Systems—DECMAS

Stefan Seelecke [1,*], Julian Neu [1], Sipontina Croce [2], Jonas Hubertus [3], Günter Schultes [3] and Gianluca Rizzello [2]

[1] Intelligent Material Systems Lab, Department of Systems Engineering, Saarland University, 66123 Saarbrücken, Germany
[2] Adaptive Polymer Systems Lab, Department of Systems Engineering, Saarland University, 66123 Saarbrücken, Germany
[3] Sensors and Thin Films Group, University of Applied Sciences of Saarland, 66117 Saarbrücken, Germany
* Correspondence: stefan.seelecke@imsl.uni-saarland.de

Abstract: This paper presents results of the first phase of "Dielectric Elastomer Cooperative Microactuator Systems" (DECMAS), a project within the German Research Foundation Priority Program 2206, "Cooperative Multistable Multistage Microactuator Systems" (KOMMMA). The goal is the development of a soft cooperative microactuator system combining high flexibility with large-stroke/high-frequency actuation and self-sensing capabilities. The softness is due to a completely polymer-based approach using dielectric elastomer membrane structures and a specific silicone bias system designed to achieve large strokes. The approach thus avoids fluidic or pneumatic components, enabling, e.g., future smart textile applications with cooperative sensing, haptics, and even acoustic features. The paper introduces design concepts and a first soft, single-actuator demonstrator along with experimental characterization, before expanding it to a 3×1 system. This system is used to experimentally study coupling effects, supported by finite element and lumped parameter simulations, which represent the basis for future cooperative control methods. Finally, the paper also introduces a new methodology to fabricate metal-based electrodes of sub-micrometer thickness with high membrane-straining capability and extremely low resistance. These electrodes will enable further miniaturization towards future microscale applications.

Keywords: dielectric elastomers; actuators; micro-system; cooperative

Citation: Seelecke, S.; Neu, J.; Croce, S.; Hubertus, J.; Schultes, G.; Rizzello, G. Dielectric Elastomer Cooperative Microactuator Systems—DECMAS. *Actuators* **2023**, *12*, 141. https://doi.org/10.3390/act12040141

Academic Editor: Kenji Uchino

Received: 16 February 2023
Revised: 23 March 2023
Accepted: 24 March 2023
Published: 27 March 2023

1. Introduction

Cooperative microsystems usually rely on silicon [1], fluidic [2], shape memory alloy [3], or piezoelectric transducers [4] to implement actuation. Despite their ease of miniaturization, those solutions generally result in small strokes, low efficiency, and stiff designs. Dielectric elastomers (DEs), consisting of thin and stretchable capacitors that expand when subjected to high voltage, appear to be a suitable alternative technology for cooperative microactuators [5–7]. They feature properties such as large deformations, inherent flexibility, high scalability, high energy density and efficiency, as well as the ability to work as actuators and capacitive sensors at the same time (self-sensing [8]). While multistability is generally introduced in actuators to save energy, in the case of DEs (which are already low-power actuators due to their capacitive nature), multistability can be used in unconventional ways. For instance, a multistable biasing mechanism allows the magnification of the system stroke by up to an order of magnitude compared to a linear bias spring [9].

2. System Design

In order to achieve large strokes in dielectric elastomer membrane systems, Hodgins et al. [9] demonstrated the usefulness of so-called negative-rate bias springs based on stainless steel (NBS). This property can also be realized using soft silicone dome structures (Figure 1a). Neu et al. [10–12] showed that, depending on the dome geometry, the negative slope in a force–displacement curve can be tuned to match the corresponding dielectric elastomer actuator (DEA) curves (Figure 1b,c). Coupled with a 50 µm silicone film (Wacker ELASTOSIL 2030) of 10 mm diameter, the resulting DEA-NBS system (Figure 1d) could be shown to generate strokes of approximately 2 mm under various inputs while being able to adjust to different curvature configurations, thus demonstrating the flexibility of the approach.

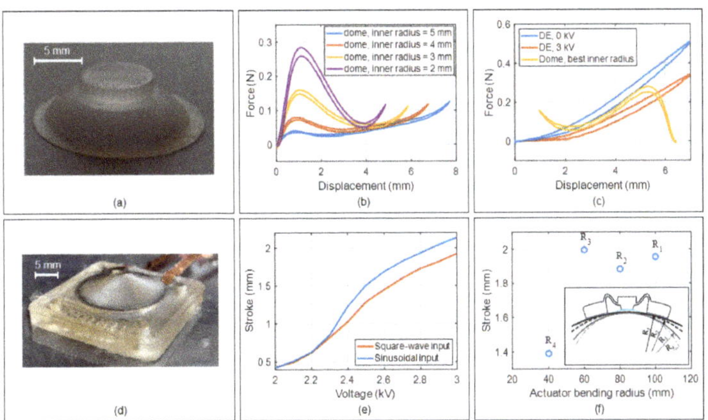

Figure 1. Silicone-based dome as negative-rate bias spring (**a**), load-deformation behavior (**b,c**), DE actuator (**d**), DEA stroke (**e**), stroke for different bending radii (**f**).

The concept was then extended to a multiactuator configuration [13–15]. In order to minimize fabrication and assembly efforts, future systems should ideally consist only of one DE membrane. This, however, introduces potential coupling effects that first need to be investigated for cooperative actuation. Figure 2a shows a 3 × 1 DEA system with carbon black (CB)/polydimethylsiloxane (PDMS) electrodes screen-printed on the same silicon membrane [16], which was subsequently characterized in a specifically designed test rig (Figure 2b). The plots in Figure 2c illustrate that the actuator performance of the center DE2 is not affected by a simultaneous displacement of its neighbors, while DE2 is still able to sense its own displacement as well as those of its neighbors through a capacitance measurement (Figure 2d).

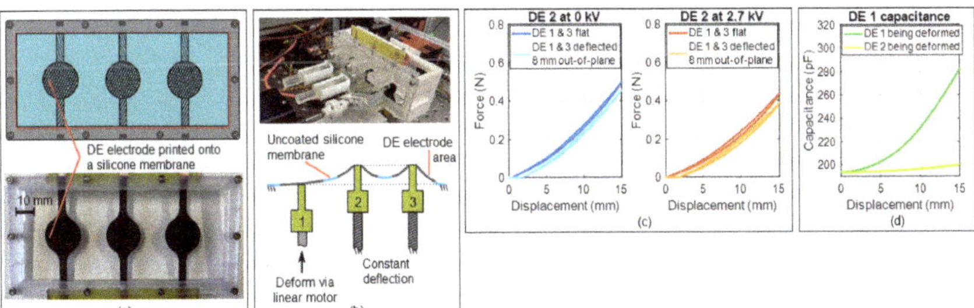

Figure 2. 3 × 1 DEA membrane w/CB/PDMS electrodes (**a**), test rig for coupling characterization (**b**), center DE actuation properties unaffected by neighbor deflection (**c**), center DE sensing properties capable of self-detection and neighbor detection (**d**).

These features demonstrate the system's potential for future cooperative behavior, as illustrated by Figure 3a,b. Figure 3c shows an example of a simple wave propagation experiment.

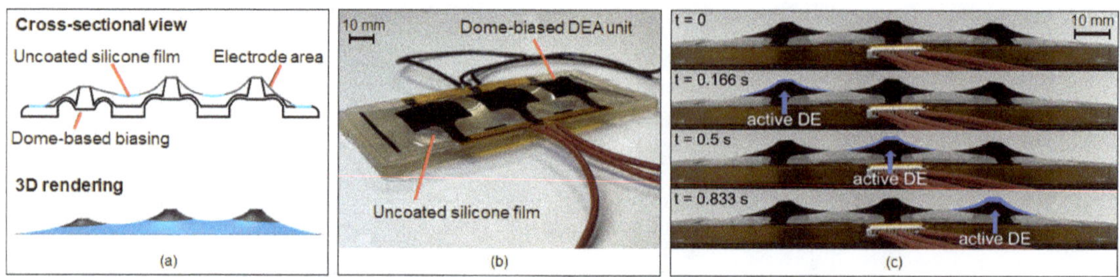

Figure 3. 3 × 1 DEA membrane w/polymeric dome bias, schematic side view (**a**), 3 × 1 DEA membrane w/ polymeric dome bias, full view (**b**), 3 × 1 DEA membrane w/polymeric dome bias, simple wave propagation (**c**).

3. Modeling and Simulation

For a suitable choice of system geometries and materials, it is crucial to rely on simulations of the mechanical and electro-mechanical behavior. Figure 4 shows a finite element simulation performed with COMSOL [17,18], identifying the dome properties for the above silicone bias system.

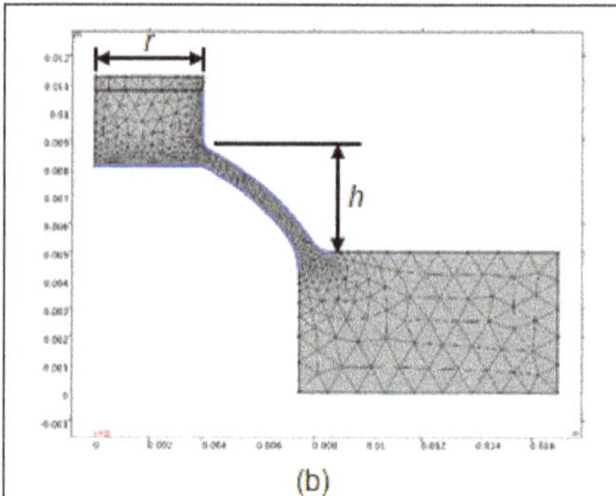

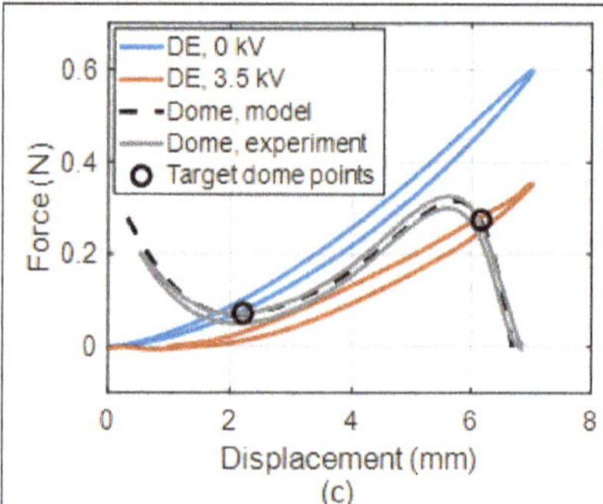

Figure 4. Single element polymeric dome bias, FE mesh (**a,b**), electro-mechanical simulation of single DEA system, experimental validation (**c**).

While Figure 4 illustrates the behavior of an individual dome-based DEA system, the plots in Figure 5 are the results of an electromechanical FE simulation of the 3 × 1 system above [19,20]. The simulation is able to reproduce the coupling behavior from Figure 2, while also giving an estimate of the effect of spacing between individual DEAs.

Based on the above COMSOL approach, Croce et al. also developed a lumped-parameter version of the model [21,22], which will allow for the computationally efficient implementation of future cooperative control strategies (see Figure 6).

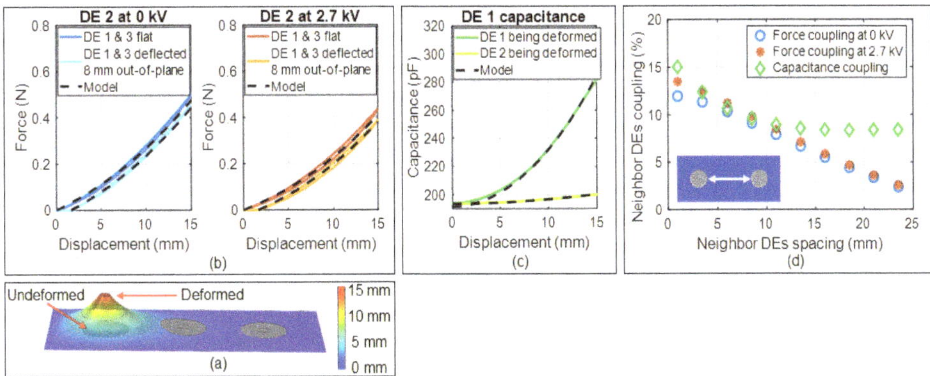

Figure 5. 3 × 1 DEA system, FE mesh (**a**), electro-mechanical simulation of 3 × 1 DEA system, experimental validation actuation (**b**), electro-mechanical simulation of 3 × 1 DEA system, experimental validation sensing (**c**), electro-mechanical coupling effects (**d**).

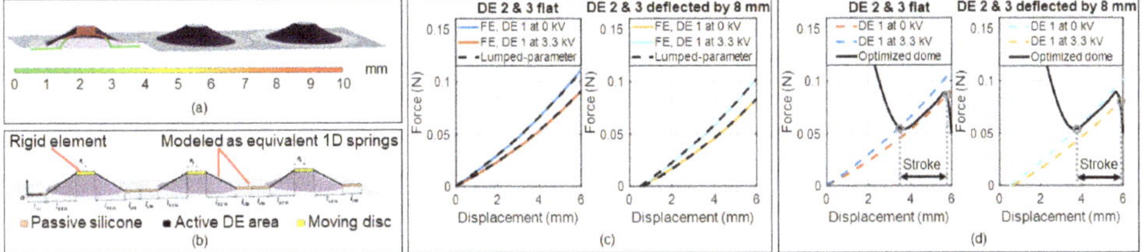

Figure 6. 3 × 1 DEA system, lumped-parameter model for computational efficiency (**a,b**), electro-mechanical simulation of 3 × 1 DEA system, experimental validation actuation (**c,d**).

4. Electrodes

While the CB/PDMS electrodes from above have been successfully used for a number of macro-scale applications, their miniaturization potential is somewhat limited due to the resolution of the underlying screen-printing technology. To address this issue, Hubertus et al. [23,24] developed a technology based on the sputter deposition of Ni-based electrodes. One of the key features of the approach is pre-stretching the PDMS membrane prior to the electrode deposition. Unloading of the membrane after sputtering then leads to a strongly wrinkled surface; see the top left of Figure 7a. Operating the DE afterwards in a strain range below the pre-stretch level unfolds the wrinkles, which, coupled with the extremely low electrode thickness of <20 nm, does not impact the stiffness of the DEA negatively (Figure 7a, top right), hence allowing for efficient actuation.

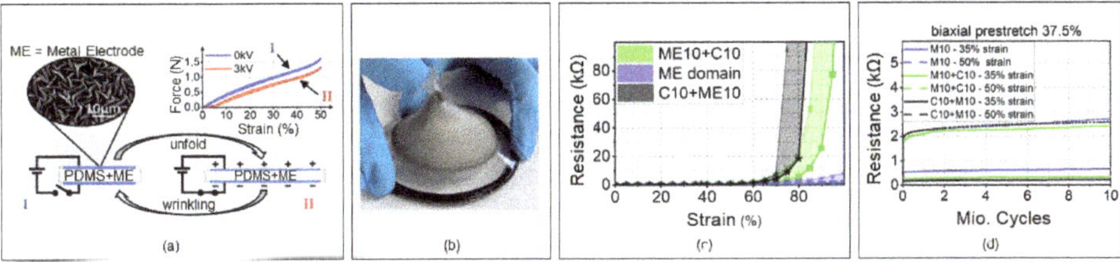

Figure 7. Single element DEA with Ni sputtered nano-scale electrode (**a**), strong adhesion between electrode and silicone membrane (**b**), low resistance over large strain range (**c**), high cycle fatigue with stable resistance behavior (**d**).

Figure 7b displays the strong adhesion between the electrode and the PDMS membrane. Furthermore, despite the low layer thickness, the electrode features a very small electric resistance in the operation range up to the pre-stretch level (Figure 7c), remaining stable over several million cycles (Figure 7d).

Additionally, the authors developed a highly efficient way of structuring the electrodes on the DEA's top and bottom sides (Figure 8a) to obtain desired shapes for a broad range of target applications [25]. Tuning the wavelength and other laser operation parameters, they could utilize metal and polymer absorption behavior (Figure 8b) such that top and bottom electrode layers could be simultaneously removed with a single laser with high precision from one side only (Figure 8c).

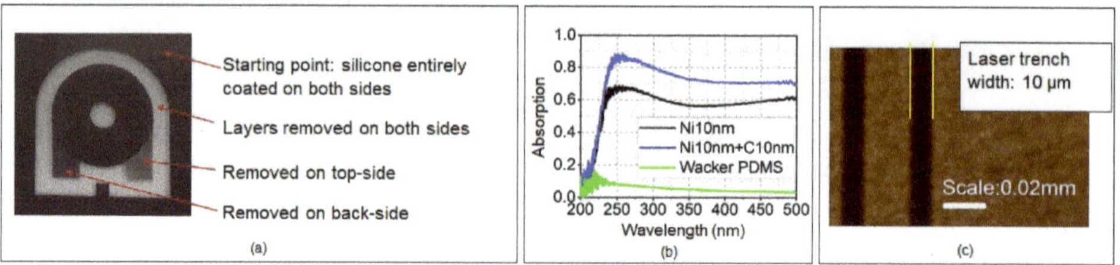

Figure 8. Simultaneous two-side electrode structuring technique (**a**), based on optimal absorption properties (**b**), high resolution suitable for future MEMS applications (**c**).

5. Conclusions and Outlook

The paper presented first steps towards future cooperative dielectric-elastomer-based microactuator systems, detailing design, simulation, and system fabrication including a novel electrode technology. Based on a fully polymeric approach without fluidics or pneumatics, these systems (Figure 9a,b) will be suitable for, e.g., wearable and other mobile applications in the future. They will be able to generate a multitude of complex cooperative motion patterns such as self-organized transport processes for microconveyor systems or wave propagation signals in wearable haptics devices, potentially combining these with hapto-acoustic features [26,27]; see Figure 9c.

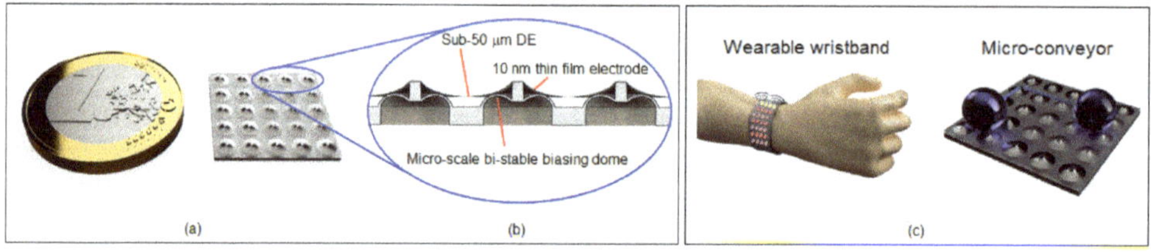

Figure 9. Future fluidics-free and highly flexible cooperative DEA-MEMS system (**a,b**), potential applications in smart textiles or self-organizing microconveyor systems (**c**).

The next steps in Phase 2 of the DECMAS project will address a systematic study on the miniaturization and subsequent fabrication of the developed system with sub-50 µm PDMS membranes and microstructured Ni-sputtered electrodes. These steps will make use of further simulations and will particularly develop novel cooperative model-based control methods. In addition, an extension to a 3×3 design is planned to generate 2D arrays such as the ones proposed in, e.g., [28].

Author Contributions: Conceptualization, S.S., G.R. and G.S.; methodology, S.S., G.R. and G.S.; simulations, S.C.; hardware and experimental validation, J.N.; electrode development, J.H.; writing—original draft preparation, S.S.; writing—review and editing, S.S.; supervision, S.S., G.R. and G.S.; project administration, S.S., G.R. and G.S.; funding acquisition, S.S., G.R. and G.S. All authors have read and agreed to the published version of the manuscript.

Funding: The German Research Foundation (DFG) through the three DECMAS projects (RI3030/2-1, SCHU1609/7-1, SE704/9-1) within the Priority Program SPP 2206—Cooperative Multistage Multistable Microactuator Systems (KOMMMA).

Acknowledgments: We would like to thank Wacker Chemie AG for providing the ELASTOSIL 2030 silicone films used in the projects.

Conflicts of Interest: The authors declare no conflict of interest.

References

1. Bohringer, K.-F.; Donald, B.R.; MacDonald, N.C. Single-crystal silicon actuator arrays for micro manipulation tasks. In Proceedings of the Ninth International Workshop on Micro Electromechanical Systems, San Diego, CA, USA, 11–15 February 1996; pp. 7–12.
2. Vandelli, N.; Wroblewski, D.; Velonis, M.; Bifano, T. Development of a MEMS Microvalve Array for Fluid Flow Control. *J. Microelectromech. Syst.* **1998**, *7*, 395–403. [CrossRef]
3. Mineta, T.; Yanatori, H.; Hiyoshi, K.; Tsuji, K.; Ono, Y.; Abe, K. Tactile display MEMS device with SU8 micro-pin and spring on SMA film actuator array. *IEEE Transducers* **2017**, *2017*, 2031–2034.
4. Hishinuma, Y.; Yang, E.-H. Piezoelectric Unimorph Microactuator Arrays for Single-Crystal Silicon Continuous-Membrane Deformable Mirror. *J. Microelectromech Syst.* **2006**, *15*, 370–379. [CrossRef]
5. Kornbluh, R.D.; Pelrine, R.; Prahlad, H.; Heydt, R. Electroactive polymers: An emerging technology for MEMS. *SPIE Mems/Moems Compon. Appl.* **2004**, *5344*, 13–27.
6. Balakrisnan, B.; Smela, E. Challenges in the microfabrication of dielectric elastomer actuators. *SPIE Eapad* **2010**, *2010*, 76420K.
7. Akbari, S.; Shea, H.R. An array of 100 μm × 100 μm dielectric elastomer actuators with 80% strain for tissue engineering applications. *Sens. Actuator A Phys.* **2012**, *186*, 236–241. [CrossRef]
8. Rizzello, G.; Naso, D.; York, A.; Seelecke, S. A Self-Sensing Approach for Dielectric Elastomer Actuators Based on Online Estimation Algorithms. *IEEE/ASME Trans. Mechatron.* **2017**, *22*, 728–738. [CrossRef]
9. Hodgins, M.; York, A.; Seelecke, S. Experimental comparison of bias elements for out-of-plane DEAP actuator system. *Smart Mater. Struct.* **2013**, *22*, 094016. [CrossRef]
10. Neu, J.; Hubertus, J.; Croce, S.; Schultes, G.; Seelecke, S.; Rizzello, G. Fully Polymeric Domes as High-Stroke Biasing System for Soft Dielectric Elastomer Actuators. *Front. Robot. AI* **2021**, *8*, 171. [CrossRef]
11. Neu, J.; Croce, S.; Hubertus, J.; Rizzello, G.; Schultes, G.; Seelecke, S. Design and characterization of polymeric domes as biasing elements for dielectric elastomer membrane actuators. *Actuator* **2021**, *2021*, 442–445.
12. Neu, J.; Croce, S.; Hubertus, J.; Schultes, G.; Rizzello, G.; Seelecke, S. Assembly and characterization of a DE actuator based on polymeric domes as biasing element. *Proceedings* **2020**, *64*, 24.
13. Neu, J.; Croce, S.; Willian, T.; Hubertus, J.; Schultes, G.; Seelecke, S.; Rizzello, G. Distributed Electro-Mechanical Coupling Effects in Dielectric Elastomer Membrane Arrays: System Design and Experimental Characterization. *Exp. Mech.* **2023**, *63*, 79–95. [CrossRef]
14. Neu, J.; Croce, S.; Hubertus, J.; Schultes, G.; Seelecke, S.; Rizzello, G. Characterization and Modeling of an Array of Dielectric Elastomer Taxels. *SPIE Eapad* **2021**, *XXIII*, 115870R.
15. Neu, J.; Croce, S.; Hubertus, J.; Schultes, G.; Seelecke, S.; Rizzello, G. Experimental characterization of the mechanical coupling in a DE-array. *SPIE Eapad* **2022**, *XXIV*, 120420H.
16. Fasolt, B.; Hodgins, M.; Rizzello, G.; Seelecke, S. Effect of screen printing parameters on sensor and actuator performance of dielectric elastomer (DE) membranes. *Sens. Actuator A Phys.* **2017**, *265*, 10–19. [CrossRef]
17. Croce, S.; Neu, J.; Hubertus, J.; Seelecke, S.; Schultes, G.; Rizzello, G. Model-Based Design Optimization of Soft Polymeric Domes Used as Nonlinear Biasing Systems for Dielectric Elastomer Actuators. *Actuators* **2021**, *10*, 209. [CrossRef]
18. Croce, S.; Neu, J.; Hubertus, J.; Rizzello, G.; Seelecke, S.; Schultes, G. Modeling and simulation of compliant biasing systems for dielectric elastomer membranes based on polymeric domes. *Actuator* **2021**, *2021*, 446–449.
19. Croce, S.; Neu, J.; Moretti, G.; Hubertus, J.; Schultes, G.; Rizzello, G. Finite element modeling and validation of a soft array of spatially coupled dielectric elastomer transducers. *Smart Mater. Struct.* **2022**, *31*, 084001. [CrossRef]
20. Croce, S.; Neu, J.; Hubertus, J.; Schultes, G.; Seelecke, S.; Rizzello, G. Finite Element Modeling and Parameter Study of a Fully-Polymeric Array of Coupled Dielectric Elastomers. *SPIE Eapad* **2022**, *XXIV*, 120420B.
21. Croce, S.; Neu, J.; Hubertus, J.; Seelecke, S.; Schultes, G.; Rizzello, G. Modeling and simulation of an array of Dielectric Elastomeric Actuator Membranes. *Proceedings* **2020**, *64*, 28.
22. Croce, S.; Moretti, G.; Neu, J.; Hubertus, J.; Seelecke, S.; Schultes, G.; Rizzello, G. Finite Element Modeling and Simulation of a Soft Array of Dielectric Elastomer Actuators. *ASME Smasis* **2021**, *85499*, V001T07/A003.

23. Hubertus, J.; Fasolt, B.; Linnebach, P.; Seelecke, S.; Schultes, G. Electromechanical evaluation of sub-micron NiCr-carbon thin films as highly conductive and compliant electrodes for dielectric elastomers. *Sens. Actuators A Phys.* **2020**, *315*, 112243. [CrossRef]

24. Hubertus, J.; Croce, S.; Neu, J.; Seelecke, S.; Rizzello, G.; Schultes, G. Laser Structuring of Thin Metal Films of Compliant Electrodes on Dielectric Elastomers. *Adv. Funct. Mater.* **2023**, 2214176. [CrossRef]

25. Hubertus, J.; Neu, J.; Croce, S.; Rizzello, G.; Seelecke, S.; Schultes, G. Nanoscale nickel-based thin films as highly conductive electrodes for dielectric elastomer applications with extremely high stretchability up to 200%. *ACS Appl. Mater. Interfaces* **2021**, *13*, 39894–39904. [CrossRef]

26. Gratz-Kelly, S.; Rizzello, G.; Fontana, M.; Seelecke, S.; Moretti, G. A Multi-Mode, Multi-Frequency Dielectric Elastomer Actuator. *Adv. Funct. Mater.* **2020**, *32*, 2201889. [CrossRef]

27. Gratz-Kelly, S.; Krüger, T.; Rizzello, G.; Seelecke, S.; Moretti, G. An audio-tactile interface based on dielectric elastomer actuators. *Smart Mater. Struct.* 2023; *accepted*. [CrossRef]

28. Chen, F.; Cao, J.; Zhang, H.; Wang, M.Y.; Zhu, J.; Zhang, Y.F. Programmable Deformations of Networked Inflated Dielectric Elastomer Actuators. *IEEE/ASME Trans. Mechatron.* **2019**, *24*, 45–55. [CrossRef]

Review

A Review of Cooperative Actuator and Sensor Systems Based on Dielectric Elastomer Transducers

Gianluca Rizzello 

Department of Systems Engineering, Saarland University, 66123 Saarbrücken, Germany; gianluca.rizzello@imsl.uni-saarland.de

Abstract: This paper presents an overview of cooperative actuator and sensor systems based on dielectric elastomer (DE) transducers. A DE consists of a flexible capacitor made of a thin layer of soft dielectric material (e.g., acrylic, silicone) surrounded with a compliant electrode, which is able to work as an actuator or as a sensor. Features such as large deformation, high compliance, flexibility, energy efficiency, lightweight, self-sensing, and low cost make DE technology particularly attractive for the realization of mechatronic systems that are capable of performance not achievable with alternative technologies. If several DEs are arranged in an array-like configuration, new concepts of cooperative actuator/sensor systems can be enabled, in which novel applications and features are made possible by the synergistic operations among nearby elements. The goal of this paper is to review recent advances in the area of cooperative DE systems technology. After summarizing the basic operating principle of DE transducers, several applications of cooperative DE actuators and sensors from the recent literature are discussed, ranging from haptic interfaces and bio-inspired robots to micro-scale devices and tactile sensors. Finally, challenges and perspectives for the future development of cooperative DE systems are discussed.

Keywords: dielectric elastomer (DE); dielectric elastomer actuator (DEA); dielectric elastomer sensor (DES); cooperative actuator; micro-actuator; array actuator; soft actuator

Citation: Rizzello, G. A Review of Cooperative Actuator and Sensor Systems Based on Dielectric Elastomer Transducers. *Actuators* **2023**, *12*, 46. https://doi.org/10.3390/act12020046

Academic Editor: Federico Carpi

Received: 23 December 2022
Revised: 10 January 2023
Accepted: 12 January 2023
Published: 18 January 2023

1. Introduction

The development of cooperative systems, in which several entities (or agents) perform a complex task by sharing information with their neighbors and coordinating in a decentralized fashion, represents an attractive and highly challenging goal for researchers working in many areas [1]. Compared to centralized system architectures, the advantages of cooperative solutions include ability to adapt to a great variety of configurations and/or environments, modularity, robustness to failures and disturbances, and a reduced amount of computational effort. The paradigm shift from centralized to cooperative architectures has been made possible by recent technological advances in miniaturized actuators and sensors [2], as well as by the development of reliable and communication paradigms [3] and distributed control algorithms [4]. At the macro-scale, the introduction of cooperative strategies has led to a number of novel concepts and applications, which range from the intelligent management of smart grids [5] and sensor networks [6] to the coordination of teams of autonomous vehicles [7], robots [8], and unmanned aerial vehicles [9], to mention few examples. At the same time, cooperative concepts have also been successfully applied to micro electro-mechanical systems (MEMS), leading to devices in which a complex global task is accomplished via the cooperation of several micro-actuator units. Some relevant examples include arrays of micro-actuators functioning as micro-conveyors [10–13], micro-manipulators [14–16], micro-fluidic systems [17], and reconfigurable structures [18,19].

In macro-scale cooperative systems, the individual behavior of each agent is generally well understood, and most of the technological issues are related to communication and control strategies, as well as energy autonomy. In case of meso- or micro scale cooperative devices, however, integration and miniaturization efforts also represent highly critical technological challenges [20]. Moreover, even though some authors succeeded in implementing advanced control paradigms in cooperative micro-actuators [21–24], most such devices are still controlled via centralized or open-loop methods, possibly due to the challenges in miniaturizing the sensing, communication, and online processing units. As a result, the full exploitation of the benefits of cooperative control in micro-actuator systems is still far from being achieved. Examples of micro-actuator application areas that may benefit from new features introduced by cooperative control methods include haptics [25,26], wearables [27], and reconfigurable displays [28].

A potential means to enhance miniaturization while keeping the desired actuation/sensing and cooperative functionalities consists of adopting highly integrated multifunctional transducers based on smart materials. This term refers to active materials capable of modifying their mechanical properties (e.g., geometry, force, stiffness) when subjected to an external stimulus of electrical, magnetic, thermal, or chemical nature [29–31]. Thanks to the so-called self-sensing operating mode, smart material transducers are able to work as actuators and sensors at the same time, thus allowing to reduce cost and size of the final device [32–34]. Even though most of the state-of-the-art cooperative micro-actuators are driven by conventional technologies (e.g., silicon actuators, microfluidic valves), a number of smart-materials-based prototypes have also been successfully developed. Some notable examples include tactile displays made of shape memory alloy [35], piezoelectric unimorph actuators [36], shape memory polymer flexile tactile displays [37], and artificial skins based of stimuli-responsive hydrogel [38].

Among the many transducers that appear as potential candidates for the design of cooperative micro-actuators, dielectric elastomers (DEs) represent a highly promising alternative [39]. A DE consists of a flexible and highly stretchable capacitor made of polymeric material. Thanks to their large deformability, design scalability, high energy density and efficiency, and dual actuator/sensor behavior, DEs have become a popular technology in various areas of mechatronics, including industrial actuators [40,41], soft robotics [42,43], wearables [44,45], and artificial muscles [46,47]. Despite most of currently developed DE-based devices are stand-alone systems that operate at the macroscopic level [48], these active materials offer many opportunities for the development of innovative small-scale cooperative applications [30,49]. In contrast to other types of cooperative systems based on micro-valves and silicon MEMS technologies, the intrinsic flexibility and large deformability of DE transducers open up new areas of applications, such as intelligent wearables, smart skins, and bio-inspired soft robots.

The goal of this review paper is to provide an overview of recently developed cooperative actuator and sensor systems based on DE technology. Throughout this work, the term cooperative is used to refer to those devices in which many DE units are arranged in an array-like layout, resulting in a fully integrated system where the single elements interact with each other to perform coordinated tasks of various complexity. Potential examples include reconfigurable haptic displays, bio-inspired robots capable of coordinated motion, conveyors of small objects, and wearable sensor surfaces for the detection of local pressure distributions. After describing the basic operating principle of DE transducers and presenting the most common types of DE actuator configurations, a wide variety of cooperative DE systems from the recent literature is presented, focusing on both actuation and sensing applications. Although most of the current cooperative DE systems are developed to operate at the macro-scale, examples of concepts that push the miniaturization limits of the technology will also be highlighted.

The remainder of this paper is organized as follows. Section 2 describes the basic operating principle of DE transducers and presents the most common types of actuator configurations used in cooperative applications. A variety of cooperative DE actuators and sensors from the literature are then reviewed in Section 3. Finally, Section 4 outlines future perspective and possible new research directions for cooperative DE systems.

2. Dielectric Elastomer Transducers

This section summarizes the basic operating principle of DE transducers. After describing the basic actuation and sensing principle of the transducer, the most common types of DE layouts encountered in cooperative applications are presented.

2.1. Dielectric Elastomer Material and Operating Principle

A DE basically consist of a thin (typically 20–100 μm) layer of highly elastic dielectric material coated with compliant electrodes on both surfaces. Common materials used as dielectric are acrylics [50], silicones [51], natural rubbers [52], synthetic rubbers [53], and polyurethane [54], while the compliant electrodes can be manufactured via carbon-based compounds [55] or thin metal films [56]. The first documented investigation of DEs is due to Röntgen, who demonstrated their operating principle back in 1880 [57]. The effect was then rediscovered more than one century after by Pelrine and coworkers [58], causing a renovated interest in the technology by the smart materials research community. Over the last two decades, DEs have generated a large amount of basic and applied research, which led to the publications of the first standards in 2015 [59].

To understand the operating principle of a DE actuator (DEA), we consider the sketch provided in Figure 1a in which an undeformed DE membrane is depicted. When an electric voltage difference is applied between the electrodes, charges of opposite signs are stored onto them. The combination of attractive electrostatic forces among charges of opposite signs (i.e., between the two electrodes), and repulsive electrostatic forces among charges on the same sign (i.e., on the same electrode) cause the DE membrane to reduce in thickness and expand in area, as shown in Figure 1b. This electro-mechanical transduction principle can be quantified via the following equation [58]:

$$\sigma_{Max} = -\epsilon_0\epsilon_r E^2 = -\epsilon_0\epsilon_r \left(\frac{v}{z}\right)^2,$$ (1)

where σ_{Max} is the Maxwell stress, i.e., the compressive stress that arises as a consequence of the applied voltage (cf. Figure 1b), ϵ_0 and ϵ_r are the vacuum and DE relative permittivity, respectively, while E is the electric field in the material and is given by the ratio between applied voltage v and membrane thickness z. In order to practically understand the effect of Maxwell stress in Equation (1) on the overall material response, Figure 1c reports an example of (in-plane) force-displacement curves of a DE membrane for different applied voltages [60]. As it can be seen, the application of the voltage results into a change of the elastic force. This fact plays a key role when designing a DEA, as it will be discussed in Section 2.2. Furthermore, it can be noticed that the DEA curves also exhibit nonlinearities and a rate-dependent hysteresis because of the complex response of the elastomer.

Other than working as an actuator, a DE membrane can also be used in sensing applications. With respect to the DE sensor (DES) layouts shown in Figure 2a,b, we can express the electrical capacitance via the usual parallel plate capacitor equation:

$$C_i = \epsilon_0\epsilon_r \frac{A_i}{z_i}, \ i = 0, 1,$$ (2)

where C_i, A_i, and z_i represent the capacitance, surface area, and thickness of the DES when being undeformed ($i = 0$, Figure 2a) and deformed by an external in-plane force ($i = 1$, Figure 2b), respectively. Since the DE material is incompressible [39], the total volume of elastomer remains constant in both states, i.e., $z_0 A_0 = z_1 A_1$. Using this fact in conjunction with Equation (2) leads to:

$$\frac{C_1}{C_0} = \left(\frac{z_0}{z_1}\right)^2 = \left(\frac{A_1}{A_0}\right)^2,$$ (3)

which implies that we can use the change in capacitance to detect geometric changes in the DES membrane thickness or area. An example of experimental capacitance-displacement curve of a DES is shown in Figure 2c, confirming the monotonic relationship between the two quantities.

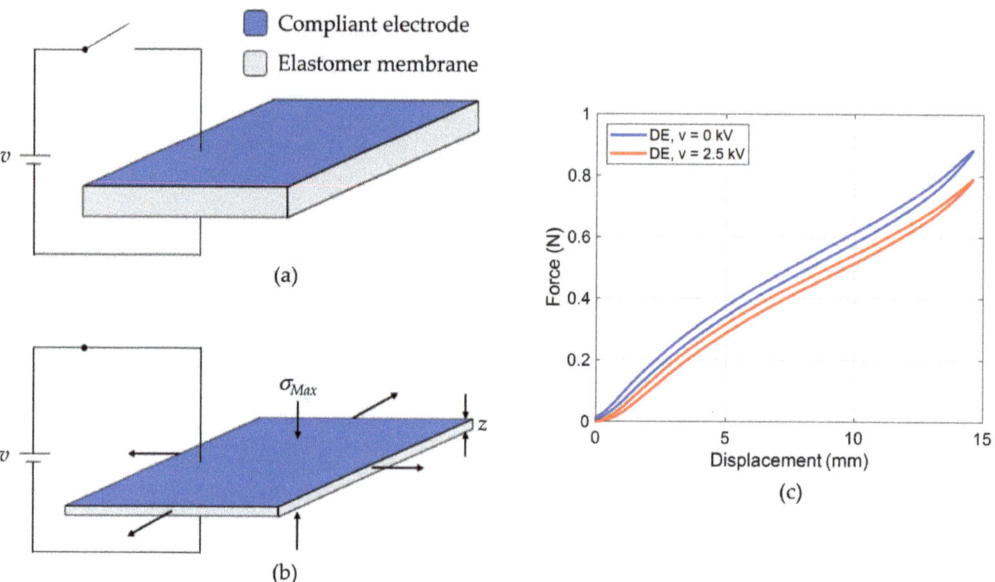

Figure 1. Operating principle of a DEA, deactivated state (**a**), activated state (**b**), and example of force-displacement characteristics for different applied voltages (**c**).

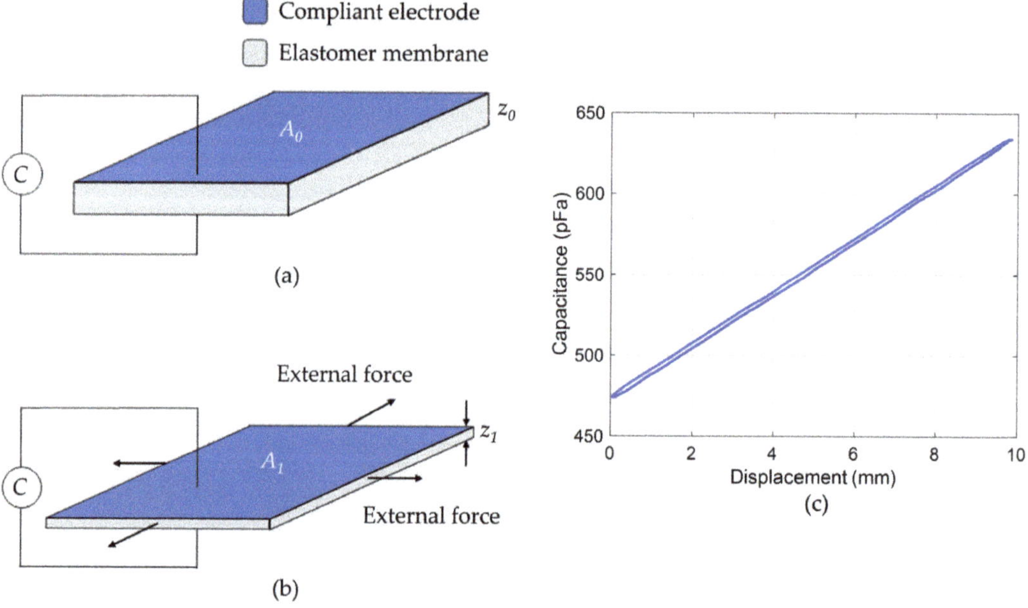

Figure 2. Operating principle of a DES, undeformed state (**a**), deformed state (**b**), and example of capacitance-displacement characteristics (**c**).

In general, the performance of a DE are strongly dependent on the adopted material. For instance, while acrylic elastomers are able to sustain reversible deformations up to 400–500% and exhibit a relatively low elastic modulus (on the order of 20 kPa), they are also subject to high electro-mechanical losses which, in turn, results in inefficient and slow operations [53,61]. On the other hand, silicone elastomers are more rigid (on the order of 100–1000 kPa) and can sustain deformations up to 100%, but are significantly faster (>50 times), exhibit smaller losses, and have a longer (>100 times) cyclic lifetime [62,63]. Another important aspect is represented by the compliant electrodes, which are used to apply a high voltage. Since the elastomer can undergo very large deformations, the electrode must also be capable of high stretchability and compliance, while exhibiting at the same time low mechanical degradation and resistivity [56,64]. Commonly, these features are achieved with carbon-based materials, such as carbon powder [65], carbon grease [66], or carbon black [67], whose manufacturing can be systematically addressed via inkjet printing [68], screen printing [55], or dip-coating processes [69]. While this process usually results in electrodes thickness on the order of 5 μm and sheet resistances of 50 kΩ/$\square$, it has recently been shown that ultrathin (10 nm) sputter-deposited metallic electrodes result in a sheet resistance on the order of 0.5 kΩ/$\square$, and are able to withstand a strain up to 200% while remaining electrically conductive [70]. Nanoscale electrodes, in conjunction with laser-structuring methods [71], open up the possibility of using DE transducers in micro-scale applications.

2.2. Dielectric Elastomer Actuators Configurations

The Maxwell stress described by Equation (1) represents the main physical principle that enables the use of DE in actuator applications. The main advantages of DEAs include large deformations, high compliance and flexibility, low power consumption (on the order of mW), high energy density and efficiency, broad bandwidth (from DC up to several kHz), lightweight, and low cost [39]. The intrinsic multi-functionalities of DE transducers makes it possible to use them as actuators and sensors at the same time, by exploiting the so-called self-sensing mode. In this way, one can reconstruct the DEA displacement [34,72–74] or force [75] via online processing of electrical quantities only (e.g., electric voltage and current), and eventually use this information to implement a sensorless control system [76–78]. Major drawbacks of DEA technology include the need to generate a high voltage on the order of several kV (for membrane thickness within the range 20–100 μm) to produce a meaningful actuation. This, in turn, results into high electric fields which are close to the breakdown strength of the material [79,80], negatively affecting the lifetime of the device. In addition, the strong nonlinearities of the material, in combination with the large deformations, makes their accurate modeling and control a highly challenging task [81,82].

The high flexibility of DEAs makes it possible to generate a variety of layouts and actuation modes [83]. It can be noted that the Maxwell stress principle shown in Figure 1 can be exploited to generate two different actuation modes, i.e., thickness contraction and membrane expansion. Those two modes are exploited to develop different types of DEAs.

The thickness reduction mode can be naturally used to develop contractile actuators. Since the absolute thickness reduction of a single DE layer is on the order of few micrometers, one generally achieves a macroscopic stroke by stacking together several layers, which are then activated simultaneously with the same high voltage signal [84,85]. A sketch showing the layout and principle of such actuators is reported in Figure 3a. Stack DEAs allow to simply tune the actuation stroke and force by changing the number of layers and the electrode surface areas, thus they appear as highly scalable. Other advantages of this topology include high compactness and energy density. In general, stack DEAs are preferred when one needs to generate higher forces rather than large stroke.

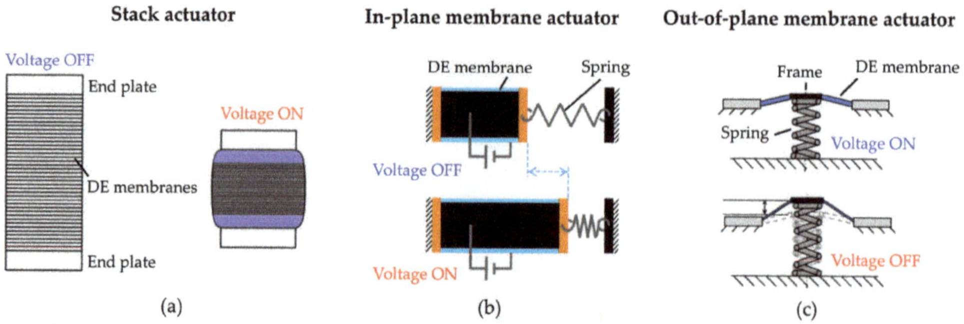

Figure 3. Different types of DEAs: stack actuator (**a**), in-plane membrane actuator (**b**), out-of-plane membrane actuator (**c**).

At the same time, the electrode area expansion mode can be also exploited as a means for an actuation. In this case, differently from the stack actuator case, the motion is an elongation rather than a contraction. Membrane DEAs can be further divided into two sub-categories, i.e., in-plane actuators and out-of-plane actuators. On the one hand, when an in-plane membrane DEA is electrically activated, its motion is always restricted to a two-dimensional plane, see Figure 3b. Examples of in-plane membrane DEAs include strips [86,87], uniaxial rolls [88,89], and lozenge-shaped [90]. On the other hand, the motion of out-of-plane membrane DEAs is not constrained to a two-dimensional plane, thanks to the adoption of design solutions that generate external forces directed orthogonally to the elastomer, see the example in Figure 3c. Examples in this category include cones [91–93], double-cones [91,94,95], and bending structures [96,97]. A wide variety of actuation patterns can be realized through membrane actuators, depending on the type of geometry and pre-loading condition to which the membrane is subjected. Membrane actuators are preferred when one needs to generate a larger stroke, possibly in combination with a complex motion pattern, even though stacking concepts can be used also in this case to multiply the amount of force [98].

In order to generate a meaningful actuation stroke, a membrane DEA is usually combined with a mechanical biasing system which provides a pre-load force. It is remarked that the role of such a biasing system is fundamental in determining the performance of the actuator [99], and can lead to an increase in stroke up to one order of magnitude given the same DE membrane layout and applied voltage [86]. To understand why, we notice that at equilibrium the force of the DE membrane and the one of the biasing system must be equal. Graphically, such equilibrium condition can be obtained by plotting the biasing system force-displacement curve on top of the DE membrane characteristics, and checking for the corresponding intersection points. In particular, the intersections between the biasing curve and the low/high voltage DE curve correspond to the equilibrium position when the DE is deactivated/activated; thus, their inspection allows to readily estimate the stroke. Figure 4 provides several examples of commonly adopted biasing systems, together with the corresponding force-displacement curves for prediction of the stroke. As it can be seen, conventional biasing solutions, such as hanging masses [88,100], linear springs [101,102], pressurized air [103,104], or a second DE membrane [105,106] leads to a relatively small actuation stroke. In contrast, the use of negative-stiffness types of springs results in a significantly larger amount of stroke, since the resulting change in slope allows the biasing system curve to fit between the two DEA characteristics (note that the slope of the biasing curve appears as positive even though the stiffness is negative, since we are plotting the bias curve with respect to the DE deformation, not the biasing one). In the literature, this type of negative-stiffness biasing has been achieved in several ways, e.g., via bi-stable beams [91,99], attracting permanent magnets [107], compliant [92] as well as spring-based mechanisms [108], or buckling polymeric domes [109].

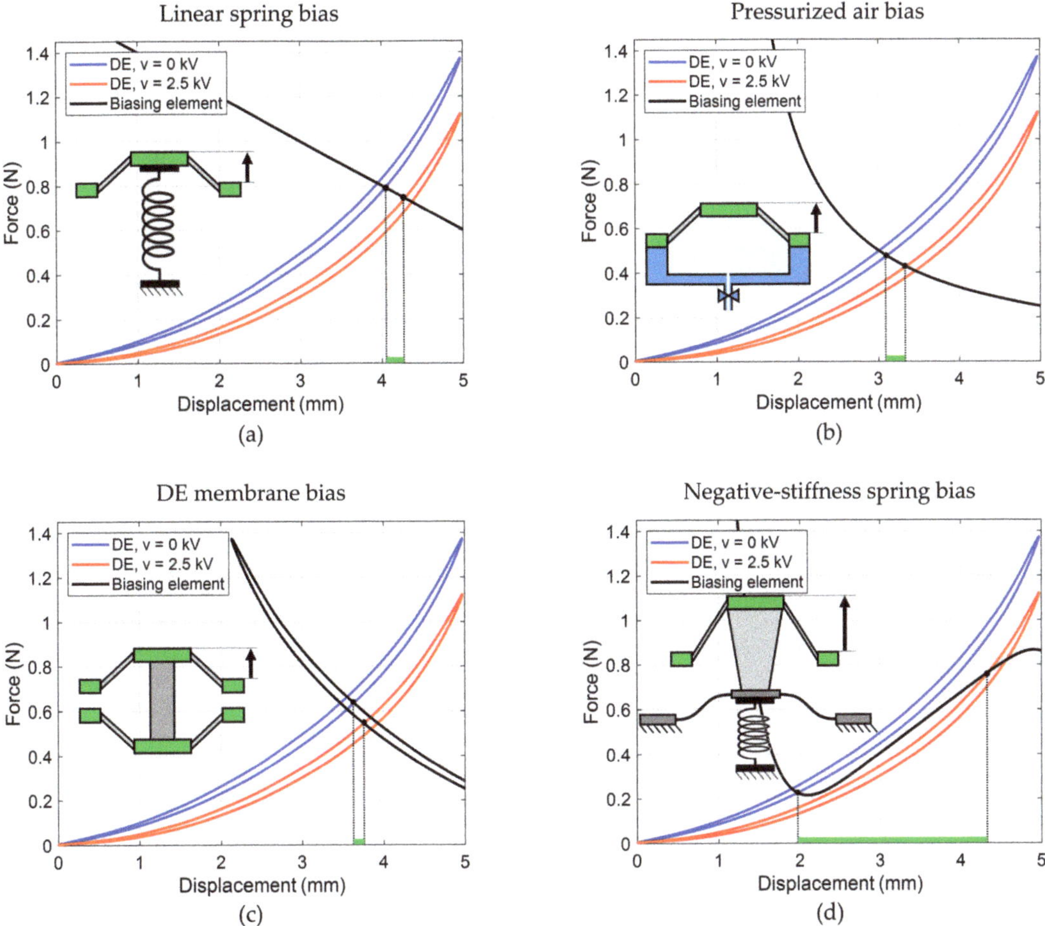

Figure 4. Examples of biasing elements used in membrane DEAs: linear spring (**a**), pressurized air (**b**), a second DE membrane (**c**), and negative-stiffness spring (**d**). In each case, the green segment quantifies the resulting actuation stroke.

3. Dielectric Elastomer Applications in Cooperative Actuator and Sensor Systems

The unique combination of large strain, scalability, high energy density, low power consumption, low cost, and self-sensing has made DE technology highly attractive for the realization of cooperative actuator and sensor systems. A prototypical example of such device is shown in Figure 5. Here, several DE elements are arranged in an array-like layout, and can be operated either as actuators or as sensors (or, eventually, as self-sensing actuators capable of executing both modes at the same time). The synergistic coordination among the various taxels enables complex tasks that are not achievable with individual actuators. Possible examples include wearable haptic interfaces capable of recognizing different touch inputs from the user and sending them wave patterns accordingly, reconfigurable displays, and micro-conveyors. Clearly, a system as the one in Figure 5 requires a synergistic combination of system design, miniaturization, microfabrication, as well as modeling and cooperative control to enable intelligent self-sensing operations.

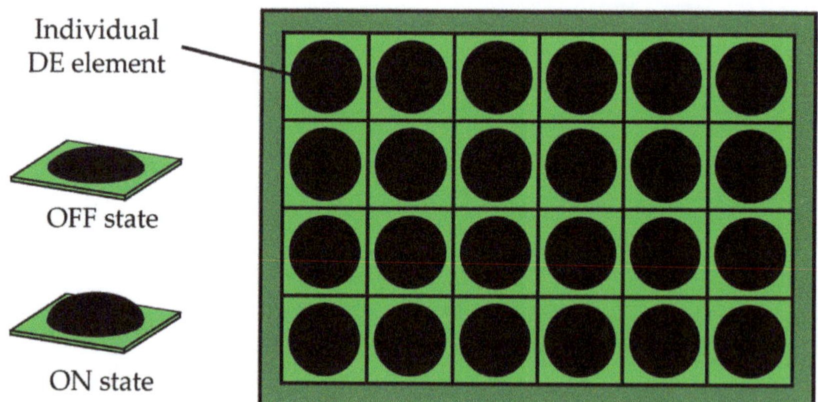

Figure 5. A prototypical example of cooperative DE actuator/sensor system.

3.1. Towards Meso- and Micro-Scale Dielectric Elastomer Actuators

Over the last two decades, progresses in DE micromachining [110–113] have made possible to systematically scale size and accuracy of DEAs to the mm and sub mm scale [114]. As an example, Carpi et al. presented in [115] a bio-inspired lens having a size similar to human eyes (7.6 mm diameter), and capable of actively adjusting its focal length of about 26%, see Figure 6a. The same group proposed in [116] a hydrostatically coupled bubble-like DEA for tactile displays, having a size of about 6 mm capable of producing an out-of-plane displacement of 18% and stresses up to 2.2 kPa under a static actuation at 2.25 kV. A similar concept was presented by Kim et al. in [117], in the context of a button-like haptic actuator application. This completely transparent DEA is about 5 mm large, and exhibits a stroke of ~1 mm when actuated via 3 kV. Hau et al. [118] presented a high-speed micro-positioning stage consisting of 2 DEAs arranged in agonist–antagonist configuration, shown in Figure 6b, which can produce a stroke of 40 μm with a maximum error of 2 μm when operating at a frequency of 60 Hz. A method to fabricate miniaturized DEAs embedded between gold electrodes was proposed by Soulimane et al. in [119], resulting in a controllable actuation of 2 μm. More DEA-based micro-fluidic actuators have also been presented in [120–123].

It has been remarked how one of the main impediments to DEA miniaturization is the high voltage needed to drive the elastomer. Equation (1) implies that, by either increasing the material permittivity or reducing the membrane thickness, the same amount of actuation stress can be obtained with a smaller input voltage [66]. By pursuing the path of thickness reduction, several authors have developed DE micro-actuators capable of operating below the 1 kV threshold. In [124], Ren et al. proposed a high-lift micro aerial robot that weighs about 150 g and achieves a high lift-to-weight ratio of 3.7, see Figure 6c. Actuation is provided by means of 20-layer DEA membrane with each layer being 10 μm thick, which is operated via 630 V at a frequency of 400 Hz. In [125], Rosset et al. developed and characterized a DEA micro-actuator based on a 30 μm thick layer with compliant electrodes based on metal ion implantation, which exhibits a displacement of ~100 μm when operated at ~500 V. A remarkable milestone in DEA miniaturization was achieved by Poulin et al. [126], who developed a pad printing method to produce 3 μm thick DEA, which, in turn, can be operated at 245 V while maintaining an actuation strain on the order of 7.5%. A picture of the resulting actuator is shown in Figure 6d.

All-polymer electrically tunable lens

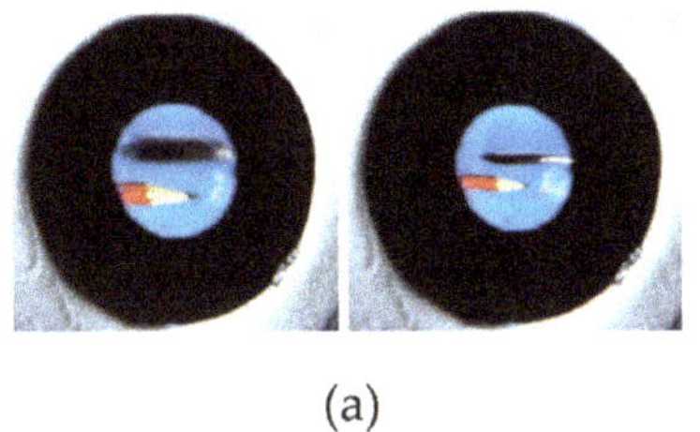

(a)

High-speed micropositioning stage

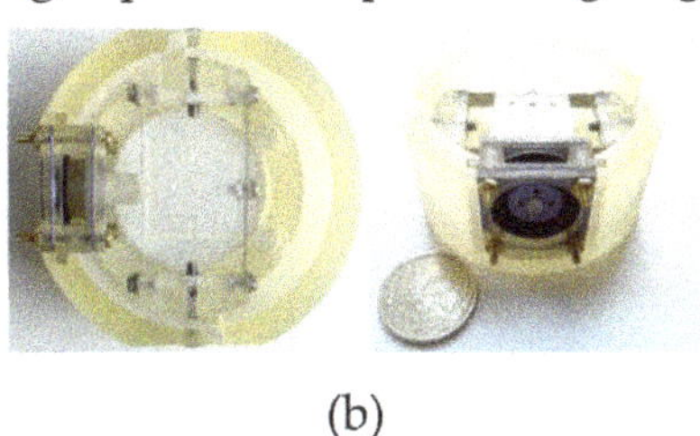

(b)

High-lift micro-aerial robot

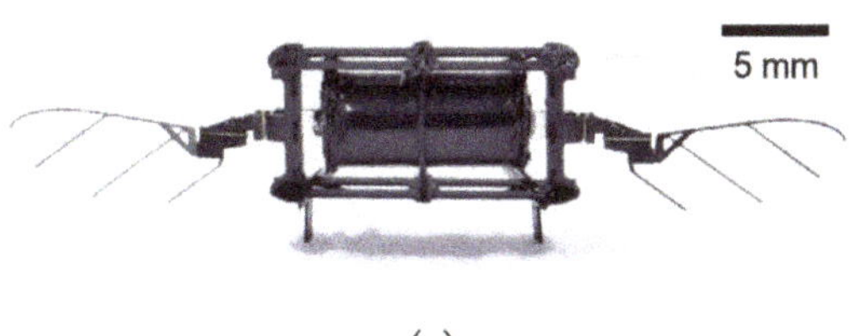

(c)

3 µm thick low-voltage actuator

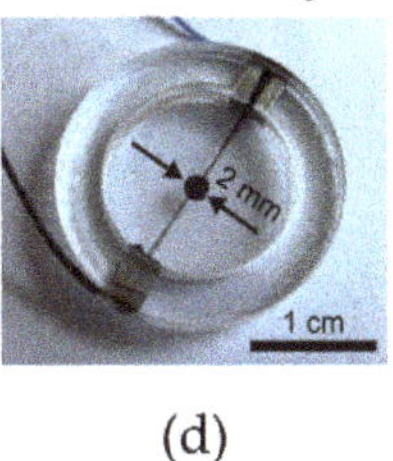

(d)

Figure 6. Examples of meso- and micro-scale DEAs: (**a**) All-polymer electrically tunable lens [115], reproduced with permission from John Wiley and Sons, copyright 2011; (**b**) High-speed micro-positioning stage [118], reproduced with permission from IEEE, copyright 2017; (**c**) High-lift micro-aerial robot [124], reproduced with permission from John Wiley and Sons, copyright 2022; (**d**) 3 µm thick low-voltage actuator [126], reproduced with permission from AIP Publishing, copyright 2015.

All the devices and micro-actuators mentioned above, however, make use of DEAs as stand-alone transducers. The subsequent sections will present cooperative systems in which multiple DEA/DES units are combined together in an array-like fashion.

3.2. Cooperative Dielectric Elastomer Actuators

3.2.1. One-Dimensional Arrays

The simplest types of cooperative DEAs are represented by macro-scale one-dimensional arrays. An example of such a system is the object of investigation of the DECMAS project, which aims at developing fully polymeric intelligent arrays of DEA elements that can be used, e.g., in intelligent wearables and soft robotics. The key elements of the DECMAS project are the multi-stable polymeric domes shown in Figure 7a, which are used as biasing elements for the individual DEA units. In this way, a large stroke can be obtained without the need to use pressurized air or metal beams, making the overall concept significantly easier to miniaturize while keeping it flexible [109]. The potential for miniaturization is also enhanced by the adoption of ultrathin (10 nm) and highly stretchable sputter-deposited metallic electrodes [70]. By means of model-based optimization, it was possible to optimize the design of the biasing domes and, in turn, achieve a fully polymeric DEA capable of a stroke on the order of 3 mm for an actuator with a radius of 20 mm [109], see Figure 7b. Current research studies focus on the characterization [127] and modeling [82] of the electro-mechanical coupling effects occurring when many of those elements are placed onto a common elastic substrate in an array configuration (cf. Figure 7c), as well as on the development of laser-structuring electrode techniques for enabling local activation of the DEAs [71]. A picture of the first prototype of 1-by-3 actuator array is shown in Figure 7d.

Bi-stable polymeric dome

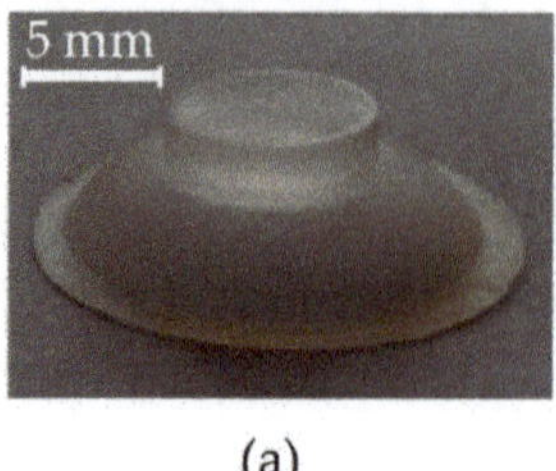

(a)

Fully-polymeric DEA unit

(b)

Membrane with a 1-by-3 array electrode layout

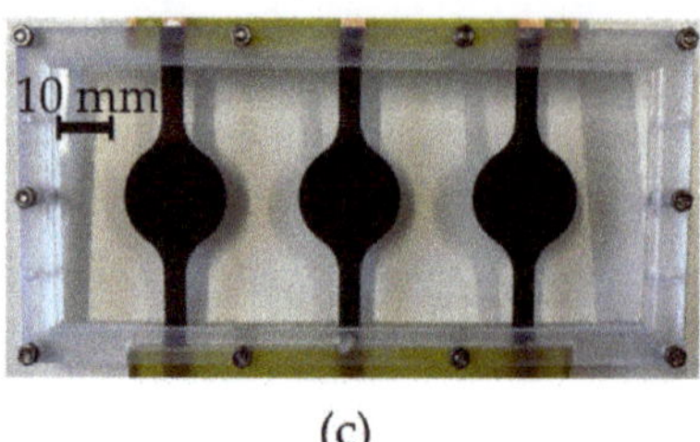

(c)

First prototype of 1-by-3 actuator array

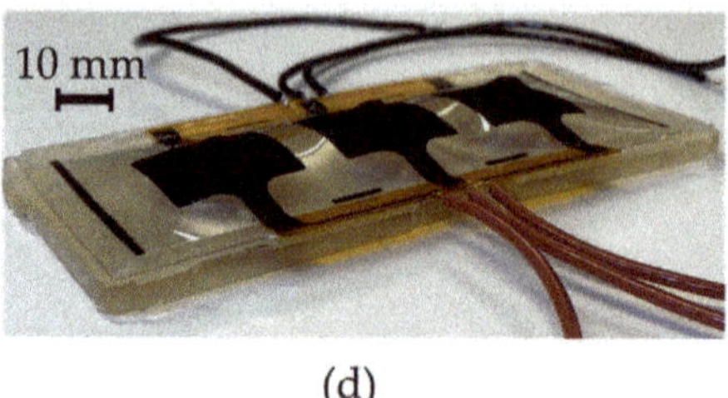

(d)

Figure 7. Fully polymeric DEA array developed in the context of the DECMAS project: (**a**) Bi-stable polymeric dome; (**b**) Fully polymeric DEA unit; (**c**) Silicone membrane with a screen-printed 1-by-3 electrode layout; (**d**) First prototype of 1-by-3 actuator array.

Another example of a 1D array of DEAs was presented by Yu et al. in [128], reported in Figure 8a. Here, a 1-by-4 array of DEAs, each one having a size of 160 mm × 135 mm and characterized by a different pre-stretch, is used to achieve a broadband tunable resonator. In particular, the amount of pre-stretch and applied DC voltage can be used to tune the natural frequency of each DEA and, in turn, the attenuation bandwidth of each resonator. The synergy among the four elements allows the combination of the attenuation bandwidth of each individual DEA, leading to an increase in the effective bandwidth by a factor of 10, with a bandwidth tunability up to 14%. An untethered 1-by-10 array of feel-through haptic elements was presented in [129], based on a 18 µm thick silicone DEA membrane, see Figure 8b. The device can be worn on the user's fingertip, and can generate perceivable vibrations up to 500 Hz when operated below 500 V. Lotz et al. presented in [113] a fabrication technique for silicone-based miniaturized stacked DEAs, and used it to develop a 1-by-8 array operating as a microfluidic pump, where every actuator contains 30 layers of 50 µm each and has a total length of 40 mm. The periodic activation of the DEAs causes the wall of the pump to undergo a periodic motion, which, in turn, propels the fluid at a flowrate of 12 µL/min under a maximum pressure of 0.4 kPa. A similar actuator was then used in [130] to develop a fluidic micro-mixer, shown in Figure 8c. Here, three 1-by-4 arrays of stacked DEAs were used to activate two pumping chambers and one mixing chamber, distributed over a 33 mm × 25 mm layout. In [131], Schlatter et al. presented a multi-material inkjet printing method for integrated and soft multifunctional machines, and used it to develop a 1 DEA flexible peristaltic pump with six integrated actuators capable of pumping up to 13.8 µL/min when operating at 1 Hz and 3.8 kV, see Figure 8d.

1-by-4 tunable resonator

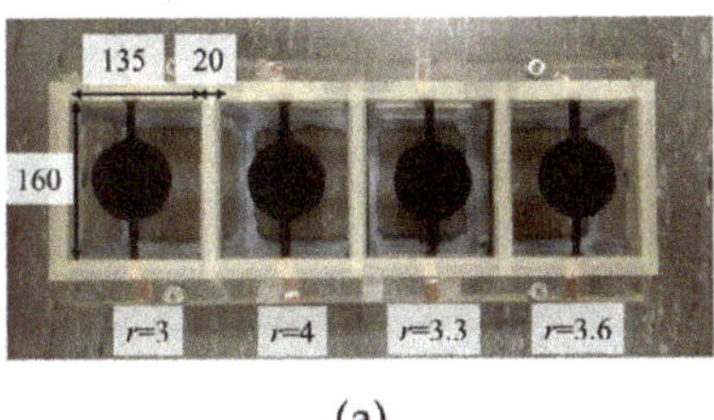

1-by-10 untethered feel-through haptic actuator

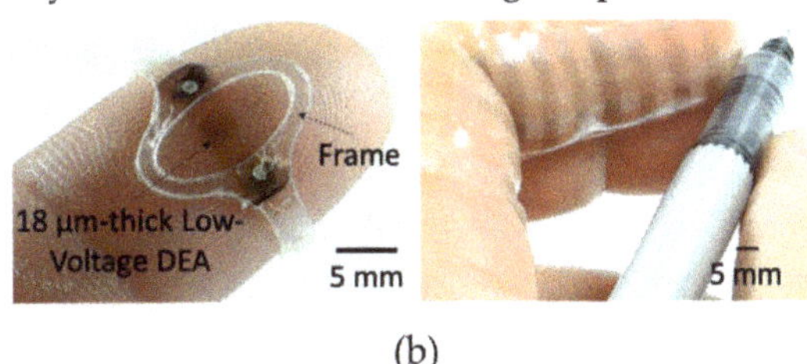

(a)

(b)

Fluidic micro-mixer based on three 1-by-4 arrays

Flexible peristaltic pump with 6 actuators

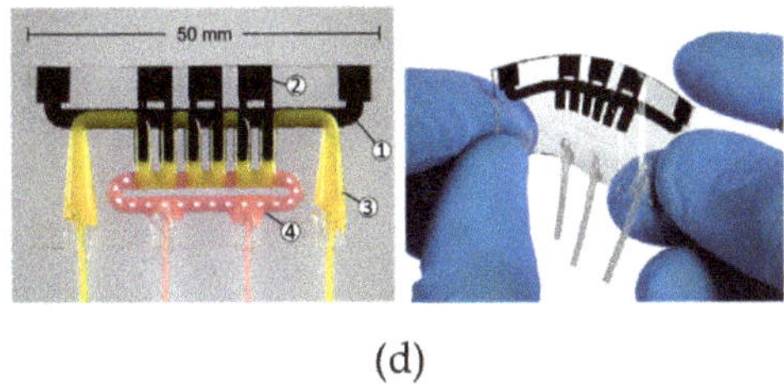

(c)

(d)

Figure 8. Examples 1D DE arrays: (**a**) 1-by-4 tunable resonator [128]; (**b**) 1-by-10 untethered fee-through haptic actuator [129], reproduced with permission from John Wiley and Sons, copyright 2020; (**c**) Fluidic micro-mixer based on three 1-by-4 arrays [130], reproduced with permission from IOP Publishing Ltd., copyright 2018; (**d**) Flexible peristaltic pump with 6 integrated actuators [131].

3.2.2. Cooperative Bio-Inspired Robots

Different types of 1D array concepts have also been proposed for the development of bio-inspired DEA conveyors. O'Brien et al. presented in [132] modeling and experimental validation of a 1-by-4 array of acrylic-based bending DEA, with the aim of replicating the generation of travelling waves generated by ctenophores. By implementing a coordinated mechano-sensitivity concept, in which the motion of one element triggers the actuation of the subsequent one via a capacitance measurement, the authors succeeded in propelling tubes of various materials and sizes, i.e., a 13.9 g and 10.11 mm diameter Teflon tube and an 18.63 g and 4.77 mm diameter brass tube. In [133], a different concept of cilia array conveyor was presented based on a DE matrix and dielectric barium titanate nanoparticles. A transportation experiment showed how the array of cilia can drive a cargo at an average speed of 30 mm/min when operating at a frequency of 2.25 Hz with a voltage of 20 kV. A compliant slug drive capable of generating travelling waves for object micro-transportation by taking inspiration from the motion of invertebrate animals was presented by Schlatter et al. in [131], see Figure 9a. The slug drive has an area of 25 cm^2 and contains 28 integrated DEAs, each one having an area of 0.126 mm^2. The authors showed how the slug drive allows conveying a 15 mm long, 0.26 g plastic object, achieving a travelling speed of 6.1 µm/s when using a 3.5 kV, 0.2 Hz square wave driving signal.

The generation of travelling waves through cooperation of several DEAs has also been used to achieve bio-inspired robotic locomotion. Zhao et al. [134] presented the design of a soft creeping robot consisting of a 1-by-4 array of elliptic frames, which can deform under high voltage application thanks to the use of DEAs. The cooperative activation of each actuation unit allows each one of the four segments to switch between an elliptical and a circular shape, leading to a forward peristaltic motion. Although the results are preliminary, they show the high potential of cooperative DEAs for the achievement of complex actuation patterns. An annelid-like peristaltic crawling robot based on a 1-by-4 array of acrylic DEAs operating in bending mode was proposed by Lu et al. in [135], shown in Figure 9b. The developed robot is 150 mm long, weights 8 g, and achieves a forward or backward motion with a maximum speed of 11.5 mm/s and a maximum speed/mass ratio of 86.23 mm/min·g thanks to the cooperative coordination of all the actuated segments. A multi-segment annular soft robot driven by acrylic DEAs was presented by Li et al. in [136]. The system layout corresponds to a circle with 6 DEA segments distributed along its circumference, as shown in Figure 9c. The resulting actuator weights less than 1 g, has a diameter of ~50 mm, and can achieve a variety of motion patterns (i.e., rolling, creeping, and peristalsis) thanks to the synergistic activation of different combinations of DEA units. Pfeil et al. [137] presented a bio-inspired worm-like crawling robot, consisting of a 1-by-3 serial connection of silicone-based cylindrical DEAs with additional textile reinforcement. Relative elongations of 2.4% and generated forces of 0.29 N were achieved, which, in turn, resulted into a locomotion speed of 28 mm/min. A bio-inspired robot capable of mimicking the crawling motion of a caterpillar was presented in [138]. A 1-by-6 array of DEAs is powered through a DE-based electronic oscillator, which automatically generated the coordinated periodic signals that were needed to set the robot in motion. A speed up to 50 mm/min is achieved with a driving voltage of 4 kV. A different concept of walking hexapod robot is presented by Nguyen et al. in [139], also reported in Figure 9d. Here, a silicone-based double-cone silicone DEA with a patterned electrode is used to generate different motions of the robot legs. In particular, by activating different combinations of electrode segments, translation along both in-plane axes as well as rotation of the robotic leg can be achieved. By properly coordinating and synchronizing those three motion modes on the six legs of the robots, the authors successfully demonstrated the ability of the robot to move at an average speed of 30 mm/s (about 12 body lengths/min).

Slug drive micro-conveyor with 28 actuators

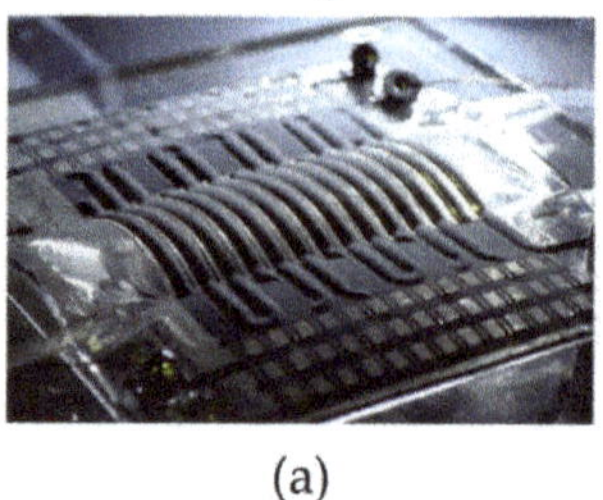

(a)

1-by-4 annelid-like crawling robot

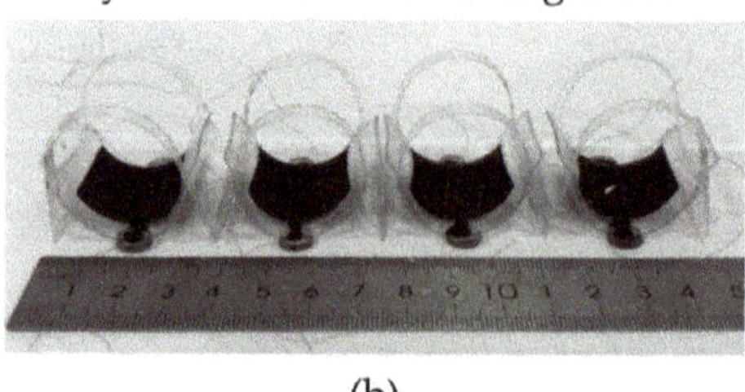

(b)

1-by-6 multi-segment annular robot

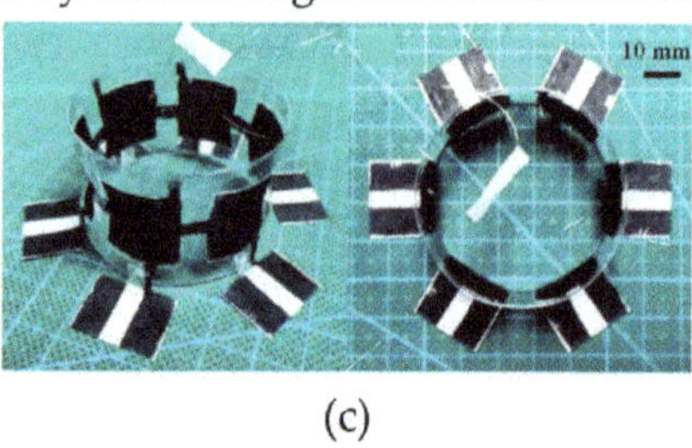

(c)

Multiple-degrees-of-freedom hexapod robot

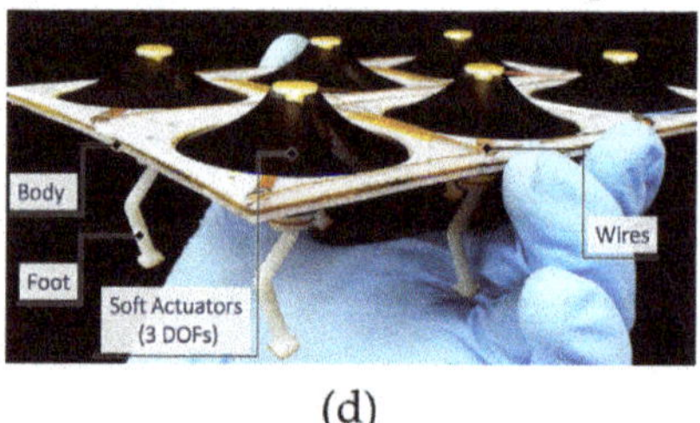

(d)

Figure 9. Examples of DEA bio-inspired robots: (**a**) Slug drive micro-conveyor with 28 integrated actuators [131]; (**b**) 1-by-4 annelid-like peristaltic crawling robot [135], reproduced with permission from IOP Publishing Ltd., copyright 2020; (**c**) 1-by-6 multi-segment annular soft robot [136], reproduced with permission from IOP Publishing Ltd., copyright 2018; (**d**) Multiple-degrees-of-freedom walking hexapod robot [139], reproduced with permission from Elsevier, copyright 2017.

3.2.3. Two-Dimensional Arrays for Out-of-Plane Actuation

Two-dimensional cooperative DEA concepts, capable of operating both out-of-plane and in-plane, have also been explored by several authors. Chen et al. [140] developed a 3-by-3 array of acrylic DEAs. Each actuator operates out of plane, is biased via pressurized air at 2 kPa, and has a circular shape with a radius of 10 mm. A model-based design technique is adopted to suppress the snap-through instability of the device, thus allowing continuous actuation of each element. The authors showed that by modulating the high voltage signals applied to each DEA, the array can be programmed to achieve different desired configurations. An optically addressable and stretchable DEA matrix was presented by Hajiesmaili and Clarke in [141], see Figure 10a. By integrating percolating networks of zinc oxide nanowires into the soft elastomer array, the authors showed the possibility of achieving a localized control of each DEA in a 6-by-6 array. A micro-actuator concept for microfluidic is discussed in [30]. Here, a 7-by-7 array of DEAs operating in enhanced thickness mode is presented, in which localized patterns of out-of-plane bumps can be created upon electrical activation. A 9-by-9 array of DEA-based varifocal micro-lenses was proposed by Wang et al. in [142]. The array, shown in Figure 10b, has a total size of is 40 mm × 40 mm, while the diameter of each actuator element is of 1 mm. The DEA is based on a 1 mm thick acrylic membrane, which is activated at 5 kV. The authors show how the device enables tunability of the focal length from 950 mm to infinity. A concept for enhancing the performance of cooperative DEAs through the use of a fractal interconnection architecture was presented by Burugupally et al. in [143], see Figure 10c. Here, the authors presented bi-dimensional arrays of various sizes in which DEAs are arranged alternatively onto a common substrate. The actuators are based on a 100 μm thick acrylic DE membrane, and undergo a radial expansion when activated. The most complex design comprises 25 actuators having a diameter of 1.8 mm each, arranged alternatingly on a square frame. In-plane displacements on the order of 0.1 mm are achieved for a voltage of about 5 kV.

6-by-6 optically addressable matrix

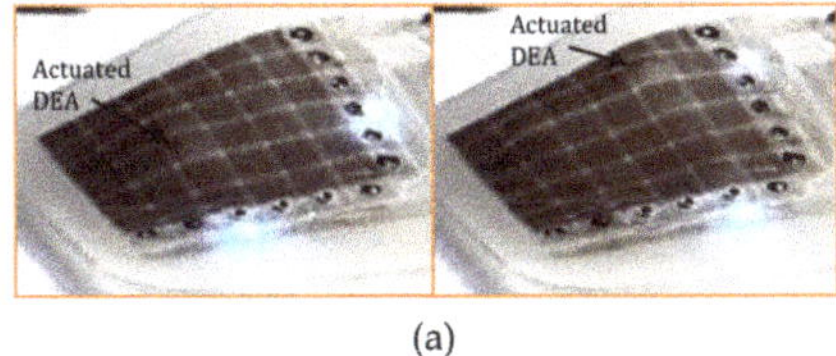

(a)

9-by-9 array of varifocal micro-lenses

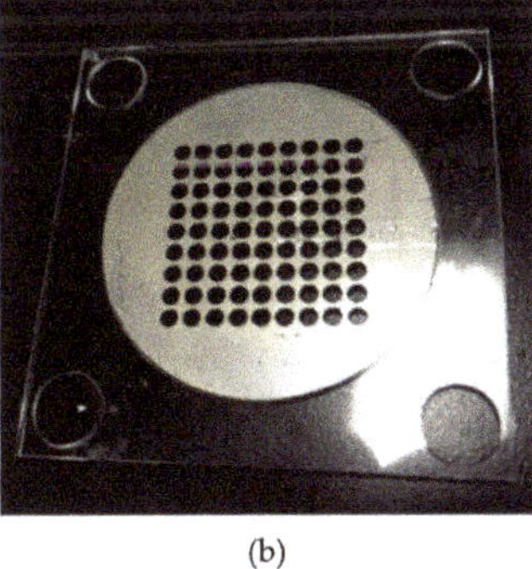

(b)

Array of 25 actuators with fractal interconnects

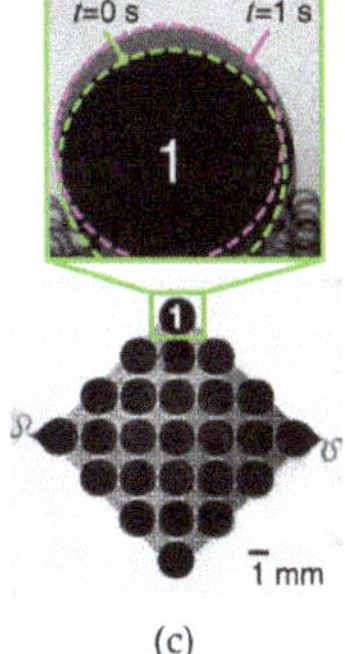

(c)

Figure 10. Examples of two-dimensional DEA arrays operating out of plane: (**a**) 6-by-6 optically addressable matrix [141], reproduced with permission from Royal Society of Chemistry, copyright 2022; (**b**) 9-by-9 array of varifocal micro-lenses [142]; (**c**) Array of distributed electrodes with fractal interconnects architecture [143], reproduced with permission from IOP Publishing Ltd., copyright 2021.

3.2.4. Two-Dimensional Arrays for in-Plane Actuation

While the previously discussed concepts of two-dimensional arrays operate out of plane, several authors also investigated the possibility of generating a cooperative in-plane actuation via DEA systems. A concept of a two-dimensional array of hollow DEA cylinders is presented in [144]. By means of finite element simulations, the authors studied the propagation of an acoustic wave, and confirmed that adjusting the DEA voltage enables an effective tuning of the acoustic band gap of the device. No experimental validation was presented.

A soft-wave handling system capable of transporting fragile and soft objects was proposed by Wang et al. in [145]. The system consists of a 4-by-4 array of hydrostatically coupled acrylic DEA units, each one having a diameter of 55 mm. By concurrent activation of different DEA elements with phase-shifted 1 Hz sinusoidal waves, the authors were able to generate a travelling wave that set a rolling ball in motion. Although the proposed systems exhibit a high potential for future DEA-based conveyor systems, the authors acknowledge the high losses of the adopted acrylic material as one of the limiting factor for the actuator performance.

Akbari and Shea investigated the use of in-plane operating micro-arrays of DEAs to provide in-plane mechanical actuation to cell cultures. The first generation of those devices, shown in Figure 11a, is based on a 30 µm thick silicone membrane consists of an 8-by-9 array of 0.1 mm × 0.2 mm DEA elements capable of a strain of 4.7% at 2.9 kV [146]. A second generation of cell stretcher, always based on the same type of DE membrane, is made of 0.1 mm × 0.1 mm DEA elements and can produce a strain of 37% when driven at 3.6 kV [147]. An expanded view of the micro-actuator grid is reported Figure 11b. Those devices represent some of the most remarkable examples of micro-scale cooperative DEA systems.

First generation cell stretcher
with 0.1 mm × 0.2 mm micro-actuators   Second generation cell stretcher
with 0.1 mm × 0.1 mm micro-actuators

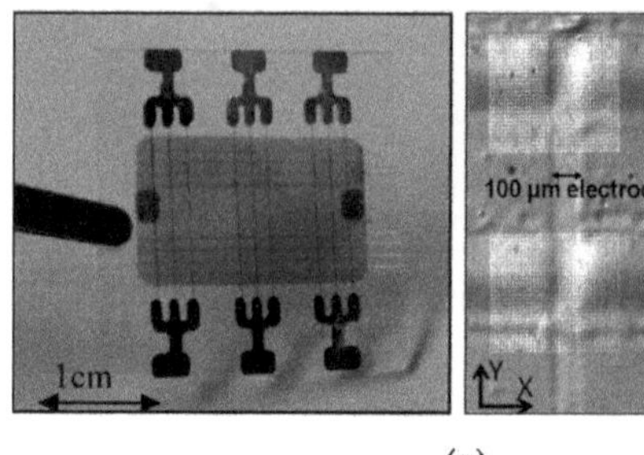

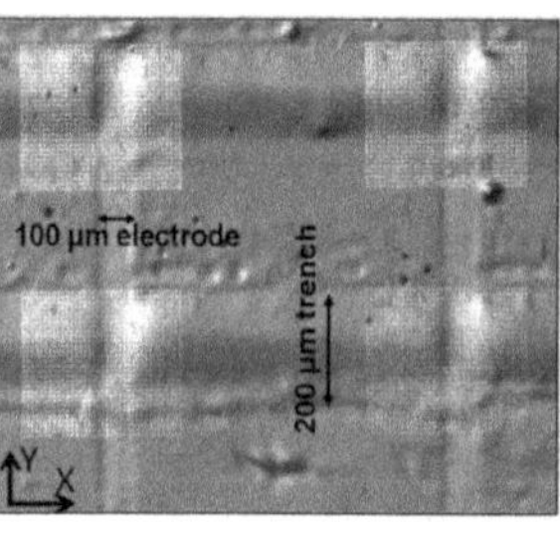

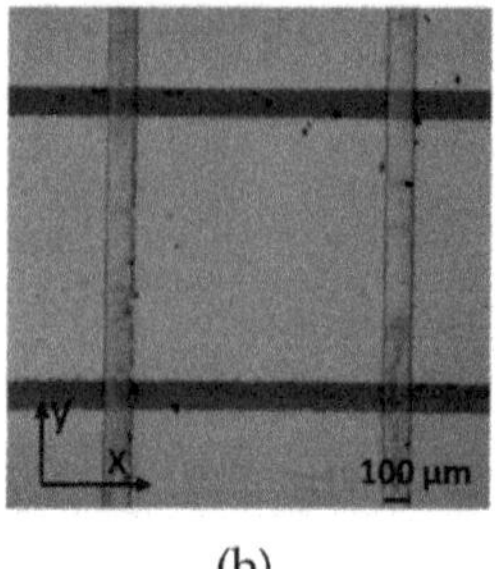

(a)   (b)

Figure 11. Examples of two-dimensional DEA arrays operating in plane: (**a**) First generation cell stretcher based on a 8-by-9 array with of 0.1 mm × 0.2 mm micro-actuators [146], reproduced with permission from IOP Publishing Ltd., copyright 2012; (**b**) Second generation cell stretcher based on 0.1 mm × 0.2 mm micro-actuators [147], reproduced with permission from Elsevier, copyright 2012.

3.2.5. Two-Dimensional Arrays for Haptics and Wearables

Two-dimensional cooperative DEAs have also found natural applications in the field of haptics. Matysek et al. [148] presented a concept for a tactile display consisting of a 3-by-3 array of 1 mm × 1 mm DEA taxels. Each actuator is based on a stack of 25 µm thick silicone elastomer layers. Marette et al. developed in [111] a 4-by-4 array of independently controllable circular DEAs, each one characterized by a diameter of 4 mm and a thickness of 17 µm. Each DEA produces a stroke of 0.25 mm when activated via a 1.4 kV driving voltage, which is switched by using high-voltage metal-oxide thin-film transistors that can be switched with a 30 V gate voltage. The tactile display operates under pneumatic bias at 50 mbar, and can still generate a stroke even when bent to a 5 mm radius of curvature, as shown in Figure 12a. A bidirectional haptic display, which combines a 4-by-4 array of DEAs with a 5-by-5 array of resistive sensors, was proposed by Phung et al. in [149]. The device is 130 mm long × 130 mm wide × 13 mm high. The design is based on rigidly coupled double-cone DEAs made of a 50 μm thick silicone elastomer, operated at a driving voltage of 3.5 kV. The device reaches frequencies up to 300 Hz, providing 0.52 mm of displacement and 0.6 N of normal force. The device can also measure normal forces up to 6 N and the position of touches, and use this information to control the corresponding tactile actuator units. A 2-by-3 tactile display is presented in [150] and also shown in Figure 12b. The device makes use of the liquid coupling between touch spot and DEA to transmit the tactile sensation to the user. The radius of each actuator is 4 mm, while the radius of the touching spot is limited to 1.5 m. A 90 μm thick silicone film is chosen to fabricate the actuators, operated up to a voltage of 7 kV. Displacements of about 240–120 μm at 3–10 Hz are achieved, with forces large enough to simulate the sensation of touch.

4-by-4 low-voltage flexible tactile display   2-by-3 tactile display with liquid coupling

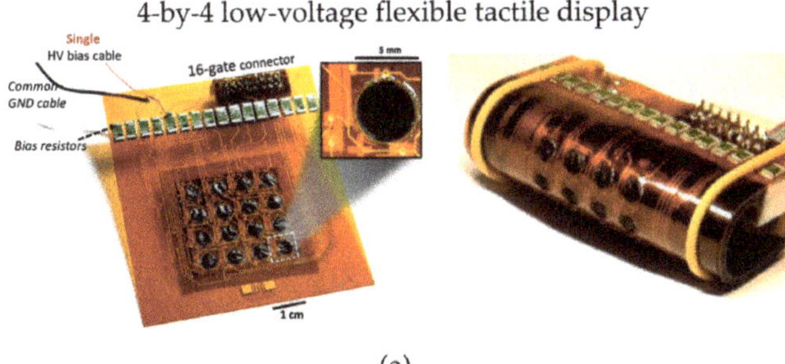

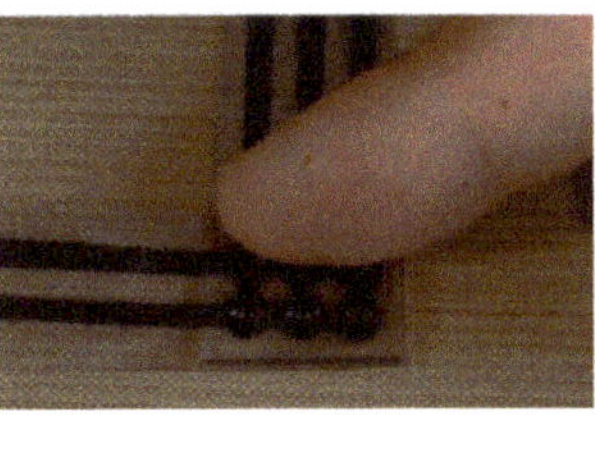

(a)   (b)

Figure 12. Examples of two-dimensional DEA arrays for tactile displays: (**a**) 4-by-4 flexible array of low-voltage actuators [111], reproduced with permission from John Wiley and Sons, copyright 2017; (**b**) 2-by-3 tactile display based on liquid coupling [150], reproduced with permission from Elsevier, copyright 2014.

In the area of wearables, Lee et al. [151] presented a textile-embedded haptic display consisting of a 3-by-5 array of circular DEAs. Each actuator unit has a diameter of 20 mm and is on the skin of the user via a cylindrical polymer indenter. The overall display has a thickness of 7 mm and a weight of 60 g; thus, can be easily worn on the user's forearm. A multi-layered actuator design permits the use of driving voltages of 1 kV only, while ensuring, at the same time, a bandwidth of 240 Hz. A wearable tactile communicator, consisting of a 2-by-2 array of circular DEAs embedded in a wearable armband, is proposed by Zhao et al. in [152]. Each actuator consists of a 10-layer silicone DEA with a hollow cylindrical shape, having an outer radius of 5 mm. A free displacement of ~60 mm and a blocking force of ~30 mN are obtained at a frequency of 300 Hz, considering a driving frequency of 300 Hz.

3.2.6. Two-Dimensional Arrays for Refreshable Braille Displays

Several authors have also investigated the use of DE technology to develop refreshable Braille displays. A 2-by-8 array concept for a refreshable Braille display was discussed in [30]. Each Braille dot has a diameter of 1.5 mm and a relative spacing distance of 2.3 mm. Actuation is provided by spring-biased circular DEAs having a diameter of 2 mm, each one resulting in forces up to 25 g. Chakraborti et al. [153] developed a 2-by-3 Braille matrix of refreshable dots. The actuation is provided by silicone-based thin-walled DEA tubes, having outer and inner diameters of 0.51 mm and 0.94 mm, respectively, and a length of 20 mm. A steady displacement of 1 mm and a response time of 0.1 s are achieved with a driving voltage of 1 kV. Qu et al. [154] also proposed a 2-by-3 refreshable Braille display based on DEAs. The device operates under pneumatic bias, adopts 0.25 mm thick acrylic membranes as elastomer. The total size of the device is of 30 mm × 22 mm × 16 mm, while the diameter of the single Braille dot is of 2 mm. An actuation displacement larger than 0.6 mm is achieved for a biasing pressure of 4 kPa and a driving voltage of 3.25 kV. The ability of this prototype in reproducing different letters is illustrated in Figure 13a. A prototype of a 2-by-4 refreshable Braille display was presented by Frediani et al. in [155], based on acrylic DEAs, see Figure 13b. Each actuator unit has the size of about 1.5 mm, and is capable of out-of-plane displacements of ~ 750 μm. In this case, electrical insulation with the user's finger is achieved via hydrostatic coupling with a passive membrane, which acts as a bias for the DEA.

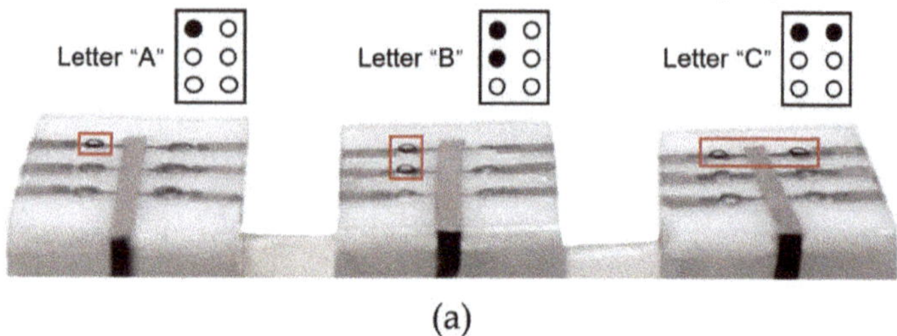

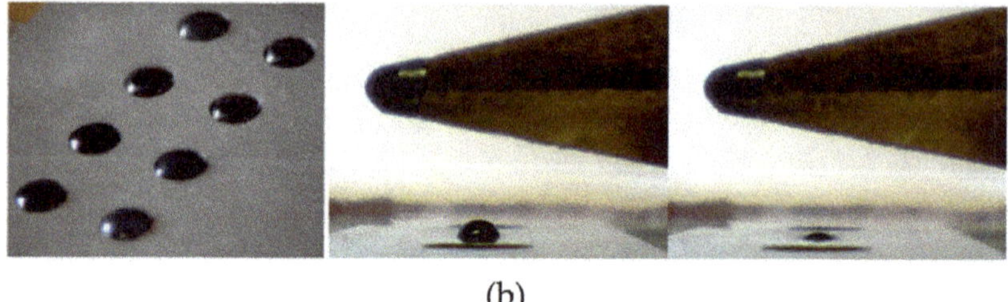

Figure 13. Examples of two-dimensional DEA arrays for refreshable Braille displays: (**a**) 2-by-3 pneumatically biased Braille display [154], reproduced with permission from John Wiley and Sons, copyright 2020; (**b**) 2-by-4 Braille display with hydrostatically coupled actuated dots [155], reproduced with permission from Elsevier, copyright 2018.

3.2.7. Three-Dimensional Reconfigurable Structures

All concepts presented so far permit the achievement of cooperative actuation by means of an array of actuators distributed on a surface, all operating either in plane or out of plane. Few authors have proposed more advanced concepts of cooperative reconfigurable structures based on DE technology. An example of reconfigurable 3D meso-structured driven by DEAs is introduced in [156]. The developed 3D meso-structures can be morphed into several distinct geometries, which are based on 100 μm thick circular DEAs made of acrylic materials. By combining theoretical and experimental studies, the authors are able to develop a large variety (~30) of actuated 3D structures, one of which shown in Figure 14a. Sun et al. presented in [157] an origami-inspired 3D folding actuator based on DEAs. The actuation is based on bending elements driven by circular (~10 mm diameter) DEAs of acrylic material, which are able to generate a bending angle of 120° and 90° when activated with 5.5 kV and 4 kV, respectively. Those two types of bending are then used to develop actuators capable of folding in pyramidal and cubic shapes, respectively, also shown in Figure 14b. The authors demonstrated how the bending actuator can be used to develop gripping and locking functions.

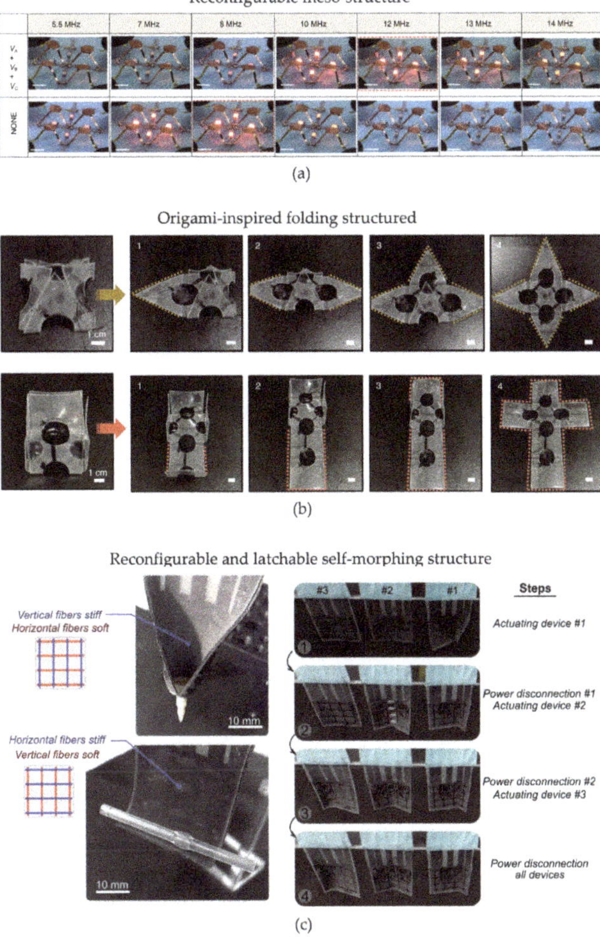

Figure 14. Examples of three-dimensional DEA reconfigurable structures: (**a**) 3D reconfigurable meso-structure [156]; (**b**) 3D origami-inspired folding actuator [157]; (**c**) Reconfigurable and latchable shape-morphing structure [158], reproduced with permission from John Wiley and Sons, copyright 2020.

The idea of combining cooperative DE actuation and bi-stability was also explored in the literature. In [158], Aksoy and Shea proposed a reconfigurable and latchable soft structure that combines layers of DEAs and shapes memory polymers. Joule heating allows reducing the stiffness of the shape memory polymers by two orders of magnitude. Actuation of DEAs allows then the structure to bend along the selected soft axis, thus allowing it to dynamically tune the orientation and location of soft and hard regions. Cooling down the shape memory polymers causes them to become stiff again, thus permitting them to lock in place the shape of the structure without requiring further DEA activation. This mechanism allows the development of several types of complex shapes, as illustrated in Figure 14c, which can be kept in place without requiring additional energy consumption. Tip-bending angles up to 300° and blocking forces over 27 mN can be achieved with a driving force of 5 kV. The authors then demonstrate how this concept allows the development of grippers for grasping objects of various shapes. A similar principle was also used by Meng et al. in [159] to develop a bending structure. Here, a bending angle of 50° is achieved with an activation voltage of 6 kV for the DEA, and a heating up to 50 °C for the shape memory polymer.

3.3. Cooperative Dielectric Elastomer Sensors

Although the main focus of this paper is on actuators, in this section we present a few relevant examples of DE applications in the area of cooperative sensors. In this context, one usually leverages on the large deformations of DE transducers to reconstruct large deformation or pressure patterns localized over a 1D or 2D region. An example of a one-dimensional DES array was presented by Xu et al. in [160], shown in Figure 15a. The device consists of a 1-by-4 array of silicone DES elements. By means of a capacitance reconstruction method built upon a multi-frequency technique, the deformed element within the array is detected using only a single-data acquisition channel.

A 2-by-2 array of silicone-based DESs was proposed by Zhang et al. [161], also reported in Figure 15b. The sensor module allows the detection of the location and magnitude of compressive forces applied on a surface area of 15 mm × 15 mm. By embedding an air chamber in the sensor, the authors achieved an increase in sensitivity by a factor of ~20, with a corresponding a detection range of 382 kPa. A further example of 2-by-2 DES array suitable for tactile sensing applications was proposed by Ham et al. in [162]. The device has a size of 10 mm × 10 mm × 3 mm and can conform to curved surfaces, such as a robotic arm or a prosthetic hand. The sensor array has a dynamic range of 500 kPa in the normal direction, can provide tactile information with a frequency up to 100 Hz, and is robust to electromagnetic interference during contacts thanks to the use of active shielding. A further example soft tactile sensor based on acrylic DE was presented by Kadooka et al. in [163]. The device comprises a DEA module, based on an unimorph-type structure that undergoes a bending upon high voltage application, and three 2-by-2 arrays of DES element. The total size of the device is 15 mm x 35 mm, while the area of each sensor is ~1.2 mm^2, and the footprint of each 2-by-2 element equals ~ 8 mm^2. Dome-shaped bumps positioned over the DES array permit the redistribution of tactile forces, allowing the proper scaling up of the magnitude of the sensed load. The recorded resolution equals 27 mN for normal forces and 67 mN for shear forces, respectively, for a maximum applied force of 5 N.

Zhu et al. [164] proposed an example of a multi-modal sensor, consisting of a 4-by-4 array of silicone DES, see Figure 15c. A multi-layer structure was used, comprising a top protection layer (0.1 mm thick), a first sensor layer (0.6 mm thick) used to measure the applied pressor, a second underlying sensor array (2.4 mm thick) enabling localization features, and a passive substrate (1.9 mm thick). A digital acquisition system was used to identify the unique location of each taxel. A 4-by-4 DES grid was presented by Meyer et al. in [165], shown in Figure 15d. The DES taxels were based on a 50 mm thick silicone membrane on which four rows and four columns of electrode lines were screen-printed to enable the eight required electrical connections. The resulting structure was a fully polymeric and highly stretchable membrane, with 16 taxels each one having a size of 10 mm × 10 mm, for a total array size of 65 mm × 65 mm. Different types of load distributions applied on the grid, corresponding to out-of-plane displacements up to 4 mm, are effectively estimated via a commercially available integrated circuit for capacitance estimation.

A similar sensor array concept was also proposed by Lee et al. [166], this time based on a 5-by-5 grid applied on a silicone DE film with micro-pores, see Figure 15e. The device size equals 33 mm × 33 mm, while the size of the single-sensing element equaled 5 mm × 4.45 mm. The device allows reconstructing localized pressures below 0.02 kPa with a high sensitivity of 1.18 kPa^{-1}, and a fast response time of 150 ms. A further example of a 5-by-5 DEs array was proposed by Kwon et al. in [167]. Due to the presence of 3D micro-pores in the DES layer, the sensitivity of the device can be dramatically increased while maintaining the large deformation features. The sensor is characterized by a sensitivity of 0.6 kPa^{-1} for pressures below 5 kPa, and has a maximum range of 130 kPa, which makes it suitable for tactile sensing applications. The effectiveness of the sensor in detecting spatially distributed pressure patterns shaped as alphabetic letters is also demonstrated. Kyaw et al. [168] proposed a 5-by-5 array for a pressure-sensitive electronic skin, shown in Figure 15f. Each electrode has a size of 3 mm × 3 mm, for a total active sensing area of ~500 mm^2. The device is able to sense pressures up to 40 kPa with a sensitivity of ~0.01 kPa^{-1}. Compared to silicone-based DES, the adopted Ecoflex material is characterized by a higher accuracy due to a higher linearity in its response. A deformable interface for human-touch recognition was developed by Larson et al. [169]. Here, a concept is developed named Orbtouch in which a 5-by-5 soft array of DES is integrated onto a deformable bubble-like structure. The overall circular membrane has a radius of 45 mm and is 2 mm thick, while each taxel has a size of 5 mm × 5 mm, and spacing among neighboring elements is also equal to 5 mm. A convolutional neutral network is used to classify the sensing information from the various channels in the array, allowing, in turn, to localize interactions with the interface with an accuracy of 97.6%. The authors showed the flexibility and effectiveness of the concept by letting users play Tetris with it.

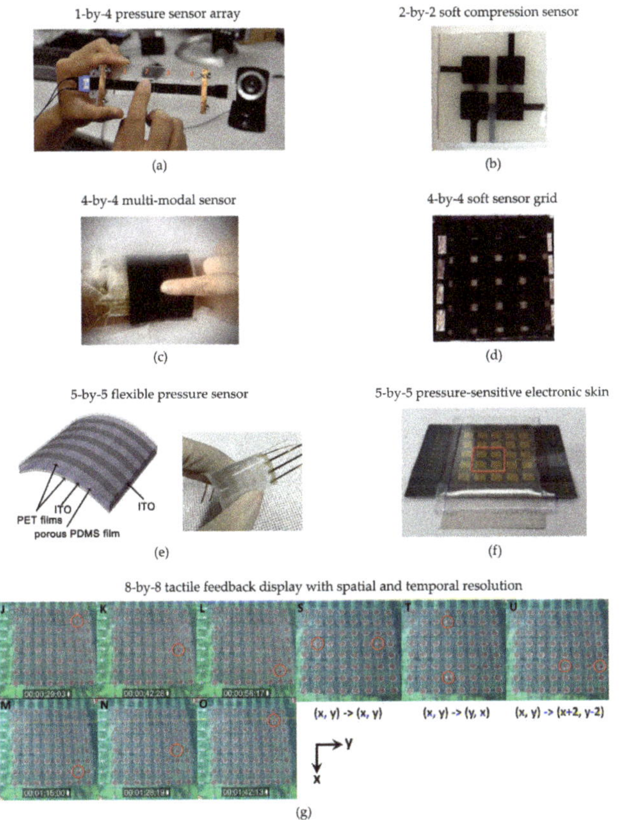

Figure 15. Examples of arrays of DES: (**a**) 1-by-4 pressure sensor array [160], reproduced with permission from John Wiley and Sons, copyright 2015; (**b**) 2-by-2 soft compression sensor [161], reproduced with permission from IOP Publishing Ltd., copyright 2016; (**c**) 4-by-4 multi-modal sensor for simultaneous detection of amplitude and location of touch pressure [164]; (**d**) 4-by-4 soft sensor grid [165], reproduced with permission from SPIE, copyright 2020; (**e**) 5-by-5 flexible pressure sensor [166], reproduced with permission from Elsevier, copyright 2016; (**f**) 5-by-5 pressure-sensitive electronic skin [168], reproduced with permission from Royal Society of Chemistry, copyright 2017; (**g**) 8-by-8 tactile feedback display with spatial and temporal resolution [170], reproduced with permission from Springer Nature, copyright 2013.

As a final example of more complex sensor layout, Vishniakou et al. [170] developed a tactile feedback display with spatial and temporal resolutions based on an 8-by-8 array of circular DES, reported in Figure 15g. The array is based on an acrylic elastomer and is capable of distributed-pressure sensing with a spatial resolution of 3 mm. The same device can also be operated as an actuator, exhibiting a maximum out-of-plane displacement of ~150 µm under an applied voltage of 4 kV and a blocking force on the order of 10 mN. The estimated sensitivity is of about 15 kPa.

4. Conclusions

This paper presented an overview of cooperative actuator and sensor systems based on soft dielectric elastomer technology. DE transducers make it possible to develop cooperative arrays of actuators with improved lightweight, flexibility, controllability, and energy consumption compared to alternative technologies (e.g., pneumatics), thus enabling applications such as wearable soft robots and sensors, haptic interfaces, or reconfigurable structures. The provided overview showed how cooperative DE technology has allowed the development of a variety of systems, which range from coordinated micropumps and bio-inspired crawling robots to bi-dimensional reconfigurable structures, soft conveyors haptic interfaces, refreshable Braille displays, and wearable tactile sensors. The surge of publications in this field over the last few years suggests that the interest in cooperative DE actuator and sensor technology will steadily increase in the near future.

Even though the presented prototypes have succeeded in showcasing the potential of DE technology for a variety of applications, many practical limitations still exist. In general, the corresponding taxel size is on the meso- (order of mm^2) rather than micro-scale, and only few prototypes succeeded in delivering actuation in the sub mm regime, e.g., [146,147]. Therefore, additional effort in both design and manufacturing is still needed to effectively scale the systems at the micro-scale level. Additionally, most current devices are based on a limited number of elements per array, each one producing a relatively small stroke compared to the potential DE large deformation. This is essentially due to limitations in current microfabrication methods for both DE membrane and electrodes, lack of bias systems suitable for miniaturization (e.g., pneumatic), as well as challenges in miniaturizing the control and sensing electronics. In this sense, the multi-stable polymeric domes presented in [109] may offer a great potential means for future miniaturization of cooperative DEAs with large stroke. Finally, it is worth noting that almost the totality of the current devices are controlled via centralized feedforward approaches rather than by means of feedback cooperative strategies, without taking advantage of the DE self-sensing functionalities. This may be partially due to the highly nonlinear and hysteretic behavior of such materials, which complicates the design of control and self-sensing algorithms. In principle, by properly exploiting self-sensing, each individual actuator unit can autonomously reconstruct information on its current electromechanical state based on simplified sensing hardware, i.e., local measurements of voltage and current, and use this information to perform a desired cooperative task in a fully autonomous and fault-tolerant way. Some examples include the generation of a specific shape, the propagation of a dynamic motion pattern (e.g., a wave as in [132]), or the transportation of an object on a path by avoiding damaged taxels. This concept, although highly attractive, is still largely unexploited in current cooperative DE systems. As a result, the full potential and intelligence offered by the technology are yet to be realized. Once those novel design and control concepts are developed, it will be possible to develop the next generation of fully autonomous, intelligent, and fault-tolerant DE cooperative micro-actuator/-sensor systems.

Funding: This research was funded by Deutsche Forschungsgemeinschaft (DFG, German Research Foundation), Priority Program SPP 2206 "Cooperative Multistage Multistable Microactuator Systems" (Projects: RI3030/2-1, SCHU1609/7-1, SE704/9-1).

Data Availability Statement: Not applicable.

Conflicts of Interest: The authors declare no conflict of interest. The funders had no role in the design of the study; in the collection, analyses, or interpretation of data; in the writing of the manuscript; or in the decision to publish the results.

References

1. Butenko, S.; Murphey, R.; Pardalos, P.M. *Cooperative Control: Models, Applications and Algorithms*; Springer: Berlin/Heidelberg, Germany, 2003.
2. Wilson, S.A.; Jourdain, R.P.J.; Zhang, Q.; Dorey, R.A.; Bowen, C.R.; Willander, M.; Wahab, Q.U.; Willander, M.; Al-hilli, S.M.; Nur, O.; et al. New materials for micro-scale sensors and actuators: An engineering review. *Mater. Sci. Eng. R Rep.* **2007**, *56*, 1–129. [CrossRef]
3. Gao, J.; Xiao, Y.; Liu, J.; Liang, W.; Chen, C.L.P. A survey of communication/networking in Smart Grids. *Futur. Gener. Comput. Syst.* **2012**, *28*, 391–404. [CrossRef]
4. Chen, F.; Ren, W. On the Control of Multi-Agent Systems: A Survey. *Found. Trends® Syst. Control* **2019**, *6*, 339–499. [CrossRef]
5. Vaccaro, A.; Velotto, G.; Zobaa, A.F. A Decentralized and Cooperative Architecture for Optimal Voltage Regulation in Smart Grids. *IEEE Trans. Ind. Electron.* **2011**, *58*, 4593–4602. [CrossRef]
6. Patwari, N.; Ash, J.N.; Kyperountas, S.; Hero, A.O.; Moses, R.L.; Correal, N.S. Locating the nodes: Cooperative localization in wireless sensor networks. *IEEE Signal Process. Mag.* **2005**, *22*, 54–69. [CrossRef]
7. Lee, J.; Park, B. Development and Evaluation of a Cooperative Vehicle Intersection Control Algorithm Under the Connected Vehicles Environment. *IEEE Trans. Intell. Transp. Syst.* **2012**, *13*, 81–90. [CrossRef]
8. Khamis, A.; Hussein, A.; Elmogy, A. *Multi-Robot Task Allocation: A Review of the State-of-the-Art*; Springer: Cham, Switzerland, 2015; pp. 31–51.
9. Chandler, P.R.; Swaroop, D.; Howlett, J.K.; Pachter, M.; Fowler, J.M. Complexity in UAV Cooperative Control. In Proceedings of the 2002 American Control Conference, Anchorage, AK, USA, 8–10 May 2002; pp. 1831–1836.
10. Laurent, G.J.; Delettre, A.; Zeggari, R.; Yahiaoui, R.; Manceau, J.-F.; Le Fort-Piat, N. Micropositioning and Fast Transport Using a Contactless Micro-Conveyor. *Micromachines* **2014**, *5*, 66–80. [CrossRef]
11. Petit, L.; Hassine, A.; Terrien, J.; Lamarque, F.; Prelle, C. Development of a Control Module for a Digital Electromagnetic Actuators Array. *IEEE Trans. Ind. Electron.* **2014**, *61*, 4788–4796. [CrossRef]
12. Ataka, M.; Legrand, B.; Buchaillot, L.; Collard, D.; Fujita, H. Design, Fabrication, and Operation of Two-Dimensional Conveyance System with Ciliary Actuator Arrays. *IEEE/ASME Trans. Mechatron.* **2009**, *14*, 119–125. [CrossRef]
13. Konishi, S.; Fujita, H. A conveyance system using air flow based on the concept of distributed micro motion systems. *J. Microelectromech. Syst.* **1994**, *3*, 54–58. [CrossRef]
14. Luntz, J.E.; Messner, W.; Choset, H. Distributed Manipulation Using Discrete Actuator Arrays. *Int. J. Robot. Res.* **2001**, *20*, 553–583. [CrossRef]
15. Yahiaoui, R.; Zeggari, R.; Malapert, J.; Manceau, J.-F. A MEMS-based pneumatic micro-conveyor for planar micromanipulation. *Mechatronics* **2012**, *22*, 515–521. [CrossRef]
16. Berlin, A.; Biegelsen, D.; Cheung, P.; Fromherz, M.; Goldberg, D.; Jackson, W.; Preas, B.; Reich, J.; Swartz, L.-E. Motion Control of Planar Objects Using Large-Area Arrays of Mems-Like Distributed Manipulators. *Micromechatronics* **2000**, 1–5.
17. Vandelli, N.; Wroblewski, D.; Velonis, M.; Bifano, T. Development of a MEMS Microvalve Array for Fluid Flow Control. *J. Microelectromech. Syst.* **1998**, *7*, 395–403. [CrossRef]
18. Tellers, M.C.; Pulskamp, J.S.; Bedair, S.S.; Rudy, R.Q.; Kierzewski, I.M.; Polcawich, R.G.; Bergbreiter, S.E. Piezoelectric actuator array for motion-enabled reconfigurable RF circuits. In Proceedings of the 2015 Transducers-2015 18th International Conference on Solid-State Sensors, Actuators and Microsystems (Transducers), Anchorage, AK, USA, 21–25 June 2015; pp. 819–822. [CrossRef]
19. El-Baz, D.; Piranda, B.; Bourgeois, J. A Distributed Algorithm for a Reconfigurable Modular Surface. In Proceedings of the 2014 IEEE International Parallel & Distributed Processing Symposium Workshops, Phoenix, AZ, USA, 19–23 May 2014; pp. 1591–1598. [CrossRef]
20. Bourgeois, J.; Goldstein, S.C. Distributed Intelligent MEMS: Progresses and Perspectives. *IEEE Syst. J.* **2015**, *9*, 1057–1068. [CrossRef]
21. Boutoustous, K.; Laurent, G.J.; Dedu, E.; Matignon, L.; Bourgeois, J.; Le Fort-Piat, N. Distributed Control Architecture for Smart Surfaces. In Proceedings of the 2010 IEEE/RSJ International Conference on Intelligent Robots and Systems, Taipei, Taiwan, 18–22 October 2010; pp. 2018–2024.
22. Konishi, S.; Fujita, H. System design for cooperative control of a microactuator array. *IEEE Trans. Ind. Electron.* **1995**, *42*, 449–454. [CrossRef]
23. Matignon, L.; Laurent, G.J.; Le Fort-Piat, N.; Chapuis, Y.-A. Designing Decentralized Controllers for Distributed-Air-Jet MEMS-Based Micromanipulators by Reinforcement Learning. *J. Intell. Robot. Syst.* **2010**, *59*, 145–166. [CrossRef]
24. Fukuta, Y.; Chapuis, Y.-A.; Mita, Y.; Fujita, H. Design, Fabrication, and Control of MEMS-Based Actuator Arrays for Air-Flow Distributed Micromanipulation. *J. Microelectromech. Syst.* **2006**, *15*, 912–926. [CrossRef]
25. Amato, M.; De Vittorio, M.; Petroni, S. Advanced MEMS Technologies for Tactile Sensing and Actuation. *MEMS Fundam. Technol. Appl.* **2013**, 351–380.

26. Chouvardas, V.G.; Miliou, A.N.; Hatalis, M.K. Tactile Display Applications: A State of the Art Survey. In Proceedings of the 2nd Balkan Conference in Informatics, Ohrid, Macedon, 26–28 September 2005; pp. 290–303.

27. Velázquez, R. Wearable Assistive Devices for the Blind. In *Wearable and Autonomous Biomedical Devices and Systems for Smart Environment*; Springer: Berlin/Heidelberg, Germany, 2010; pp. 331–349.

28. Wilhelm, E.; Schwarz, T.; Jaworek, G.; Voigt, A.; Rapp, B.E. *Towards Displaying Graphics on a Cheap, Large-Scale Braille Display*; Springer: Cham, Switzerland, 2014; pp. 662–669.

29. Munasinghe, K.C.; Bowatta, B.G.C.T.; Abayarathne, H.Y.R.; Kumararathna, N.; Maduwantha, L.K.A.H.; Arachchige, N.M.P.; Amarasinghe, Y.W.R. New MEMS based micro gripper using SMA for micro level object manipulation and assembling. In Proceedings of the 2016 Moratuwa Engineering Research Conference (MERCon), Moratuwa, Sri Lanka, 5–6 April 2016; pp. 36–41.

30. Kornbluh, R.D.; Pelrine, R.; Prahlad, H.; Heydt, R. *Electroactive Polymers: An Emerging Technology for MEMS*; Janson, S.W., Henning, A.K., Eds.; International Society for Optics and Photonics: Bellingham, WA, USA, 2004; Volume 5344, pp. 13–27.

31. Liu, H.; Jui Tay, C.; Quan, C.; Kobayashi, T.; Lee, C. Piezoelectric MEMS Energy Harvester for Low-Frequency Vibrations With Wideband Operation Range and Steadily Increased Output Power. *Artic. J. Microelectromech. Syst.* **2011**, *20*, 1131–1142. [CrossRef]

32. Dhanalakshmi, K. Demonstration of self-sensing in Shape Memory Alloy actuated gripper. In Proceedings of the IEEE International Symposium on Intelligent Control (ISIC), Hyderabad, India, 28–30 August 2013; pp. 218–222.

33. Islam, M.N.; Seethaler, R.J. Sensorless Position Control for Piezoelectric Actuators Using A Hybrid Position Observer. *IEEE/ASME Trans. Mechatron.* **2014**, *19*, 667–675. [CrossRef]

34. Rizzello, G.; Naso, D.; York, A.; Seelecke, S. A Self-Sensing Approach for Dielectric Elastomer Actuators Based on Online Estimation Algorithms. *IEEE/ASME Trans. Mechatron.* **2017**, *22*, 728–738. [CrossRef]

35. Mineta, T.; Yanatori, H.; Hiyoshi, K.; Tsuji, K.; Ono, Y.; Abe, K. Tactile display MEMS device with SU8 micro-pin and spring on SMA film actuator array. In Proceedings of the 2017 19th International Conference on Solid-State Sensors, Actuators and Microsystems (TRANSDUCERS), Kaohsiung, Taiwan, 18–22 June 2017; pp. 2031–2034.

36. Hishinuma, Y.; Yang, E.-H. Piezoelectric Unimorph Microactuator Arrays for Single-Crystal Silicon Continuous-Membrane Deformable Mirror. *J. Microelectromech. Syst.* **2006**, *15*, 370–379. [CrossRef]

37. Besse, N.; Zarate, J.J.; Rosset, S.; Shea, H.R. Flexible haptic display with 768 independently controllable shape memory polymers taxels. In Proceedings of the 2017 19th International Conference on Solid-State Sensors, Actuators and Microsystems (TRANSDUCERS), Kaohsiung, Taiwan, 18–22 June 2017; pp. 323–326.

38. Richter, A.; Paschew, G. Optoelectrothermic Control of Highly Integrated Polymer-Based MEMS Applied in an Artificial Skin. *Adv. Mater.* **2009**, *21*, 979–983. [CrossRef]

39. Carpi, F.; De Rossi, D.; Kornbluh, R.; Pelrine, R.E.; Sommer-Larsen, P. *Dielectric Elastomers as Electromechanical Transducers: Fundamentals, Materials, Devices, Models and Applications of an Emerging Electroactive Polymer Technology*; Elsevier: Amsterdam, The Netherlands, 2011.

40. Hill, M.; Rizzello, G.; Seelecke, S. Development and Experimental Characterization of a Pneumatic Valve Actuated by a Dielectric Elastomer Membrane. *Smart Mater. Struct.* **2017**, *26*, 085023. [CrossRef]

41. Loverich, J.J.; Kanno, I.; Kotera, H. Concepts for a new class of all-polymer micropumps. *Lab Chip* **2006**, *6*, 1147–1154. [CrossRef]

42. Gu, G.-Y.; Zhu, J.; Zhu, L.-M.; Zhu, X. A survey on dielectric elastomer actuators for soft robots. *Bioinspir. Biomim.* **2017**, *12*, 011003. [CrossRef] [PubMed]

43. Guo, Y.; Liu, L.; Liu, Y.; Leng, J. Review of Dielectric Elastomer Actuators and their Applications in Soft Robots. *Adv. Intell. Syst.* **2021**, *3*, 2000282. [CrossRef]

44. Huang, B.; Li, M.; Mei, T.; McCoul, D.; Qin, S.; Zhao, Z.; Zhao, J. Wearable stretch sensors for motion measurement of the wrist joint based on dielectric elastomers. *Sensors* **2017**, *17*, 2708. [CrossRef]

45. Kelley, C.R.; Kauffman, J.L. Towards wearable tremor suppression using dielectric elastomer stack actuators. *Smart Mater. Struct.* **2020**, *30*, 025006. [CrossRef]

46. Kovacs, G.; Lochmatter, P.; Wissler, M. An arm wrestling robot driven by dielectric elastomer actuators. *Smart Mater. Struct.* **2007**, *16*, S306–S317. [CrossRef]

47. Pelrine, R.; Kornbluh, R.D.; Pei, Q.; Stanford, S.; Oh, S.; Eckerle, J.; Full, R.J.; Rosenthal, M.A.; Meijer, K. Smart Structures and Materials. In Proceedings of the SPIE'S 9th Annual International Symposium on Smart Structures and Materials, San Diego, CA, USA, 17–21 March 2002; Volume 4695, pp. 126–137.

48. Wang, N.F.; Cui, C.Y.; Guo, H.; Chen, B.C.; Zhang, X.M. Advances in dielectric elastomer actuation technology. *Sci. China Technol. Sci.* **2018**, *61*, 1512–1527. [CrossRef]

49. Balakrisnan, B.; Smela, E. *Challenges in the Microfabrication of Dielectric Elastomer Actuators*; Bar-Cohen, Y., Ed.; International Society for Optics and Photonics: Bellingham, WA, USA, 2010; Volume 7642, pp. 141–150.

50. Patrick, L.; Gabor, K.; Silvain, M. Characterization of dielectric elastomer actuators based on a hyperelastic film model. *Sens. Actuators A Phys.* **2007**, *135*, 748–757. [CrossRef]

51. York, A.; Dunn, J.; Seelecke, S. Experimental characterization of the hysteretic and rate-dependent electromechanical behavior of dielectric electro-active polymer actuators. *Smart Mater. Struct.* **2010**, *19*, 094014. [CrossRef]

52. Kaltseis, R.; Keplinger, C.; Koh, S.J.A.; Baumgartner, R.; Goh, Y.F.; Ng, W.H.; Kogler, A.; Tröls, A.; Foo, C.C.; Suo, Z.; et al. Natural rubber for sustainable high-power electrical energy generation. *RSC Adv.* **2014**, *4*, 27905–27913. [CrossRef]

53. Chen, Y.; Agostini, L.; Moretti, G.; Fontana, M.; Vertechy, R. Dielectric elastomer materials for large-strain actuation and energy harvesting: A comparison between styrenic rubber, natural rubber and acrylic elastomer. *Smart Mater. Struct.* **2019**, *28*, 114001. [CrossRef]

54. Chen, T.; Qiu, J.; Zhu, K.; Li, J. Electro-mechanical performance of polyurethane dielectric elastomer flexible micro-actuator composite modified with titanium dioxide-graphene hybrid fillers. *Mater. Des.* **2016**, *90*, 1069–1076. [CrossRef]

55. Fasolt, B.; Hodgins, M.; Rizzello, G.; Seelecke, S. Effect of screen printing parameters on sensor and actuator performance of dielectric elastomer (DE) membranes. *Sens. Actuators A Phys.* **2017**, *265*, 10–19. [CrossRef]

56. Rosset, S.; Shea, H.R. Flexible and stretchable electrodes for dielectric elastomer actuators. *Appl. Phys. A* **2012**, *110*, 281–307. [CrossRef]

57. Röntgen, W.C. Ueber die durch Electricität bewirkten Form- und Volumenänderungen von dielectrischen Körpern. *Ann. Der Phys. Und Chem.* **1880**, *247*, 771–786. [CrossRef]

58. Pelrine, R.; Kornbluh, R.; Joseph, J. Electrostriction of polymer dielectrics with compliant electrodes as a means of actuation. *Sens. Actuators A Phys.* **1998**, *64*, 77–85. [CrossRef]

59. Carpi, F.; Anderson, I.; Bauer, S.; Frediani, G.; Gallone, G.; Gei, M.; Graaf, C.; Jean-Mistral, C.; Kaal, W.; Kofod, G.; et al. Standards for dielectric elastomer transducers. *Smart Mater. Struct.* **2015**, *24*, 105025. [CrossRef]

60. Rizzello, G.; Loew, P.; Agostini, L.; Fontana, M.; Seelecke, S. A lumped parameter model for strip-shaped dielectric elastomer membrane transducers with arbitrary aspect ratio. *Smart Mater. Struct.* **2020**, *29*, 115030. [CrossRef]

61. Shian, S.; Huang, J.; Zhu, S.; Clarke, D.R. Optimizing the Electrical Energy Conversion Cycle of Dielectric Elastomer Generators. *Adv. Mater.* **2014**, *26*, 6617–6621. [CrossRef]

62. Michel, S.; Zhang, X.Q.; Wissler, M.; Löwe, C.; Kovacs, G. A comparison between silicone and acrylic elastomers as dielectric materials in electroactive polymer actuators. *Polym. Int.* **2009**, *59*, 391–399. [CrossRef]

63. Chen, Y.; Agostini, L.; Moretti, G.; Berselli, G.; Fontana, M.; Vertechy, R. Fatigue life performances of silicone elastomer membranes for dielectric elastomer transducers: Preliminary results. In Proceedings of the SPIE-The International Society for Optical Engineering, Denver, CO, USA, 3–7 March 2019; Volume 10966, pp. 158–167.

64. Youn, J.-H.; Jeong, S.M.; Hwang, G.; Kim, H.; Hyeon, K.; Park, J.; Kyung, K.-U. Dielectric Elastomer Actuator for Soft Robotics Applications and Challenges. *Appl. Sci.* **2020**, *10*, 640. [CrossRef]

65. Pelrine, R.; Kornbluh, R.; Joseph, J.; Heydt, R.; Pei, Q.; Chiba, S. High-field deformation of elastomeric dielectrics for actuators. *Mater. Sci. Eng. C* **2000**, *11*, 89–100. [CrossRef]

66. Romasanta, L.J.; Lopez-Manchado, M.A.; Verdejo, R. Increasing the performance of dielectric elastomer actuators: A review from the materials perspective. *Prog. Polym. Sci.* **2015**, *51*, 188–211. [CrossRef]

67. Hodgins, M.; Seelecke, S. Systematic experimental study of pure shear type dielectric elastomer membranes with different electrode and film thicknesses. *Smart Mater. Struct.* **2016**, *25*, 095001. [CrossRef]

68. Schlatter, S.; Rosset, S.; Shea, H. Inkjet Printing of Carbon Black Electrodes for Dielectric Elastomer Actuators. In *SPIE Electroactive Polymer Actuators and Devices*; SPIE: Bellingham, WA, USA, 2017; Volume 10163, pp. 177–185.

69. Klug, F.; Solano-Arana, S.; Hoffmann, N.J.; Schlaak, H.F. Multilayer dielectric elastomer tubular transducers for soft robotic applications. *Smart Mater. Struct.* **2019**, *28*, 104004. [CrossRef]

70. Hubertus, J.; Neu, J.; Croce, S.; Rizzello, G.; Seelecke, S.; Schultes, G. Nanoscale Nickel-Based Thin Films as Highly Conductive Electrodes for Dielectric Elastomer Applications with Extremely High Stretchability up to 200%. *ACS Appl. Mater. Interfaces* **2021**, *13*, 39894–39904. [CrossRef] [PubMed]

71. Hubertus, J.; Croce, S.; Neu, J.; Rizzello, G.; Seelecke, S.; Schultes, G. Electromechanical characterization and laser structuring of Ni-based sputtered metallic compliant electrodes for DE applications. In Proceedings of the International Conference and Exhibition on New Actuator Systems and Applications, online, 17–19 February 2021; pp. 321–324.

72. Jung, K.; Kim, K.J.; Choi, H.R. A self-sensing dielectric elastomer actuator. *Sens. Actuators A Phys.* **2008**, *143*, 343–351. [CrossRef]

73. Gisby, T.A.; O'Brien, B.M.; Anderson, I.A. Self sensing feedback for dielectric elastomer actuators. *Appl. Phys. Lett.* **2013**, *102*, 193703. [CrossRef]

74. Hoffstadt, T.; Griese, M.; Maas, J. Online identification algorithms for integrated dielectric electroactive polymer sensors and self-sensing concepts. *Smart Mater. Struct.* **2014**, *23*, 104007. [CrossRef]

75. Rizzello, G.; Fugaro, F.; Naso, D.; Seelecke, S. Simultaneous Self-Sensing of Displacement and Force for Soft Dielectric Elastomer Actuators. *IEEE Robot. Autom. Lett.* **2018**, *3*, 1230–1236. [CrossRef]

76. Rizzello, G.; Naso, D.; York, A.; Seelecke, S. Closed loop control of dielectric elastomer actuators based on self-sensing displacement feedback. *Smart Mater. Struct.* **2016**, *25*, 035034. [CrossRef]

77. Hoffstadt, T.; Maas, J. Sensorless force control for dielectric elastomer transducers. *J. Intell. Mater. Syst. Struct.* **2018**, *30*, 1419–1434. [CrossRef]

78. Rizzello, G.; Serafino, P.; Naso, D.; Seelecke, S. Towards Sensorless Soft Robotics: Self-Sensing Stiffness Control of Dielectric Elastomer Actuators. *IEEE Trans. Robot.* **2020**, *36*, 174–188. [CrossRef]

79. Gatti, D.; Haus, H.; Matysek, M.; Frohnapfel, B.; Tropea, C.; Schlaak, H.F. The dielectric breakdown limit of silicone dielectric elastomer actuators. *Appl. Phys. Lett.* **2014**, *104*, 052905. [CrossRef]

80. Fasolt, B.; Welsch, F.; Jank, M.; Seelecke, S. Effect of actuation parameters and environment on the breakdown voltage of silicone dielectric elastomer films. *Smart Mater. Struct.* **2019**, *28*, 094002. [CrossRef]

81. Suo, Z. Theory of dielectric elastomers. *Acta Mech. Solida Sin.* **2010**, *23*, 549–578. [CrossRef]

82. Croce, S.; Neu, J.; Moretti, G.; Hubertus, J.; Schultes, G.; Rizzello, G. Finite element modeling and validation of a soft array of spatially coupled dielectric elastomer transducers. *Smart Mater. Struct.* **2022**, *31*, 084001. [CrossRef]

83. Hajiesmaili, E.; Clarke, D.R. Dielectric elastomer actuators. *J. Appl. Phys.* **2021**, *129*, 151102. [CrossRef]

84. Kovacs, G.; Düring, L.; Michel, S.; Terrasi, G. Stacked dielectric elastomer actuator for tensile force transmission. *Sens. Actuators A Phys.* **2009**, *155*, 299–307. [CrossRef]

85. Maas, J.; Tepel, D.; Hoffstadt, T. Actuator design and automated manufacturing process for DEAP-based multilayer stack-actuators. *Meccanica* **2015**, *50*, 2839–2854. [CrossRef]

86. Hau, S.; Bruch, D.; Rizzello, G.; Motzki, P.; Seelecke, S. Silicone based dielectric elastomer strip actuators coupled with nonlinear biasing elements for large actuation strains. *Smart Mater. Struct.* **2018**, *27*, 074003. [CrossRef]

87. Kwak, J.W.; Chi, H.J..; Jung, K.M.; Koo, J.C.; Jeon, J.W.; Lee, Y.; Nam, J.-D.; Ryew, Y.; Choi, H.-R. A Face Robot Actuated With Artificial Muscle Based on Dielectric Elastomer. *J. Mech. Sci. Technol.* **2005**, *19*, 578–588. [CrossRef]

88. Kunze, J.; Prechtl, J.; Bruch, D.; Fasolt, B.; Nalbach, S.; Motzki, P.; Seelecke, S.; Rizzello, G. Design, Manufacturing, and Characterization of Thin, Core-Free, Rolled Dielectric Elastomer Actuators. *Actuators* **2021**, *10*, 69. [CrossRef]

89. Rajamani, A.; Grissom, M.; Rahn, C.; Ma, Y.; Zhang, Q. Wound roll dielectric elastomer actuators: Fabrication, analysis and experiments. In Proceedings of the 2005 IEEE/RSJ International Conference on Intelligent Robots and Systems, IROS, Edmonton, AB, Canada, 2–6 August 2005; Volume 13, pp. 2587–2592.

90. Moretti, G.; Sarina, L.; Agostini, L.; Vertechy, R.; Berselli, G.; Fontana, M. Styrenic-rubber dielectric elastomer actuator with inherent stiffness compensation. *Actuators* **2020**, *9*, 44. [CrossRef]

91. Follador, M.; Cianchetti, M.; Mazzolai, B. Design of a compact bistable mechanism based on dielectric elastomer actuators. *Meccanica* **2015**, *50*, 2741–2749. [CrossRef]

92. Berselli, G.; Vertechy, R.; Vassura, G.; Parenti-Castelli, V. Optimal Synthesis of Conically Shaped Dielectric Elastomer Linear Actuators: Design Methodology and Experimental Validation. *IEEE/ASME Trans. Mechatron.* **2011**, *16*, 67–79. [CrossRef]

93. Rizzello, G.; Hodgins, M.; Naso, D.; York, A.; Seelecke, S. Modeling of the effects of the electrical dynamics on the electromechanical response of a DEAP circular actuator with a mass-spring load. *Smart Mater. Struct.* **2015**, *24*, 094003. [CrossRef]

94. Nalbach, S.; Banda, R.M.; Croce, S.; Rizzello, G.; Naso, D.; Seelecke, S. Modeling and Design Optimization of a Rotational Soft Robotic System Driven by Double Cone Dielectric Elastomer Actuators. *Front. Robot. AI* **2020**, *6*, 150. [CrossRef] [PubMed]

95. Cao, C.; Chen, L.; Li, B.; Chen, G.; Nie, Z.; Wang, L.; Gao, X. Toward broad optimal output bandwidth dielectric elastomer actuators. *Sci. China Technol. Sci.* **2022**, *65*, 1137–1148. [CrossRef]

96. Kofod, G.; Wirges, W.; Paajanen, M.; Bauer, S. Energy minimization for self-organized structure formation and actuation. *Appl. Phys. Lett.* **2007**, *90*, 081916. [CrossRef]

97. McGough, K.; Ahmed, S.; Frecker, M.; Ounaies, Z. Finite element analysis and validation of dielectric elastomer actuators used for active origami. *Smart Mater. Struct.* **2014**, *23*, 094002. [CrossRef]

98. Hau, S.; Rizzello, G.; Seelecke, S. A novel dielectric elastomer membrane actuator concept for high-force applications. *Extrem. Mech. Lett.* **2018**, *23*, 24–28. [CrossRef]

99. Hodgins, M.; York, A.; Seelecke, S. Experimental comparison of bias elements for out-of-plane DEAP actuator system. *Smart Mater. Struct.* **2013**, *22*, 094016. [CrossRef]

100. Wang, Y.; Zhu, J. Artificial muscles for jaw movements. *Extrem. Mech. Lett.* **2016**, *6*, 88–95. [CrossRef]

101. Zhang, R.; Lochmatter, P.; Kunz, A.; Kovacs, G. Spring roll dielectric elastomer actuators for a portable force feedback glove. In Proceedings of the Smart Structures and Materials 2006: Electroactive Polymer Actuators and Devices (EAPAD), San Diego, CA, USA, 26 February–2 March 2006; Volume 6168, pp. 505–516.

102. He, T.; Cui, L.; Chen, C.; Suo, Z. Nonlinear deformation analysis of a dielectric elastomer membrane–spring system. *Smart Mater. Struct.* **2010**, *19*, 085017. [CrossRef]

103. Keplinger, C.; Li, T.; Baumgartner, R.; Suo, Z.; Bauer, S. Harnessing snap-through instability in soft dielectrics to achieve giant voltage-triggered deformation. *Soft Matter* **2012**, *8*, 285. [CrossRef]

104. Li, T.; Keplinger, C.; Baumgartner, R.; Bauer, S.; Yang, W.; Suo, Z. Giant voltage-induced deformation in dielectric elastomers near the verge of snap-through instability. *J. Mech. Phys. Solids* **2013**, *61*, 611–628. [CrossRef]

105. Jordi, C.; Michel, S.; Kovacs, G.; Ermanni, P. Scaling of planar dielectric elastomer actuators in an agonist-antagonist configuration. *Sens. Actuators A Phys.* **2010**, *161*, 182–190. [CrossRef]

106. Carpi, F.; Frediani, G.; De Rossi, D. Hydrostatically Coupled Dielectric Elastomer Actuators. *IEEE/ASME Trans. Mechatron.* **2010**, *15*, 308–315. [CrossRef]

107. Loew, P.; Rizzello, G.; Seelecke, S. A novel biasing mechanism for circular out-of-plane dielectric actuators based on permanent magnets. *Mechatronics* **2018**, *56*, 48–57. [CrossRef]

108. Cao, C.; Chen, L.; Hill, T.L.; Wang, L.; Gao, X. Exploiting Bistability for High-Performance Dielectric Elastomer Resonators. *IEEE/ASME Trans. Mechatron.* **2022**, *27*, 5994–6005. [CrossRef]

109. Neu, J.; Hubertus, J.; Croce, S.; Schultes, G.; Seelecke, S.; Rizzello, G. Fully Polymeric Domes as High-Stroke Biasing System for Soft Dielectric Elastomer Actuators. *Front. Robot. AI* **2021**, *8*, 695918. [CrossRef]

110. Dubois, P.; Rosset, S.; Koster, S.; Stauffer, J.; Mikhaïlovc, S.; Dadras, M.; de Rooij, N.-F.; Shea, H. Microactuators based on ion implanted dielectric electroactive polymer (EAP) membranes. *Sens. Actuators A Phys.* **2006**, *131*, 147–154. [CrossRef]

111. Marette, A.; Poulin, A.; Besse, N.; Rosset, S.; Briand, D.; Shea, H. Thin Film Transistors: Flexible Zinc-Tin Oxide Thin Film Transistors Operating at 1 kV for Integrated Switching of Dielectric Elastomer Actuators Arrays. *Adv. Mater.* **2017**, *29*, 1700880. [CrossRef]
112. Poulin, A.; Rosset, S.; Shea, H. Fully printed 3 microns thick dielectric elastomer actuator. In Proceedings of the SPIE Smart Structures and Materials + Nondestructive Evaluation and Health Monitoring, Las Vegas, NV, USA, 20–24 March 2016; Volume 9798, pp. 36–46.
113. Lotz, P.; Matysek, M.; Schlaak, H.F. Fabrication and Application of Miniaturized Dielectric Elastomer Stack Actuators. *IEEE/ASME Trans. Mechatron.* **2011**, *16*, 58–66. [CrossRef]
114. Wang, K.; Ouyang, G.; Chen, X.; Jakobsen, H. Engineering Electroactive Dielectric Elastomers for Miniature Electromechanical Transducers. *Polym. Rev.* **2017**, *57*, 369–396. [CrossRef]
115. Carpi, F.; Frediani, G.; Turco, S.; De Rossi, D. Bioinspired Tunable Lens with Muscle-Like Electroactive Elastomers. *Adv. Funct. Mater.* **2011**, *21*, 4152–4158. [CrossRef]
116. Carpi, F.; Frediani, G.; Tarantino, S.; De Rossi, D. Millimetre-scale bubble-like dielectric elastomer actuators. *Polym. Int.* **2009**, *59*, 407–414. [CrossRef]
117. Kim, U.; Kang, J.; Lee, C.; Kwon, H.Y.; Hwang, S.; Moon, H.; Koo, J.C.; Nam, J.-D.; Hong, B.H.; Choi, J.-B.; et al. A transparent and stretchable graphene-based actuator for tactile display. *Nanotechnology* **2013**, *24*, 145501. [CrossRef]
118. Hau, S.; Rizzello, G.; Hodgins, M.; York, A.; Seelecke, S. Design and control of a high-speed positioning system based on dielectric elastomer membrane actuators. *IEEE/ASME Trans. Mechatron.* **2017**, *22*, 1259–1267. [CrossRef]
119. Soulimane, S.; Pinon, S.; Shih, W.P.; Camon, H. Dielectric Elastomer Micro Actuator Made In Micromachining Technology: Finite Element Modelling and Deformation Measurement. *Procedia Eng.* **2011**, *25*, 479–482. [CrossRef]
120. Pimpin, A.; Suzuki, Y.; Kasagi, N. Microelectrostrictive actuator with large out-of-plane deformation for flow-control application. *J. Microelectromech. Syst.* **2007**, *16*, 753–764. [CrossRef]
121. Murray, C.; McCoul, D.; Sollier, E.; Ruggiero, T.; Niu, X.; Pei, Q.; Di Carlo, D. Electro-adaptive microfluidics for active tuning of channel geometry using polymer actuators. *Microfluid. Nanofluidics* **2013**, *14*, 345–358. [CrossRef]
122. Tanaka, Y.; Fujikawa, T.; Kazoe, Y.; Kitamori, T. An active valve incorporated into a microchip using a high strain electroactive polymer. *Sens. Actuators B Chem.* **2013**, *184*, 163–169. [CrossRef]
123. Mohd Ghazali, F.A.; Mah, C.K.; AbuZaiter, A.; Chee, P.S.; Mohamed Ali, M.S. Soft dielectric elastomer actuator micropump. *Sens. Actuators A Phys.* **2017**, *263*, 276–284. [CrossRef]
124. Ren, Z.; Kim, S.; Ji, X.; Zhu, W.; Niroui, F.; Kong, J.; Chen, Y. A High-Lift Micro-Aerial-Robot Powered by Low-Voltage and Long-Endurance Dielectric Elastomer Actuators. *Adv. Mater.* **2022**, *34*, 2106757. [CrossRef]
125. Rosset, S.; Niklaus, M.; Dubois, P.; Shea, H.R. Mechanical characterization of a dielectric elastomer microactuator with ion-implanted electrodes. *Sens. Actuators A Phys.* **2008**, *144*, 185–193. [CrossRef]
126. Poulin, A.; Rosset, S.; Shea, H.R. Printing low-voltage dielectric elastomer actuators. *Appl. Phys. Lett.* **2015**, *107*, 244104. [CrossRef]
127. Neu, J.; Croce, S.; Willian, T.; Hubertus, J.; Schultes, G.; Seelecke, S.; Rizzello, G. Distributed Electro-Mechanical Coupling Effects in a Dielectric Elastomer Membrane Array. *Exp. Mech.* **2022**, *63*, 79–95. [CrossRef]
128. Yu, X.; Lu, Z.; Cui, F.; Cheng, L.; Cui, Y. Tunable acoustic metamaterial with an array of resonators actuated by dielectric elastomer. *Extrem. Mech. Lett.* **2017**, *12*, 37–40. [CrossRef]
129. Ji, X.; Liu, X.; Cacucciolo, V.; Civet, Y.; El Haitami, A.; Cantin, S.; Perriard, Y.; Shea, H. Untethered Feel-Through Haptics Using 18-µm Thick Dielectric Elastomer Actuators. *Adv. Funct. Mater.* **2021**, *31*, 2006639. [CrossRef]
130. Solano-Arana, S.; Klug, F.; Mößinger, H.; Förster-Zügel, F.; Schlaak, H.F. A novel application of dielectric stack actuators: A pumping micromixer. *Smart Mater. Struct.* **2018**, *27*, 074008. [CrossRef]
131. Schlatter, S.; Grasso, G.; Rosset, S.; Shea, H. Inkjet Printing of Complex Soft Machines with Densely Integrated Electrostatic Actuators. *Adv. Intell. Syst.* **2020**, *2*, 2000136. [CrossRef]
132. O'Brien, B.; Gisby, T.; Calius, E.; Xie, S.; Anderson, I. FEA of Dielectric Elastomer Minimum Energy Structures as a Tool for Biomimetic Design. In *SPIE Electroactive Polymer Actuators and Devices*; SPIE: Bellingham, WA, USA, 2009; Volume 7287, pp. 61–71.
133. Dai, B.; Li, S.; Xu, T.; Wang, Y.; Zhang, F.; Gu, Z.; Wang, S. Artificial Asymmetric Cilia Array of Dielectric Elastomer for Cargo Transportation. *ACS Appl. Mater. Interfaces* **2018**, *10*, 42979–42984. [CrossRef]
134. Zhao, J.; Niu, J.; Liu, L.; Yu, J. A Soft Creeping Robot Actuated by Dielectric Elastomer. In *SPIE Electroactive Polymer Actuators and Devices*; SPIE: Bellingham, WA, USA, 2014; Volume 9056, pp. 40–45.
135. Lu, X.J.; Wang, K.; Hu, T.T. Development of an annelid-like peristaltic crawling soft robot using dielectric elastomer actuators. *Bioinspir. Biomim.* **2020**, *15*, 046012. [CrossRef] [PubMed]
136. Li, W.B.; Zhang, W.M.; Zou, H.X.; Peng, Z.K.; Meng, G. Multisegment annular dielectric elastomer actuators for soft robots. *Smart Mater. Struct.* **2018**, *27*, 115024. [CrossRef]
137. Pfeil, S.; Henke, M.; Katzer, K.; Zimmermann, M.; Gerlach, G. A Worm-Like Biomimetic Crawling Robot Based on Cylindrical Dielectric Elastomer Actuators. *Front. Robot. AI* **2020**, *7*, 9. [CrossRef]
138. Henke, E.-F.M.; Schlatter, S.; Anderson, I.A. Soft Dielectric Elastomer Oscillators Driving Bioinspired Robots. *Soft Robot.* **2017**, *4*, 353–366. [CrossRef] [PubMed]
139. Nguyen, C.T.; Phung, H.; Nguyen, T.D.; Jung, H.; Choi, H.R. Multiple-degrees-of-freedom dielectric elastomer actuators for soft printable hexapod robot. *Sens. Actuators A Phys.* **2017**, *267*, 505–516. [CrossRef]

140. Chen, F.; Cao, J.; Zhang, H.; Wang, M.Y.; Zhu, J.; Zhang, Y.F. Programmable Deformations of Networked Inflated Dielectric Elastomer Actuators. *IEEE/ASME Trans. Mechatron.* **2019**, *24*, 45–55. [CrossRef]
141. Hajiesmaili, E.; Clarke, D.R. Optically addressable dielectric elastomer actuator arrays using embedded percolative networks of zinc oxide nanowires. *Mater. Horiz.* **2022**, *9*, 3110–3117. [CrossRef]
142. Wang, L.; Hayakawa, T.; Ishikawa, M. Dielectric-elastomer-based fabrication method for varifocal microlens array. *Opt. Express* **2017**, *25*, 31708. [CrossRef]
143. Burugupally, S.P.; Koppolu, B.; Danesh, N.; Lee, Y.; Indeewari, V.; Li, B. Enhancing the performance of dielectric elastomer actuators through the approach of distributed electrode array with fractal interconnects architecture. *J. Micromech. Microeng.* **2021**, *31*, 064002. [CrossRef]
144. Yang, W.-P.; Chen, L.-W.; McGough, K.; Ahmed, S. The tunable acoustic band gaps of two-dimensional phononic crystals with a dielectricelastomer cylindrical actuator. *Smart Mater. Struct.* **2007**, *17*, 015011. [CrossRef]
145. Wang, T.; Zhang, J.; Hong, J.; Wang, M.Y. Dielectric Elastomer Actuators for Soft Wave-Handling Systems. *Soft Robot.* **2017**, *4*, 61–69. [CrossRef]
146. Akbari, S.; Shea, H.R. Microfabrication and characterization of an array of dielectric elastomer actuators generating uniaxial strain to stretch individual cells. *J. Micromech. Microeng.* **2012**, *22*, 045020. [CrossRef]
147. Akbari, S.; Shea, H.R. An array of 100 m × 100 m dielectric elastomer actuators with 80% strain for tissue engineering applications. *Sens. Actuators A* **2012**, *186*, 236–241. [CrossRef]
148. Matysek, M.; Lotz, P.; Winterstein, T.; Schlaak, H.F. Dielectric elastomer actuators for tactile displays. In Proceedings of the World Haptics 2009-Third Joint EuroHaptics conference and Symposium on Haptic Interfaces for Virtual Environment and Teleoperator Systems, Online, 18–20 March 2009; pp. 290–295.
149. Phung, H.; Hoang, P.T.; Jung, H.; Nguyen, T.D.; Nguyen, C.T.; Choi, H.R. Haptic Display Responsive to Touch Driven by Soft Actuator and Soft Sensor. *IEEE/ASME Trans. Mechatron.* **2021**, *26*, 2495–2505. [CrossRef]
150. Lee, H.S.; Phung, H.; Lee, D.-H.; Kim, U.K.; Nguyen, C.T.; Moon, H.; Koo, J.C.; Choi, H.R. Design analysis and fabrication of arrayed tactile display based on dielectric elastomer actuator. *Sens. Actuators A Phys.* **2014**, *205*, 191–198. [CrossRef]
151. Lee, D.-Y.; Jeong, S.H.; Cohen, A.J.; Vogt, D.M.; Kollosche, M.; Lansberry, G.; Mengüç, Y.; Israr, A.; Clarke, D.R.; Wood, R.J. A Wearable Textile-Embedded Dielectric Elastomer Actuator Haptic Display. *Soft Robot.* **2022**, *9*, 1186–1197. [CrossRef]
152. Zhao, H.; Hussain, A.M.; Israr, A.; Vogt, D.M.; Duduta, M.; Clarke, D.R.; Wood, R.J. A Wearable Soft Haptic Communicator Based on Dielectric Elastomer Actuators. *Soft Robot.* **2020**, *7*, 451–461. [CrossRef]
153. Chakraborti, P.; Toprakci, H.A.K.; Yang, P.; Di Spigna, N.; Franzon, P.; Ghosh, T. A compact dielectric elastomer tubular actuator for refreshable Braille displays. *Sens. Actuators A Phys.* **2012**, *179*, 151–157. [CrossRef]
154. Qu, X.; Ma, X.; Shi, B.; Li, H.; Zheng, L.; Wang, C.; Liu, Z.; Fan, Y.; Chen, X.; Li, Z.; et al. Refreshable Braille Display System Based on Triboelectric Nanogenerator and Dielectric Elastomer. *Adv. Funct. Mater.* **2020**, *31*, 2006612. [CrossRef]
155. Frediani, G.; Busfield, J.; Carpi, F. Enabling portable multiple-line refreshable Braille displays with electroactive elastomers. *Med. Eng. Phys.* **2018**, *60*, 86–93. [CrossRef]
156. Pang, W.; Cheng, X.; Zhao, H.; Guo, X.; Ji, Z.; Li, G.; Liang, Y.; Xue, Z.; Song, H.; Zhang, F.; et al. Electro-mechanically controlled assembly of reconfigurable 3D mesostructures and electronic devices based on dielectric elastomer platforms. *Natl. Sci. Rev.* **2020**, *7*, 342–354. [CrossRef]
157. Sun, Y.; Li, D.; Wu, M.; Yang, Y.; Su, J.; Wong, T.; Xu, K.; Li, Y.; Li, L.; Yu, X.; et al. Origami-inspired folding assembly of dielectric elastomers for programmable soft robots. *Microsyst. Nanoeng.* **2022**, *8*, 1–11. [CrossRef] [PubMed]
158. Aksoy, B.; Shea, H. Reconfigurable and Latchable Shape-Morphing Dielectric Elastomers Based on Local Stiffness Modulation. *Adv. Funct. Mater.* **2020**, *30*, 2001597. [CrossRef]
159. Meng, J.; Qiu, Y.; Hou, C.; Zhang, Q.; Li, Y.; Wang, H. Bistable dielectric elastomer actuator with directional motion. *Sens. Actuators A Phys.* **2021**, *330*, 112889. [CrossRef]
160. Xu, D.; Tairych, A.; Anderson, I.A. Where the rubber meets the hand: Unlocking the sensing potential of dielectric elastomers. *J. Polym. Sci. Part B Polym. Phys.* **2016**, *54*, 465–472. [CrossRef]
161. Zhang, H.; Wang, M.Y.; Li, J.; Zhu, J. A soft compressive sensor using dielectric elastomers. *Smart Mater. Struct.* **2016**, *25*, 035045. [CrossRef]
162. Ham, J.; Huh, T.M.; Kim, J.; Kim, J.-O.; Park, S.; Cutkosky, M.R.; Bao, Z. Porous Dielectric Elastomer Based Flexible Multiaxial Tactile Sensor for Dexterous Robotic or Prosthetic Hands. *Adv. Mater. Technol.* **2022**, *in press*. [CrossRef]
163. Kadooka, K.; Imamura, H.; Taya, M. Tactile Sensor Integrated Dielectric Elastomer Actuator for Simultaneous Actuation and Sensing. In *Electroactive Polymer Actuators and Devices*; SPIE: Bellingham, WA, USA, 2016; Volume 9798, pp. 489–498.
164. Zhu, Y.; Giffney, T.; Aw, K. A Dielectric Elastomer-Based Multimodal Capacitive Sensor. *Sensors* **2022**, *22*, 622. [CrossRef] [PubMed]
165. Meyer, A.; Lenz, S.; Gratz-Kelly, S.; Motzki, P.; Nalbach, S.; Seelecke, S.; Rizzello, G. Experimental Characterization of a Smart Dielectric Elastomer Multi-Sensor Grid. In *SPIE Electroactive Polymer Actuators and Devices (EAPAD) XXII*; SPIE: Bellingham, WA, USA, 2020; Volume 11375, pp. 262–268.
166. Lee, B.Y.; Kim, J.; Kim, H.; Kim, C.; Lee, S.D. Low-cost flexible pressure sensor based on dielectric elastomer film with micro-pores. *Sens. Actuators A Phys.* **2016**, *240*, 103–109. [CrossRef]

167. Kwon, D.; Lee, T.-I.; Shim, J.; Ryu, S.; Kim, M.S.; Kim, S.; Kim, T.-S.; Park, I. Highly Sensitive, Flexible, and Wearable Pressure Sensor Based on a Giant Piezocapacitive Effect of Three-Dimensional Microporous Elastomeric Dielectric Layer. *ACS Appl. Mater. Interfaces* **2016**, *8*, 16922–16931. [CrossRef]
168. Kyaw, A.K.K.; Loh, H.H.C.; Yan, F.; Xu, J. A polymer transistor array with a pressure-sensitive elastomer for electronic skin. *J. Mater. Chem. C* **2017**, *5*, 12039–12043. [CrossRef]
169. Larson, C.; Spjut, J.; Knepper, R.; Shepherd, R. A Deformable Interface for Human Touch Recognition Using Stretchable Carbon Nanotube Dielectric Elastomer Sensors and Deep Neural Networks. *Soft Robot.* **2019**, *6*, 611–620. [CrossRef] [PubMed]
170. Vishniakou, S.; Lewis, B.W.; Niu, X.; Kargar, A.; Sun, K.; Kalajian, M.; Park, N.; Yang, M.; Jing, Y.; Brochu, P.; et al. Tactile Feedback Display with Spatial and Temporal Resolutions. *Sci. Rep.* **2013**, *3*, 2521. [CrossRef] [PubMed]

MDPI

St. Alban-Anlage 66

4052 Basel

Switzerland

www.mdpi.com

Actuators Editorial Office

E-mail: actuators@mdpi.com

www.mdpi.com/journal/actuators